DIMENSIONS OF HIMALAYAN GEOLOGY

DIMENSIONS OF HIMALAYAN GEOLOGY

A.K. Biyani
Department of Geology
DBS College, Dehradun (India)

SSPH

SATISH SERIAL PUBLISHING HOUSE
403, Express Tower, Commercial Complex
Azadpur, Delhi-110033 (India)
Telefax : 011 - 47073040
e-mail : info@satishserial.com, hkjain1975@yahoo.com
Website : www.satishserial.com

Published by :

SATISH SERIAL PUBLISHING HOUSE
403, Express Tower, Commercial Complex, Azadpur, Delhi-110033 (INDIA)
Telefax : 011 – 47073040
E-mail : info@satishserial.com, hkjain1975@yahoo.com

Print Again : 2026

ISBN : 81-89304-15-1

Composed, Designed & Printed in India

About the Author

After learning the alphabets of geology and higher education in Indore, Sagar and New Delhi, author joined the academic profession in Department of Geology, DBS College, Dehradun. He is having 28 years of teaching experience in structural and metamorphic geology. Dehradun being a hub of research in Himalayan geology, author is also lured by Himalayan charm. His doctoral work in the geologically under investigated Higher Himalayan Yamuna Valley 'a geological marvel' exposed him to larger geological canvass. His 'generalist' approach in research spreading over many branches of geology kept him abreast with a fair view about current status of development in major fields of geology. So far he has published 42 research papers from geothermics to geochemistry and from mass movement to metallogeny in leading Indian journals and in proceeding volumes and dozens of articles on the academic, societal and scientific issues. He is fellow or life member of 20 geological societies. As convenor, author has organized two seminars and worked as member in executive committees of several others. Credit for concepts like social geology, natural disasters safety index, safety maps for reduction of disaster impact goes to author. Under his guidance post graduate students compiled possibly the largest Geological Map of Himalaya on 1 cm to 12 km scale by incorporating large number of published small and large-scale maps. In addition to this Tectonic Zonation Map of Himalaya and Geological Map of Uttaranchal were also compiled in his guidance. Author has successfully completed minor and major research projects funded by the UGC and by B.P. Koirala Foundation and a pre-proposal project of DST. He is also actively associated with Geofraternity- a departmental society of students and teachers for promotion of academic and social activities and the Indian Science Writers' Association for scientific thrust in society.

Preface

End of writing is time of satisfaction and contemplation for an author. Satisfaction because he has no more work to do, despite full awareness of the reality that justification to available knowledge is beyond his capacity. Particularly when such summaries are attempted. Contemplation arises from the anxiety of acceptability of the work by fellow colleagues and readers, yet attempts continue.

Irresistible divine fascination of Himalaya is not without a challenge so is the case with scanning of available geological literature of Himalayan dimensions. My 'generalist' approach in the research field has helped me in this work. Turmoil of many tens of millions of years in Himalayan terrain is fully reflected in contrasting geological interpretations of Himalaya by different investigators that too spread in very diverse type of sources (books, journals, reports etc.) that made extremely difficult situation for me to accommodate each and every viewpoint. The written text is just selection of few types of rubbles from highly specialized myriad aspects from the mammoth mountain of knowledge and is aimed at to apprise the post-graduate students of the earth sciences about the current developments in Himalayan Geology. Framing of chapters is slightly unusual, contents of some chapters like paleontology, petrology and geochronology could have been easily incorporated in other chapters. This would result into diffusion of focus or thrust. For the sake of smoothness in presentation and also dealing with large number of interrelated topics repetitions of basic facts at some places have occurred. Many important references have been omitted because of their non-availability and because of this the text suffers to that extent. A locality is described by different spellings, in those cases popular spelling was adopted. My own ideas such as social geology, safety maps and remobilized cratonic nature of the Higher Himalaya are also included in the text at relevant places. Several new references have been added in the text during proof reading stage.

Idea to publish a note on Himalayan Geology for students in the Quarterly Journal of the Geological Association & Research Centre was proposed to editor Dr. S Saxena. He accepted this but suggested instead note I should write a book. His advice ultimately culminated in the form of this book. Beneficial discussions on Himalayan geology with Dr K.P. Juyal, Dr A.K. Dubey and Dr. A.C. Nanda and Dr. V.C. Thakur, senior scientists, of Wadia Institute of Himalayan Geology (WIHG), Dehradun is gratefully acknowledged. I am thankful to my colleagues Dr. M.N. Joshi, Dr. B. K. Pandey, Dr M. S. Anantharaman, Dr. P.C. Bahukhandi and Prof. J. D. Badhe for over the years of discussions on Himalayan Geology. Dr. R. K. Pande, principal, deserves special mention for promoting academic excellence in the campus.

I am grateful to directors of WIHG for allowing me to avail facilities of their institute. Appetite for knowledge is satisfied in the library of WIHG, I am thankful to librarian Mr. Saeed Ahmad and to Mr. S. S. Bhandari for sustained cooperation.

Creativity and enthusiasm of post-graduate students are reflected in compilation of Geological Map of Himalaya (one cm to twelve km scale, perhaps largest one), Tectonic Zonation Map of Himalaya (one cm to forty km scale) and Geological Map of Uttaranchal (one cm to five km scale). These maps involved labor of 4000 man-hours, yet need some improvement. Namita Jindal, Sharat Dutta, Neetu Chauhan, Jagwant Kaur, Rashmi Pradhan, Arpita Sinha, Preetika Bahuguna, Ruchika Sharma and Kuldeep Singh of 2005 batch have compiled these maps in Department of Geology. I acknowledge their meticulous efforts gratefully. Love to own creation tempted me to incorporate miniature versions of these maps in the text. I am thankful to Ms. Namita Jindal, Ms. Divya Dudeja and Mr. Suraj Prasad Bhatta for their support during writing of this book. I am thankful to Mr. Hemant Jain of Satish Serial Publishing House for prompt publication of this book.

December 22, 2005 **A. K. Biyani**

Contents

DIMENSIONS OF HIMALAYAN GEOLOGY
A. K. Biyani (2006) : pp.1-9
Satish Serial Publishing House, New Delhi
E-mail: ssp@ndf.vsnl.net.in, hkjain1975@yahoo.com; website: www.satishserial.com

Mayadhyaksena prakrtih suyate sacharacharam (Geeta, 9.10)
I preside over all the animate and inanimate physical actions of the nature

CHAPTER - I

INTRODUCTION

The majestic mountain Himalaya (Sanskrit: Himalaya i.e. house of snow) is a glorious creation of nature. The Himalaya mountain commands unique reverence by the Indians due to its religious significance particularly its snow clad peaks houses the large number of Hindu deities and innumerable number of mythological lores are written in its backdrops. Holy river Ganga famed for imperishable water originates from its lap and above all the mountain has inspired and continuously inspiring many writers and poets by its serene charm. Besides its cultural, religious and literary significance, the mountain provides effective guard against invasions by man and chilly Siberian wind. Curiosity and courage of Man forced him to explore the giant, relentlessly from time immemorial to unearth haloes of mysteries woven around it. No doubt many of them have sacrificed themselves in the pursuit of their tasks. Same sprit also guided the geologists, geographers and nature explorers to write its story on the scientific pedestals. Despite being one of the extensively and thoroughly probed regions, yet new modifications of the existing thoughts are the order of the days. However, incessative activities of wide ranging investigations provided a fairly precise geological account of this supreme Mountain Range.

1.1 Wider Geographical Frame of the Himalaya

The Himalaya is a part of the Alpine- Himalayan mountain chain. The Alpine-Himalayan mountain range is nearly 10000 km long and encompasses nearly 3.0 M sq km area. This curvilinear range is known by several regional names between its western end in Spain in Europe to eastern end in Indonesia in Asia; it is called Baltic's in Spain, Alps in Switzerland, Dinarides and Carpathians in Poland, Atlas in Greece, Antolides- Zagros in Iran, Sulaiman in Afghanistan and Pakistan, Himalaya in Pakistan, India, Nepal, Bhutan, Arakan in eastern India and Myanmar, Andaman-Nicobar in India and Indonesian islands.

The Alpine-Himalayan Range has a very long, of the order of nearly 200 Million annum (Ma), and complex tectonic and evolutionary history. Its different segments evolved in different periods of time and are intimately linked with right from the time of break up of Pangaea to present. The geological history is much older for example in case of Himalaya this dates back to Archaean-Proterozoic boundary or even to Late Archaean. Some parts of this chain have been assembled and fragmented several times in the past.

1.2 Geographical Frame of the Himalaya

The classical geographical and geological texts divide Indian sub-continent into three parts the southern part is called peninsular India, middle flat part is called Indo-Gangetic Plain and the northern one is called extra-peninsular India or Himalaya (Fig. 1.1). These three divisions show marked contrast in their geological (stratigraphical, structural, petrological, geochronological and tectonic characters) and geographical (geomorphic and topographic) settings (Krishnan, 1982). The major part of the peninsular India is comprised of the polyphase deformed and metamorphosed rocks of the Archaean and early Proterozoic age and younger sedimentary rocks mainly of Cuddapah, Vindhyan and Gondwana supergroups and smaller groups/ formations and vast Deccan Volcanic Province- the largest volcanic province in the world. The equally vast, largest molassic sedimentary province of the world and relatively flat surface Indo – Gangetic Plain is of Quaternary origin and to the north of this stands out latest orogenic product the Himalaya.

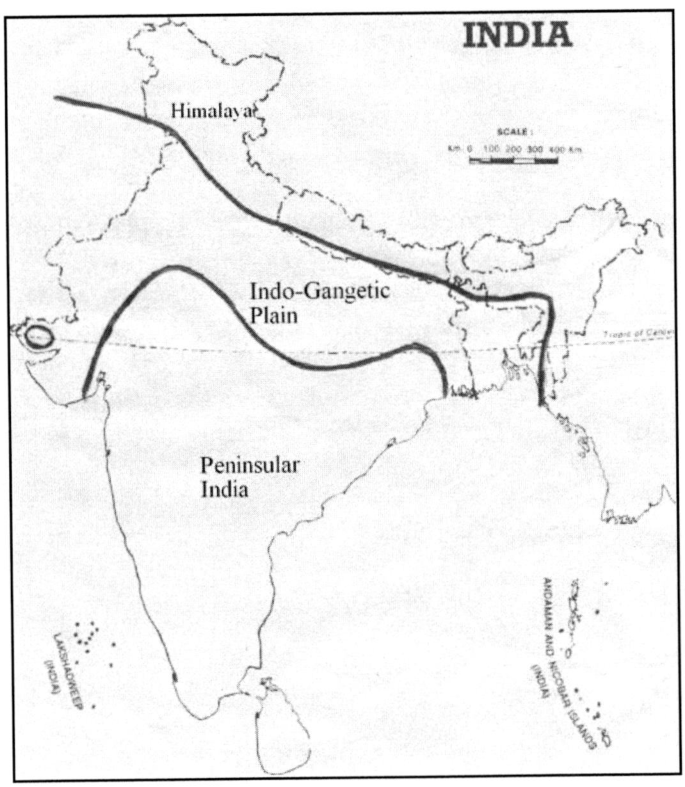

Figure 1.1: Tripartite divisions of Indian subcontinent

Geographically, the mountain divides the largest continent Asia into two unequal parts called North Asia and South Asia. The term Himalaya is geologically restricted to mountainous terrain situated between the river Brahmaputra in the east and river Indus or Sindhu in the west. Its southern and northern frontiers are also geomorphologically driven i.e. plain of Indo-Gangetic Alluvium and Tibetan Plateau respectively. The NW-SE trending Himalayan terrain is nearly 2500 km long (96°E to 70°E longitude) 240-320 km wide (29°55' to 33°20' in E to 33°-30° in W latitude). Himalaya is a tectonic mountain and to draw its boundary on the geomorphological basis is not worthwhile as tectonic elements associated with Himalayan orogeny also present across the above mentioned geomorphic realms. The mountain range occupies nearly 0.7 million (M) square kilometer or 7.0 lakh (one lakh equals to one tenth of a million) land area. The *sensu stricto* mountain range runs through countries like India, Bhutan, Nepal, Pakistan and Tibet (China).

1.3 Divisions of the Himalaya

The vast Himalayan territory for the sake of simplification is classified into divisions and sub-divisions. This exercise is attempted through geographical or latitudinal and geological or longitudinal basis.

1.3.1 Latitudinal Divisions

Burrard and Hayden (1907) divided the Himalaya latitudinally into following four divisions:

1. Assam Himalaya

The term Assam Himalaya defines a 720 km long region confined between the river Brahmaputra in the east and river Tista in the west. The name is adopted from the erstwhile eastern state of Assam of India. However, the present day boundaries of the Assam state slightly touches the Himalaya because large number of smaller states have been carved out from this once sprawling state after independence. Tsangpo River of Tibet flowing in west to east direction takes a sharp swing around the mountain Namcha Barwa in Tibet and flows in the southwest direction in Indian Territory; the river on entering into India is rechristined as the river Brahmaputra. The Assam Himalaya includes Arunachal Pradesh Himalaya, Sikkim Himalaya Darjeeling Himalaya of West Bengal state and Bhutan Himalaya. Several smaller regions on the basis of sharp geographical identity like Afa Hills, Daphla Hills, Miri Hills, and Abor Hills etc. have been recognized within the Assam Himalaya.

2. Nepal Himalaya

This 800 km long region situated between the river Tista in the east and the river Kali (Mahakali) in the west, covering the entire geographical area of the Nepal. Following the administrative set up of Nepal, the Himalaya in Nepal is divided into Eastern Nepal Himalaya, Central-, Western-, Midwestern- and Farwestern Nepal Himalayas.

3. Kumaun Himalaya

Area confined between the rivers Kali and Satluj is designated after Kumaun division of Uttaranchal state of India as Kumaun Himalaya. Three major regions namely the Kumaun and Garhwal of Uttaranchal and northern fringes of Uttar Pradesh and a part of the state of Himachal Pradesh fall in this division. This division is 320 km long and is geologically best-explored one and the longitudinal divisions show their optimal development in this sector.

4. *Punjab Himalaya*

It is 560 km long region bounded by the rivers Satluj and the Indus, in the west. Though Himalayan rocks and structure continues beyond the Indus also. Therefore, small territory situated across the Indus River, in west, is also included. The name was adopted from the Punjab state of before independence of India in 1947. However, the present day Punjab province of India and Pakistan have small Himalayan fringe in their political boundaries. The smaller regions in this sector known as Himachal Himalaya, Jammu Himalaya, Kashmir Himalaya, Ladakh Himalaya, Karakoram Himalaya, Pakistan Himalaya, Salt Range, Potwar Plateau etc.

Himalaya takes sharp turns on its eastern and western extremities toward south and south-southwest respectively. These sharp bends are called Eastern and Western syntaxes or bends.

Using the nearly central position of Nepal, Himalaya is also divided into three major regions namely the Western Himalaya i.e. Kumaun and Punjab Himalaya, Central Himalaya i.e. Nepal Himalaya and the Eastern Himalaya.

Himalaya has developed across several countries and in each country Himalaya passes through several large states which are further divided into number of regional units for examples in Pakistan Himalaya is present in Azad Kashmir or Pakistan Occupied Kashmir, Kohistan, Potwar Plateau; Kashmir, Jammu, Ladakh and Karakoram in Jammu and Kashmir; Garhwal and Kumaun in Uttaranchal, in Farwestern, Midwestern, Western, Central and Eastern Himalaya in Nepal; Darjeeling Himalaya in West Bengal.

1.3.2 Geological / Tectonic or Longitudinal Classification

This classification is widely followed as it is based on several distinctive geological characters such as structure, tectonics, stratigraphy, petrology, geomorphology and evolutionary history of the Himalaya. The general strike direction of the Himalaya is northwest-southeast in the west and east northeast-west southwest. Several linear belts of contrasting geological characters are developed roughly parallel to the general strike direction. These divisions or belts (Plate 1.1) frequently pinches and swells but in over all setup they show rough parallelism with respect to each other and are separated through major tectonic dislocations or faults. Different workers have identified three to eight divisions in the Himalaya. The commonly accepted classification is as follows:

Trans Himalaya

——————Main Mantle Thrust or Karakoram Thrust——————

Indus- Tsangpo and Shyok Suture Zones

—————— Indus Tsangpo Thrust ——————

Tethyan Himalaya

—————— Tethyan Thrust/Unconformity/Gradational Contact ——————

Higher Himalaya

—————— Main Central Thrust ——————-

Lesser Himalaya
——— Main Boundary Thrust ———
Outer Himalaya
——— Main Frontal Thrust ———
Indo- Gangetic Plain

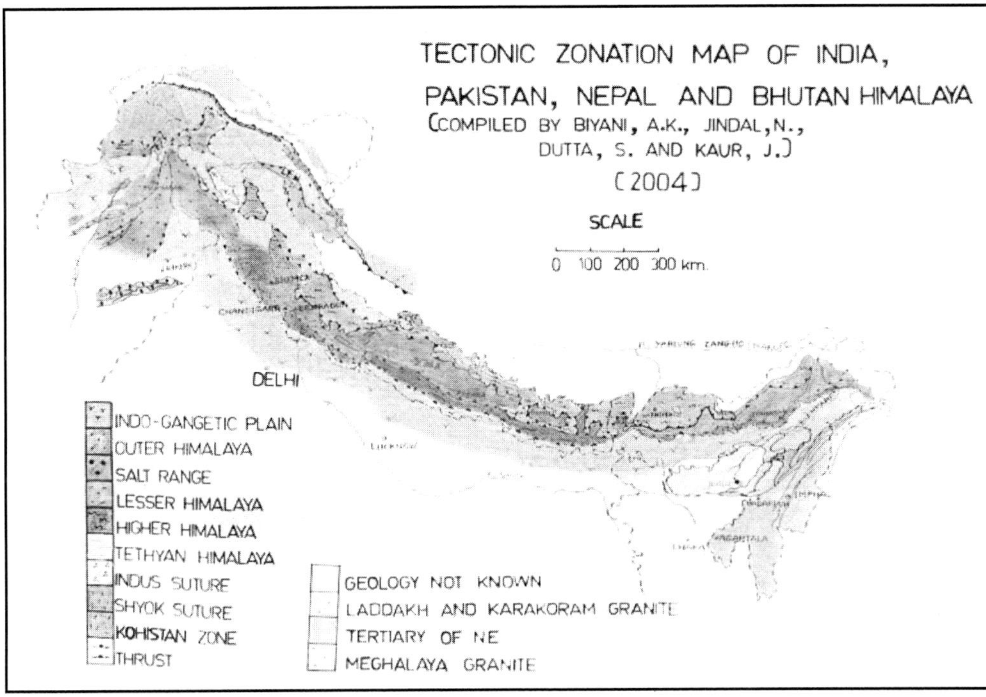

Plate 1.1. Tectonic zones or belts of Himalaya

Relatively small region of northern Pakistan and Ladakh of India across the river Indus in the west that belongs to southern part of the Eurasian plate exhibits Himalayan characters and is designated as Trans Himalaya and remaining all divisions are the part of Indian plate.

These longitudinal divisions of Himalaya are well developed in the east of Kashmir but in Kashmir and also in Pakistan some divisions disappear, or thinly developed and even in the western Kashmir alluvium of Indo-Gangetic Plain has come in contact with Tethyan Himalayan sequence.

In practice, geological workers for clear identity of a region combine together the longitudinal and latitudinal divisions. For examples, Outer Arunachal Himalaya, or Higher Kumaun Himalaya or Himachal Tethys Himalaya. Some terms also used to describe location with specific and non-specific sense e.g. Garhwal Himalaya covers entire Garhwal Division of Uttaranchal and Shimla Himalaya means region around the Shimla in Himachal Pradesh.

The Himalaya can be best described on the longitudinal basis as good geological similarity exist between one end to other end of a division and is adopted for subsequent structural, tectonic and stratigraphical descriptions of the Himalaya.

A brief account of important geological characters of these divisions is as follows:

1. Indo-Gangetic Plain

It is nearly 2700 km long and approximately 200 km wide in the east but gradually widens to 450 km in the west and covers nears of 8.0 lakh sq km area in Bangladesh, India, Nepal and Pakistan. It is a vast, featureless, highly fertile plain formed by deposition of Quaternary alluvium or molassic sediments brought mainly by the Himalayan Rivers from the north.

2. Outer Himalaya

This 10-50 km wide division is also known as Sub- or Siwalik Himalaya. It is formed due to upliftment of a part of foreland basin filled up with Neogene Quaternary molassic sediments brought by the Himalayan Rivers. The Himalayan Frontal Thrust and the Main Boundary Thrust delineate the southern boundary of Outer Himalaya with the Indo-Gangetic Plain and northern boundary with the Lesser Himalaya respectively.

3. Lesser Himalaya

It is also known as Lower Himalaya and is 50-80 km wide. It is mainly made up of poorly fossiliferous, thoroughly deformed rocks of Middle Proterozoic to Cambrian age. These sedimentary and low-grade metamorphic rocks with fair presence of intrusive and extrusive igneous rocks are of autochthonous, para-autochthonous and of allochthonous nature. The Lesser Himalaya also has localized presence of Paleozoic (Permian) and Palaeogene rocks. The Main Boundary Fault in the south and Main Central Thrust (MCT) in the north limit this division.

4. Higher Himalaya

It is also known as Central (Greisbach, 1891), Great or Inner Himalaya. This remobilized cratonic mass is of more than 20 km in thickness and its maximum width in Western Himalaya is 52 km. It is mainly made up of low to medium grade metamorphosed, green schist to upper amphibolite facies, sedimentary and igneous rocks of Proterozoic age. Higher Himalayan domain is confined between MCT in south and the Tethyan Thrust in the north. Three metamorphosed thrusts sheets namely the Chail (low grade), Jutogh (medium grade) and Vaikrita (high grade) are identified in it. These rocks show inverted disposition of metamorphic isograds and giving distinction to the Higher Himalaya of world's largest terrain of inverted nature.

5. Tethyan Himalaya

This segment is also referred as Tibetan Himalaya and is named after once existing, now disappeared, intercontinental Tethyan Sea. It is mainly made of nearly 10-12 km thick fossiliferous sedimentary rocks of Late Proterozoic to Late Mesozoic/ Eocene age. Due to tectonic overhauling, this supposed to be continuous belt is dismembered into separate blocks i.e. of Kashmir, Lahaul-Spiti in Himachal Pradesh, Kumaun, Nepal, Bhutan and Sikkim and is also present in Tibet.

6. Indus Tsangpo Suture Zone

This division joins the Indian and Eurasian Plates. Indus-Tsangpo Thrust marks its contact with the Tethyan Himalaya. Ophiolites, high-pressure metamorphic rocks, molassic sediments and intrusive and extrusive igneous rocks constitute the suture zone. It is mainly developed in Tibet, however, in India and Pakistan it is present in the Ladakh and Karakoram regions where it is branches out into two sutures zones namely the southern Indus Suture and the northern Shyok Suture.

7. Trans Himalaya

This zone is a part of southern Eurasian Plate and in India and Pakistan it is mainly developed in northern Ladakh (an administrative division of state of Jammu &Kashmir) and northern most Pakistan. It comprises of Kohistan area, Shyok Suture of Ladakh and Karakoram Zones of Western Himalaya; workers like Thakur (1992) included the suture zones in the Trans Himalaya. This zone also continues in the Tibet where ophiolites, flysch and molasse are grouped into Kailash Range, Yarlung-Tsangpo Belt, further in the east it is represented by Lhasa Block. The Indus and Shyok Sutures run parallel to each other in northern Kashmir, pointing to complex evolutionary processes operated in the western Himalaya. Ophiolites in the suture zones represent oceanic floor of Mesozoic age. Karakoram Zone is situated in further north of Shyok Suture. The contact between Karakoram Zone and Shyok Suture is designated as Main Karakoram Thrust (MKT). The Karakoram Zone has basement of metamorphic rocks over which 5 km thick sedimentary rocks of Paleozoic and Mesozoic occur with granitic intrusives.

Pamir Knot

Another interesting feature with Himalaya is that it is among one of the several mountains chain radiating out from the Pamir Knot, located little south of the center of Asia. The Himalaya is situated to the east of the Pamir Knot, and to the north of the Himalaya, Karakoram Range occurs. In the further north, the Kunlun mountain range in China making border of Tibetan Plateau originates from the Pamir Knot. The Altyn Mountains extend northeastwards from the Kunlun; the Tien Shan is the northern most branches of the Pamir Knot and the Altai Mountains of the further north. To the west of the Pamir Knot the Sulaiman range occurs in the south-west direction which continues further west as the Zagros Mountains along the coast of Iran and the Hindukush runs in the westward direction, the Hindukush is called Elburz Mountains in the further west along the border of the Caspian Sea.

Another knot also exists to northwest of the Pamir Knot. It is known as Armenian Knot and is located near the Black Sea. Elburz and Zagros Mountains are part of this knot.

1.4 Himalayan Putras

Himalayas has fascinated the men since time immemorial and has been mentioned in the oldest Hindu scriptures like Vedas, Smritis', Upanishads, Ramayana, Mahabharat and literary work of ancient writers like Kalidas and other s. Incidentally, the Mahabharat provides the first classification of Himalaya into the Outer Lesser, Higher and Trans Himalaya perfectly corresponding with the geographical/ geological classifications.

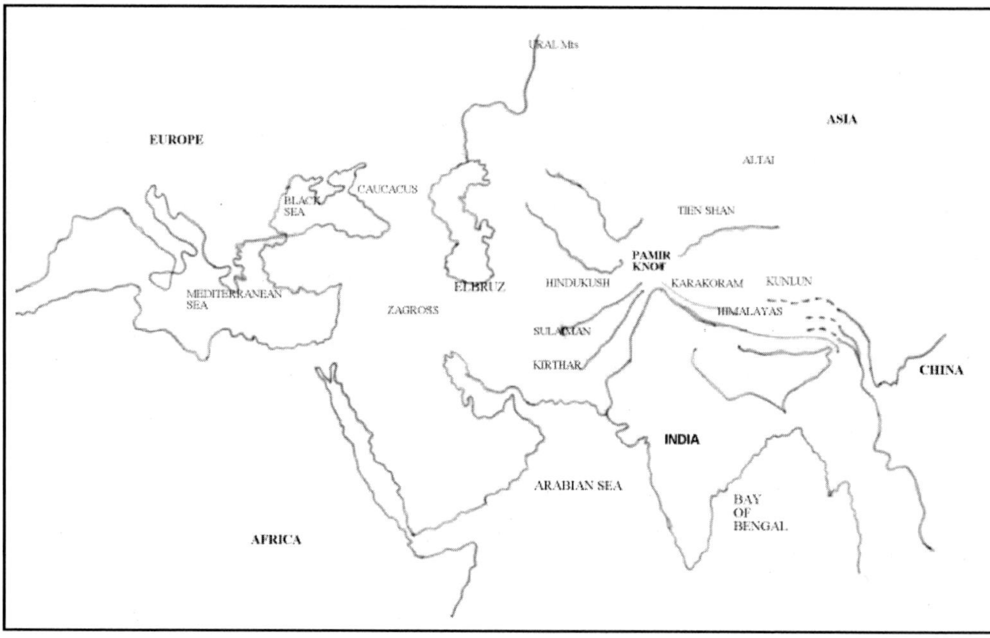

Figure 1.2. Pamir Knot

The scientific investigation of the Himalaya started comparatively late. The initial scientific investigations of the 17th and 18th century had limited objectives i.e. to acquire topographical and geographical details for trade purpose of the East India Company. Sorkhabi (2005) has ably summarized the history of geological and geographical investigations in the Himalaya. This chronological account of geographical and geological survey in Himalaya is mainly compiled from Sorkhabi's article.

Earliest geographical information of various regions of the Himalaya are provided by Keay (1781), Goft (1819, 1825), George Trebeck (1819-1825), Henry Starchy (1846) and Robert Starchy (1848), Thomas Thompson (1847-48), Andrew Fleming (1848-52), Colebrook (1867), B. Hodgson (1849), Leadflow (1817-21), G. Everest (1848-50), Voysey (1818-23) and Dangerfield (1820-23). Many among these workers were associated with the Survey of India.

P. Cautley (1830s) and H. Falconer initiated geological survey in Himalaya. They described fossils of Siwalik range. J.D. Hebberg (1842) provided on account of area between Kali and Satluj rivers and divided the rocks into primary and secondary formations. This was followed by Golden Age (1850-1890) of the Himalayan Geology, notable contributions in this period is made by Stuffwood, Delabeche, D. H. William, J. McClelland, R.D. Oldham, John, G., Henry, B., W. T. Medlicott, H.B. Medlicott, W. T. Blanford, O. Feistmental and R. Lydikker. Work of these pioneers is followed in detailed area specific investigations. C. L. Greisbach (1891) described the Central Himalaya, Middlemiss (1910) carried out detailed survey of Kangra earthquake. R.D. Oldham (1899) first time observed P and S waves, S. G. Burrard and H. H. Hayden (1907) described Himalayan physiography, G. E. Pilgrim and W.

D. West (1928) described geology of the Shimla area, Wadia (1931) provided an account of Kashmir Himalaya, J. B. Auden (1934-37) carried out investigations in Garhwal Kumaun region, A Heim and A. Gansser (1939) wrote a classical book on Central Himalaya, G. Dainelli (1922, 59), A. Desio (1930-60), P. Bordet (1960), D. Krummenacher (1956), A. Lombard, P. Misch, H.J. Schneider (1956) H. H. Hayden and T. Hagen described the geology of Nepal. Contributions of J. Coggin Brown, C. Diner, E.R. Gee, W. D. Gill, N.E. Odell and H. DeTerra are also highly praised worthy.

In the post - 1960 period boost to geological investigation in Himalaya is given by the Geological Survey of India, exploration activities of Oil and Natural Gas Corporation Ltd., Wadia Institute of Himalayan Geology and by the various academic and research institutions of India and abroad. Combined efforts of these organizations yielded enormous volume of data and better appreciation of geology. Several books, monographs, memoirs and records have been written in this period and several national and international journals have brought out specially focused issues on Himalaya and many issues of journals are exclusively devoted to Himalayan geology and few journals exclusively deal with the Himalayan Geology.

Among important books written on Himalaya geology are: Central Himalaya by A. Heim and A. Gansser (1939), Geology of Himalayas by A. Gansser (1964), Geology of Kumaun Lesser Himalaya by K. S. Valdiya (1980), Geology of Bhutan Himalaya by A. Gansser (1983), Geology of Higher Central Himalaya by A. K. Sinha (1989), Geology of Western Himalaya by V. C. Thakur (1992), Geology of Lower (Garhwal) Himalaya by P.S. Saklani (1993). Geology of Nepal Himalaya and Adjacent Countries by C. K. Sharma (1990) Tectonic and Metamorphic Investigations of Kumaun-Garhwal-Himachal Lesser Himalaya by I. C. Pande (1991), Geology of Arunachal Pradesh by G. Kumar (1997), Dynamic Himalaya by K. S. Valdiya (1998), Geology of Himachal Pradesh by S. V. Srikantia and O.N. Bhargava, Foreland sedimentation in Himalayan tectonic regime: a relook at orogenic process by V. Raiverman (2002), Geology of Uttar Pradesh and Uttaranchal by G. Kumar (2005) and research contributions in many volumes are edited by P.S. Saklani since 1978 and also by other workers engaged in Himalaya. The Geological Society of India has started a textbook program on geology of Indian states and in this series three textbooks pertain to geology of Uttar Pradesh and Uttaranchal, Arunachal Pradesh and Himachal Pradesh states have been published so far.

The Government of India duly recognizes dedication of many of these workers by conferring various honors on them. The Royal Society of England conferred honor of fellowship to Prof. D. N. Wadia for his outstanding geological investigations in Himalaya. However, positive contributions of thousands of workers have also witnessed questionable research by a few, for example, fossils reporting by V. J. Gupta.

DIMENSIONS OF HIMALAYAN GEOLOGY
A. K. Biyani (2006) : pp.11-17
Satish Serial Publishing House, New Delhi
E-mail: ssp@ndf.vsnl.net.in, hkjain1975@yahoo.com; website: www.satishserial.com

Annadbhavanti bhutani parjanyadannasambhavah
Yajnadbhavati parjanyo yajnah karmasamudbhavah (Geeta 3.10)
All beings ar evolved from food and production of food is dependent upon the rain;
rain ensues from scrifice and sacrifice is rooted in action.

CHAPTER - 2

INDO-GANGETIC PLAIN

The Indo-Gangetic Plain (IGP) is a vast, featureless, monotonous fertile plain of Quaternary origin having nearly 2700 km length from east to west (74°-88° E) and width of 200 km in east to 450 km in west (28°-32° N in West and 25°-27° in east), occupying nearly 0.8 M sq km area. This plain is largest molassic region in the world and is formed by alluvial sediments deposited by the Indus, Ganga and Brahmaputra rivers and their tributaries in a foreland basin, a product of Himalayan evolution. The Plain makes an unconformable contact in south with the sedimentary, metamorphic and igneous rocks of varying ages of the peninsular India and tectonic contact in the north along intermittently exposed thrust named Himalayan Frontal Fault (HFF); with the Outer Himalayan Siwalik rocks of the Neogene age. These rocks are also thrust over, in some areas, over the Plain. At many places in the western region, the contact is normal, unfaulted, associated with thrust related folds, imbricate thrusts and with the strike-slip faults (Raiverman et al., 1993).

The 100-1000 m thick alluvium comprises of gravel, sand and clay with occasional remains of animals and plants. The alluvium is divided into older and younger. The older alluvium or bhangar of the Middle to Upper Pleistocene in Ganga valley is dark gray in color and contains small to large size concretions of calcium carbonate, which are locally known as *kankar*. The older alluvium contains fossils of Rhinoceros, Hippopotamus, Elephas, and Eqqus etc. The newer alluvium also called *khaddar* is light colored and has much less presence of calcareous concretions. It is of Upper Pleistocene to Recent in age (Krishnan, 1982).

The IGP geomorphologically comprised of fan or cone and interfan areas. Fans (cones) are generally triangular shaped in the plan having apex at the mouth of the gorge, convex in form and having steep gradient (20 cm/km) and the interfan areas are opposite to fan with base parallel to mountain front and apex in the plain, concave edges and have gradient <10 cm/km. The term megafan in humid region is used for those fan having 100-200 km width and 100-150 km length. Area adjacent to the length of the mountain is called piedmont zone. Interfan areas considered as inactive regions with subdued fluvial activities. In Ganga plains, Yamuna, Ganga, Sarda, Gandak and Kosi megafans are identified and are separated by interfans or inter cones. Coalescing fans form the piedmont zones and these are 10-50 km wide zones. The gentle slope, low drainage density, gently undulating topography, characterizes the piedmont surface. Piedmont fan is younger than megafan (Shukla and Bora, 2003). Detached piedmont zones developed adjacent to mountain front in the south has been identified. These piedmont surfaces are formed by discrete coalescing fans and are 10-50 km wide. These surfaces show gentle 7.8 m /km slope in the bhabar region that decreases in the terai region whose distal part is sandy and swampy. Gandak megafan is 350 km long and 60-150 km wide and has 32000 sq km area. Its central part is ~10 m higher than the margin. Its apex at Tribeni is at 120 m height and the river Gandak makes a valley of 2-5 km wide just beyond the gorge. The fan is subdivided into proximal, middle and distal fans.

2.1 Surface Divisions of the Indo-Gangetic Plain

The IGP from north to south is divided into:

1. *Bhabar Belt* - It is nearly 30 km wide belt of steep slopes having ephemeral rivers of graveliferous sediments.
2. *Terai Belt* - It is 10-50 km wide belt once consisting of swamps and ponds and small sandy rivers. The swampy area has been modified now into fertile agricultural tract. Bhabar and Terai belts are also known as piedmont zone.
3. *Central Alluvial Plain* - This occupies the major part of the IGP and rivers mostly flow in SE direction.

   Sediments of the Himalayan Rivers form these three regions.
4. *Marginal Alluvial Plain* - This is north sloping area and had received sediments from meandering peninsular rivers flowing in NE direction.

IGP on the basis of surface characters also divided into three transverse areas namely Indus Plain in the west, Ganga Plain in central region and Brahmaputra Plain in the east.

2.2 Subsurface Divisions

On the basis of sub-surface characters the IGP is divided into basins (depression where sediment was deposited) and sub-basins. The important basins are:

1. Indus Basin

This basin mainly lies in the Pakistan and partly in the Rajasthan. Tertiary and Mesozoic rocks are present below alluvium cover. Thickness of sediments decreases in Rajasthan and also further in north towards subsurface Delhi-Lahore-Sargodha Ridge (DLSR).

2. Punjab Basin

The geographical extent of this basin is confined between the DLSR and Kalsi Spur (small ridge) near Dehradun. The basin configuration changes from NW-SE to ESE to EW. Four depressions or sub-basins namely the Naushera, Ram Nagar, Kangra and Subathu separated by the Chenab-Ravi, Ambala, Nalagarh and Kalsi spurs respectively are reported in the Punjab basin.

3. Ganga Basin

It is situated between Kalsi Spur in the west and Monghyar Saharsa Ridge (MRS) in the east covering Uttaranchal, Uttar Pradesh and Bihar states of India. The basin contains the Dehradun-Kalsi-Hardwar-Rishikesh, Pellagra and Kansan-Tanakpur spurs and is separated by Ramganga, Sarda and Gandak depressions respectively.

4. Ganga-Brahmaputra Basin

Ganga-Brahmaputra Basin is situated in the Bengal and Bangladesh. This basin shows altogether different geological and morphological setting, particularly characterized by passive marginal nature. It may be the remnant of Bengal bay, which shows oceanic characters.

5. Brahmaputra Basin

It mainly covers Assam state and is divided into the western and eastern parts. The western part is lying between Shillong and Mikir Hills and is known as northern Shillong Shelf. The eastern part is situated between Naga Hills and Himalaya and is known as Upper Assam Shelf.

2.3 Basement and Sub-surface Characteristics

The sub-surface information of the IGP is obtained through gravity and seismic techniques and also by deep drillings for petroleum exploration. These studies suggest the Precambrian metamorphic rocks of peninsular India constitute basement for IGP. The depth of basement is variable but it gradually increases towards north. In the central part depth of the Proterozoic basement varies between 4000 to 6000 m and is deepest at Kathgodam and Raxaul where this basement is recorded at 7.8 km depth. Metamorphosed carbonate basement is recorded at 4790 m depth at Janauri in Himachal Pradesh. In southern Nepal total thickness of Tertiary and Pre-tertiary sediment is estimated around 11 km. The maximum thickness of Pleistocene and Recent sediments in IGP 1000 m and rest thickness is constituted by pre-Quaternary to Vindhyan sedimentary rocks including Siwaliks, Subathu, Deccan Trap, Gondwana and Mesozoic/Tertiaries of Rajasthan and others. In western Uttar Pradesh shelf Vindhyan is recorded at 1720 m depth. Fig. No. 2.1 shows subsurface cross section of Mathra- Puranpur area.

Seismic study reveals that average thickness of crust is 40 km in the IGP region and has on top a 3.4 km thick layer of sedimentary rocks of low velocity. The crust is followed by mantle of low velocity (4.5 km/s). In 10-20 km depth S-wave velocity 3.80 km/sec suggesting thickening of lower basaltic layer.

The uneven metamorphic basement is made up a two types of tectonic elements one is the ridges or culminations or positive features and the second is basins or depressions or negative features. In the former category arches, domes, anticlines, upward swells and broad noses are present and in the later category downward sags, depressions, synclines and troughs are main elements. The major sub surface tectonic elements are (Verma, 1991)

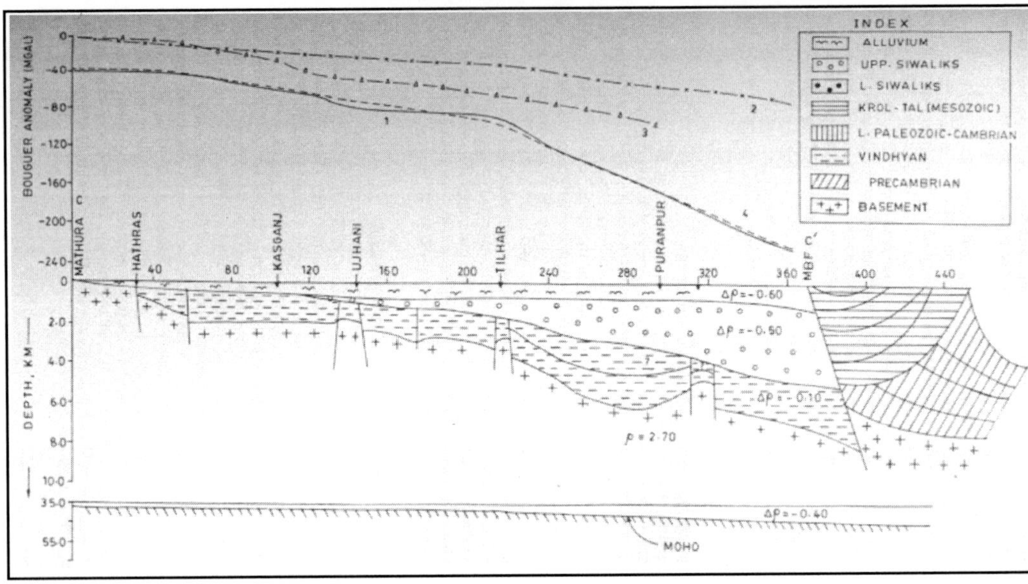

Figure 2.1. Subsurface cross section of the Indo Gangetic Plain in the Mathura- Puranpur area, abbreviations: (1) observed Bouguer anomaly, (2) gravity effect of sediments, (3) Airy anomaly, (4) computed gravity anomaly (after Verma, 1991).

1. Delhi-Lahore-Sargodha Ridge

This NW-SE trending ridge merging with Jhelum syntaxial bends. Chenab and Ravi are two important major spurs and several minor digitations are fan out from the Ridge.

2. Aravalli Horst

It is also known as Delhi-Saharanpur Ridge. It is NE-SW trending ridge; extending between the rivers Satluj to Sarda. Ambala-Nalagarh, Kalsi, Hardwar-Rishikesh, Pawalgarh and Kasganj-Tanakpur are important spurs present in the ridge.

3. Faizabad Ridge

It is northeastward extension of Bundelkhand massif. In the north, Faizabad Ridge develops two digitations one along the Ghagra River trending in N-S direction and other one is trending in NE-SW direction towards Faizabad.

4. Monghyar-Saharsa Ridge

This NW-SE trending ridge is extension of Chhota Nagpur plateau of peninsular India.

5. Dudhwa Ridge

It is an isolated ridge present near western boundary of Nepal and has E-W trend.

6. Rajmahal-Garo Gap

This is a shallow fault bounded ridge present between Rajmahal Hills and Garo Hills. The ridge is plunging in N or NE direction.

Many subsurface structural highs (similar to anticlines and are inferred by gravity survey) are also observed and are named after town or city under which they are present e.g. Mirjapur-Ghazipur High, Raxaul High, Bahraich High, Paranpur High etc.

Among important lows or depressions (similar to synclines and are inferred by gravity survey) are Ramganga depression, Sarda depression, Gandak depression (having rocks of Satpura orogeny), and Purnea half graben.

Large number of wrench and extensional faults are present in the IGP, many of them without surface expressions. Wrench faults are characterized by symmetrical or asymmetrical and half or full positive or negative flower structures. Normal Faults are present in pre-Tertiary sequences. The important subsurface faults present in the IGP are Delhi- Dehradun Fault, Moradabad Fault, Madhupur-Dhubri Fault, Lucknow Fault, Bardi Fault, Patna Fault and Malda Fault.

These faults, high and basins were active during Late Proterozoic but have not affected Siwalik and younger sediments. Some of these faults show occasional activities at present e.g. Moradabad fault.

2.4 Sedimentation in Megafans

Sands dominate fans. Megafans or cones received their sediments from braided rivers and are having gravel at the apex and clay and silt at the foot. In Kosi and Ganga megafans four zones are identified. In Kosi megafan, Zone I comprised of gravelly- sandy sediments with braided channel pattern; Zone II is predominantly sandy with braided drainage developed in 95 km length; Zone III comprised of fine sand and mud 40 km long course of river is straight and meandering Zone IV is 160 km long and characterized by fine sand and mud. In the Ganga megafan, the zones are braided, sandy braided plain, anastomosing channel plain and meandering channel zone. In the 65 km long Gandak megafan clay lenses are embedded in sandy matrix. The top 2 to 50 m thickness in the interfan areas in different fans are comprised of muddy sequences followed by basal coarse sand horizons. The rate of sediment deposition in flood plains vary between 0.05 –1.5 mm/yr measured over 1000-10000 yrs (Jain and Sinha, 2003).

Sedimentologically, distal western Ganga alluvium is composed of sand (2.1-18.3%), silt (77.7-93.5%) and clay (2.8-9.3%). Smectite (37-88%), illite (7-48%) and chlorite and kaolinite (4-19%) are the chief clay minerals. During 40-30 ka period, the rate of sediment deposition was ~ 1cm/yr (Singh et al., 2003).

2.5 Origin and Evolution of Indo-gangetic Plain

Indo-Gangetic Plain now flat area was once upon a time giant depression. Krishnan (1982) described this depression as a sag in the crust formed due to the upliftment of Himalaya.

Suess (1893) proposed and now supported by many that IGP was a foreland basin while Burrard described it as a great rift valley and Molnar et al. (1977) considered the foreland basin as deep sea trench associated with island arc structure. The IGP basin is a part of large foreland basin, which evolved in three stages. This foreland basin is formed in late stage of continents collision. Collision is resulted into evolution of Himalaya. Flexuring of such type caused by under thrusting of lithospheric plate is called peripheral foreland basin by Dickinson (1974). The foreland basin is outcome of compressional stress acted on the underthrusting plate and by the vertical load of thrust-fold belt by elastic or visco elastic behavior of deformed crust involved in the Himalayan orogenic movement. In the first stage of the foreland basin, Subathu, Dagshai, Dharamsala, Kasauli and Murree rocks were deposited. Due to deformation in Middle Miocene another foreland basin was further extended in the south of these formations in which Siwaliks were deposited. Upliftment of Siwalik in Early to Middle Pleistocene extended further in south of this foreland basin in which sediments of IGP were deposited. The IGP shows all evidences of foreland basin i.e. rising orogen in front, deformed and uplifted basin adjacent to orogen, a depositional basin and peripheral cratonic bulge. Subsidence occurred in south near the site of thrusting and further over loading and also upward deflection in further south. This gravity high or bulge is called peripheral or fore bulge. Flexural rigidity or mechanical strength of lithosphere controls the width of foreland basin. An interesting situation is developed in the IGP that Ganga basin representing topographic low between Himalaya (thrust belt) and Bundelkhand Plateau (peripheral bulge) is an under filled basin while Haryana Punjab Plains represents to an overfilled basin as the fore bulge is buried under sedimentary sequence and have no topographic history. Basin for Indo–Gangetic Plain is poly historic, pericratonic, foreland basin of early Proterozoic origin in extensional regime and has multi episodic evolutionary history.

Detailed geological study of Ganga basin revealed that basin started developing as very narrow and in displaced form in Early Miocene as Subathu rocks of Early Eocene age is not recorded in foothill bore holes data. The source of sediment was mainly Himalaya as it was rising and also being eroded away by energetic rivers (Pant and Sharma, 1993). In Middle Miocene substantial subsidence occurred due to flexuring of Indian plate and full-grown foreland basin developed. The supply of sediment even at this time was in excess, as sediments find way into Bay of Bengal where it was deposited as Bengal Fan. This was without massive transgression. During Middle Miocene to Middle Pleistocene, Siwalik time, the orogen side of Ganga plain uplifted and thrusted in the south. The last major thrusting took place along HFF about 0.5 Million Annum (Ma) ago causing upliftment of the Upper Siwalik. During the deposition of Siwalik sediment, Ganga basin migrated in the south direction due to development of fold thrust belt causing loading in the Himalaya. A southward of expansion of nearly 100 km since Middle Pleistocene of the Ganga Plain has been observed between Kanpur and Hamirpur area. In Holocene time, four loess deposits are recorded from Son-Ballen river basin. The oldest loess deposit is dated to be of 26-18 kilo (=1000) annum (Ka) age while top three are only 3-5 Ka old, they are separated by gravel beds. According to Pant and Sharma (1993) in the Ganga Plain initial sedimentation took place in shallow marine and depositional conditions gradually changed into estuarine deltaic and finally fluviatile prominent mega fans are observed adjacent to HFF.

The rivers in IGP are notorious for their changing course e.g. Kosi river between 1738-1964 has shifted 167 km in west direction and shift in Ganga and Yamuna river is estimated between 10-40 km in the east since Mahabharat (3000 year) time. The palaeocurrent pattern indicates filling of largest parabolic prism might be transverse in the beginning. This is also supported by basement topography characterized by transverse narrow depressions. This foreland basin is still inferred to be deepening at Zira in Punjab as a crystalline (granitic) basement rock is recorded at 600 m depth in a borehole. This granitic rock probably represents a blind thrust, which may emerge on the surface in future.

DIMENSIONS OF HIMALAYAN GEOLOGY
A. K. Biyani (2006) : pp.19-38
Satish Serial Publishing House, New Delhi
E-mail: ssp@ndf.vsnl.net.in, hkjain1975@yahoo.com; website: www.satishserial.com

Jnanam jneyam parijnata trividha karma chodana (Geeta,18.18)
The doer, the action and the organ of action- these are
three fold constituents of action.

CHAPTER - 3

GEOMORPHOLOGY OF HIMALAYA

Himalaya is known for its youthful geomorphic features because it is still in growing phase. The rapid upliftment of the terrain in Quaternary has adorned this orogenic belt with wide variety of landforms and some of them are justifiably described in superlative terms.

3.1 Mountains

The Himalayan region is comprised of the large number of alternate hills and valleys of different lengths and reliefs. Himalaya is famed for its towering hills and mountains. In Himalayan region the words 'hills' (low height) and 'mountains' (greater heights) are used interchangeably without any specific sense. Hills abruptly rise from the plains present in the south. Himalayan Mountain prevent the flow of northern cold winds to the south, influences the climate are repository of Indian culture.

The general characters of the mountains in the Himalaya are:

Heights in general increase in towards the north. Siwaliks comprise of hills of 1000 m height from the mean sea level. The Lesser Himalayan hills are 3500 m high while the Higher Himalayan Mountains are 3000-8000 m high; Tethyan Himalayan Mountains are 4000-8848 m high. Peaks of the inner region are covered with the snow/ glaciers.

The range of relief also varies enormously in the Himalaya. With increase in height relief difference also increases. In Karakoram region, 5900 m relief is present in an 8 km distance from the Hunza Valley (1800 m) to Rakaposhi Peak (7772 m). A relief of 7000 m in a horizontal distance of 21 km is present between top of the Nanga Parbat and floor of the Indus gorge in Pakistan.

Mountain slopes in Himalaya are uniform or straight, convex and concave. Uniform or straight slopes indicate constant intensity of erosion i.e. state of balance between upliftment and denudational processes while convex and concave slopes indicate less upliftment and more effective denudational processes and upliftment is faster than the denudation respectively. Following this interpretation then different regions would have different tectonic upliftment histories in Himalaya. In many regions slope profiles exhibit combinations of above two elements, in such cases areas exhibit break in tectonic movements. Hilltops are pointedly sharp to broad and flats or even may have depressions converted into lakes of varying sizes or are remnant of vanished glaciers.

Mountain slopes are moderate to steep and at places they are vertical. The heights of the cliffs are even a km or more. The fluvial, glacial, mass wasting and tectonic processes modify slopes.

Mountains have acquired main heights in the Pliocene- Pleistocene time and are continuously acquiring heights because of the ongoing subsurface tectonic movements. In last 10 Ma on regional level of the Himalaya is uplifted between 3-6 km while locally it is as high as 10 km. Nanga Parbat has uplifted with a rate of 0.5 to several mm/yr in the last 7 million years, however, the average rate was 2.5-5 mm/yr in last 2 million years. In some sections, the present rate of upliftment is as high as ~7-8 mm/yr and present uplift is at least six times greater than the denudation rate. This suggests that the Himalayan terrain should have been six times higher then present level. The present lower height of the terrain is attributed to discontinuous pulses of rapid upliftment, long period of quiescence, and variable rate of erosion in different regions and slope processes of erosion.

Foot regions of the mountains guide the flow directions of many rivers and the mountain slopes are mass wasting prone.

The general trend of main mountain ranges in Himalaya is NW-SE. This trend is shown by the northern most Karakoram, Ladakh, Kankar, Per Penal, Dholes Dear ranges and in south by the Siwalik Hills along with important ranges of Himachal, Kumaun, Nepal and northeastern Himalaya. While some minor ranges are trending in NE-SW, E-W and N-S directions. These minor ranges show straight, curved branching patterns. In Bhutan N-S ranges are prominent. These mountain ranges show linear shape and their length to breath ratio vary between 4:1 to 30:1. Himalaya is crowned by very large number of peaks of more than 7000 m height. Heights of the importantant peaks (in meters above mean sea level) from west to east are: Isotora North (7398), Tirich Mir (7690) in Pakistan; Kampiri Diwar (7143), Dastegil Sar (7885), Rakaposhi (7788), Kanjutsar (7760), Haramosh (7397), Nanga Parbat (8126), K2 (8611), Gasherbrum (8066), Masherbrum (7821), Saltoro Kangri (7742), Nun (7135), Saser Kangri (7672), Kamet (7750), Badrinath (7138), Dunagiri (7066), Trishul (7140), Nanda Devi (7817) in Sikkim India; Api (7132), Churen Limal (7363), Dhaulagiri (8172), Annapurna (8078), Manaslu (8156), Gauri Shankar (7144), Makalu (8481), Mt. Everest (8848) in Nepal; Gosainath (8013) and Cho Oyu (8153), in Tibet; Kanchanjunga (8598), Jonsang (7065), Chomo Yomo(6828). Pauhunri (7128) in India; Chomohhari (7327), Masang Kang (6710), Kanker Phunsum (7561) and Kula Kangri (7554) in Bhutan, Kangarneto (7090), Gyala Peri (7151) in Arunachal Pradesh in India and Namcha Barwa (7756) in Tibet.In Karakoram Himalaya five peaks are over 8000 m height and thirty six peaks are over over 7000 m height (Biyani 2003).

Mount Everest- The height of this highest peak of the world is matter of intense curiosity and is repeatedly measured. Mount Everest is also known as Sagarmatha in Nepal and Chomolangma or 'Mother of the Universe' in Tibet. Its height (30000 feet) was first determined in 1852 by mathematician Radhanath Sikdar of the Survey of India and named Peak XV. In 1856 height was recalculated at 29000 ft but declared 29002 ft to avoid the impression of rounding off and was renamed Mount Everest by Sir Andrew Waugh after the name of Sir George Everest. The 1950 survey gave the height 8848 m or 29028 ft, in 1998 American Everest Expedition proposed 8850 m height on the basis of GPS measurement and in the October, 2005 China announced that actual height of rocks in Everest is 8844.43 m and rest is snow. These height estimates are based from the mean sea level, however, Mauna Kea peak (4170 m msl) in Hawaii is above 9 km in height when it is measured from base of mid-ocean floor and the top of Chimborazo (6310 m above msl) in Ecuador is 2168 m higher than the Everest if the height is measured from the center of the earth at the equator (6384.4 m) while beneath the Everest, center of the earth is situated at 6382.3 km.

Mount Kailash (6714 m high), locally called Kangrinboq or Fery, being abode of Lord Shiva situated in Tibet is a sacred mountain for Hindus. It is an isolated remnant of volcanic debris placed over 40-60 Ma old batholith. Kailash deposits are 20-40 Ma old and are equivalent to Indus molasse.

The high mountain ranges are frequented with many depressions or passes through which easier movement of man and animals across the ranges are taking place. The important passes are Mintak, Parpik, Aghil, Marpo Karakoram, Dawan, Lanak La, Dimjor La in Kashmir; Shipki La, Bara Lacha La Jara La, Khimokul La, Muling La, Mana, Niti in Himachal Pradesh and Garhwal -Kumaun of Uttaranchal and Namja Bhanjyang, Pindu, Gya, Jelep La, Nathu La Bumla, Diphu, Hipunjan Chaukan, Donkya La (6000 m in Sikkim) in Nepal, Bhutan and in NE India. Important passes of Ladakh are Marsemik (6050 m), Digar La (6000 m), Chang La (5800 m), Khardong La (5800 m), Suru (5700 m), Burzil (5700 m), Lachulung (5500) and Polokonka (5200 m). The highest pass is Kyung zing La located at 6410 m altitude.

Complex tectonic movements are responsible for origin of mountain ranges i.e. by faulting, thrusting, folding, and upwarping of rocks along with general upliftment of area. Mountains of Ladakh are formed by upliftment of batholithic intrusions of huge dimensions and they are less deformed in comparison to other areas. Moderate to vertical slopes are the characteristic of the mountains.

In mountainous terrain, the river forms deep gorges with or without river terraces and in plain area the valley becomes broad and shallow with wide flood plains. The high mountain peaks are covered with glaciers. The glacial erosion at many places plained off the mountaintops and produced 'thatch' or plain areas or high altitude meadows of considerable areas. These 'thatch' are ideally suited for snow sports activities.

Abnormal Ridges

A common feature of orogen is that trends of some ridges are either oblique or transverse to the general trend. These obliquely or transversely trending ridges are generally formed by the complex tectonic movement that is superposed type of folding, by erosion and in some cases by upliftment along fault planes and are described as abnormal ridges. In Himalaya several abnormal ridges have been recognized and notable among these are the Singatila Ridge, making watershed between the drainage basins of Tista and Kosi rivers; Chola Ridge that separates the drainage basins of the Tista and Raidela rivers and the Narkanda-Mahasu Ridge, trending in NE-SE direction and forms the watershed between the Yamuna and Satluj rivers.

3.2 Rivers

Thousands of the rivers of varying length and orders drain Himalayan terrain. Many of these rivers are considered as holy rivers and revered by people. Many major rivers are international rivers as they flow in more than one country and some rivers draw international boundaries between two or three countries. Three river systems namely the Brahmaputra, the Ganga and the Indus or Sindhu rivers drain the Himalaya.

3.2.1 Major River Systems

Brahmaputra

The literary meaning of Brahmaputra is House of Brahma – the creator of the universe and in Tibet the river is called as Tsangpo or Yarlungzanbo that means the purifier. It is one of the mighty river of the world draining nearly 9,24,000 sq km area and has 2880 km length of which 1600 km lies in Tibet and 918 km in India with average width of 80 km in the Assam and rest in Bangladesh. The river rises from Kailash range in Tibet and flows west to is nearly straight course in the southern Tibet. The river flows in the east and takes sudden U turns around Namcha Barwa Mountain in the eastern Tibet and enters into Arunachal Pradesh (India); there it is locally named as Dishing or Sang. In India and Bangladesh the river and its tributaries are arranged in closely-knit dendritic pattern. In Assam, 640 km long river Brahmaputra enters at Kobo and leaves at Dhubri joined by 20 important tributaries on its north bank and 13 on its left bank Its important Tibetan tributaries are the Kul. Tsangpo Chema, Yungdung, Maryum, Chu, Nyang, Rang; Disang, Dihang, Siang, Subarnsiri, Dhanisiri, Torsa and Tista of Arunachal Pradesh and Assam; Raidok, Amochu, Sankash and Manas originating from Bhutan and Yamuna, Meghna, Feni and Surma of Bangladesh. 48.83 million hectare meters (M ha m) annual runoff is estimated for the Brahmaputra River basin having average annual precipitation 2551 milli meters (mm). The mean annual rainfall over sub catchments of the river varies between 2590 mm in Dihang catchment to 1735 mm in Kopili sub- catchment.

Ganga

This 2490 km long Holy River Ganga originates at Gaumukh, situated 19 km upstream of the Gangotri, from the snout of Gangotri glacier at 3900 m height. In between Gaumukh and Devprayag, the river is called Bhagirathi. At Devprayag, Bhagirathi joins with the more watery river Alaknanda and after this confluence the river is known as Ganga. The river Ganga is also called as Padma in Bangladesh. The important tributaries of the Ganga river are Yamuna and Ramganga in Uttaranchal; Kali or Mahakali drawing boundary between India and Nepal and is called Sarda in plain, Seti and Karnali renamed as Ghagra in plain,

Rapti, Kali Gandaki, and Spat Koshi (Sun Kosi, Dudh Kosi, Arun and Tamur) called Kosi in plain and Kankai draining from Nepal and the Chambal, Sind, Betwa, Ken and Tons of the peninsular India. The above mentioned lower order rivers of the Ganga system are very powerful rivers in themselves and are joined with large number of smaller rivers and their tributaries, for example, Alaknanda's tributaries are Dhauliganga, Vishnuganga, Pindar, Mandakini and Nandakini while Yamuna's tributaries are Untagad, Rishiganga, Hanumanganga, Saurigad, Tons in Himalaya and its peninsular tributaries are Chambal, Sind, Betwa, Ken, Tons, Son and Mahananda and same is true for other rivers also. Important rivers draining the Bhutan are Amo chu, Ha Paro Wangchu, Manas, Kuruchu (antecedent river), Bosari and Dhansari.

In Himalaya, the Ganga and its network of tributaries are antecedent in nature and in some segments they follow the path of tectonic weakness such as faults. V Shaped valley profiles at many places suddenly turned slit like gorge, one gorge near Gangotri is nearly 180 m deep. The runoff and rainfall of the Ganga valley in the Himalayan region is 58.93 M ha m and 1116 mm/yr. It is estimated that contribution of the Himalayan rivers in the runoff of the Ganga River is 60% and the rest 40% comes from the peninsular region.

River Yamuna is an important tributary of the Ganga system, however, in early Holocene it was a part of the Indus system. Due to neotectonic movements, subsidence and flattening of the water divide southwesterly flowing proto-Yamuna changed its course to the east and joined the Ganga system. River Yamuna also pirated the water of now disappeared *Vedic* Saraswati River (Roy and Jakhar, 2001).

Indus

Indus is corrupt name of the Sindhu of Sanskrit language. The river originates from Bokharchu glacier (4164 m) situated east of Mt. Kailash. The river has a length 2900 km and drains nearly 9,66,000 sq. km area of Tibet, India and Pakistan. In Tibet, for 1280 km the river flows in flat area in west to east direction and takes a curved path along the Nanga Parbat and then flows in the SW direction. Important tributaries of the Indus are Jhelum, Chenab, Ravi, Beas, Satluj, Zanskar, Dras, Shyok, Gartang, Shigar, Kabul, Gilgit and few tributaries of Indus also originating from the Sulaiman Range. It enters into the plains of Punjab province of Pakistan and flow through desert area of the Sindh to meet Arabian Sea. The estimated runoff and rainfall in Indian part of the Indus valleys are 18.5 M ha m and 1114 mm respectively. Indus is an excellent example of antecedent river and posses largest relief, of more than 5 km, in Ladakh in India and Pakistan. This may probably be maximum relief in the world.

Siwalik or Indobrahm River

Workers like Oldham, Pilgrim, Pasco and others in the early years of last century visualized a twin gigantic river system- Siwalik or Indobrahm and Tsangpo-Indus-Amu Dariya- originating in the eastern Tibetan highland. These rivers had drained the Himalayan region before the present river system came into existence and proposed that sediments for outer Himalayan Siwalik Supergroup were deposited by the far mighty river Siwalik or Indobrahm that entered in the Sindh estuary from the northwest, This river vanished after deposition and upliftment of Siwalik sediments. Later geomorphological and sedimentalogical (palaeocurrents) findings rejected this hypothesis. Partial capturing and drainage reversal of Tsangpo-Indus-Amu Dariya were hypothesized for present drainage configuration.

3.2.2 General Characteristics of Himalayan Rivers

Rivers like Indus and Brahmaputra originate from either side of the Kailash Mountain in Tibet, 100-150 km north of the main Himalayan watershed and flow roughly in the straight course for more than a thousand kilometers and after that they take abrupt turn towards the south. The other mighty rivers of the Himalaya rise in the north of the highest peaks of the Himalaya and cut deep narrow transverse gorges in the mountain ranges. The turbulent Himalayan Rivers are perennial in nature and fed by snowmelt, springs, lakes and rainwaters. Rivers display youth to mature states of valley development. The youthful character is amply demonstrated in field by the presence of deep gorges having steep to vertical slopes, narrow V-shaped valleys, relief of many kilometers particularly in Higher-, Tethyan- and Trans Himalaya, presence of waterfalls and rapids in longitudinal profile, very steep gradient (>50 m/km) near their source region. The deep narrow valleys are ideal for dams' construction. With in the Himalaya, depending upon geological conditions, rivers either flow in straight or in entrenched or ingrown meandering valleys. Himalayan Rivers also exhibit braided pattern with in the Himalaya as well as they flow out of the Himalaya. This pattern is resulting from high river power, excess production of sediments by river erosion and frequent change in the river course. In river Tawi near Jammu, the Nagrota braided bar is 1.79 km long, 0.29 km wide and having more than 5 m height; the Jammu braided bar has a length of 714 m length and 143 m width (Singh et al., 2006). Ganga River near Hardwar also shows spectacular braided pattern,

Tectonically, majority of rivers shows antecedent characters. Initially their valleys were consequent in nature. Gradually consequent nature of valleys turned into subsequent, as rivers have adjusted with geological realities. The length of the large rivers in the Himalayan territory is proportional to geological divisions in which they are originating. Rivers originating in the Tibet are generally the longest one, followed by the originated in Tethyan, Higher, Lesser Himalaya and the least length is shown by the rivers originating in the Outer Himalaya. So is the case of geological and topographical adjustment; most adjusted rivers are of Tibetan origin and least adjusted are of Outer Himalaya. Even with in the Himalaya, particularly in Lesser Himalaya, rivers form braided pattern with development of inland deltas in their wide valleys. All main rivers and their important tributaries are of pre-Quaternary origin and their erosive power outpaced the rate of terrain upliftment (~1 mm/year), therefore, they are antecedent in nature. These rivers have steeper gradient in Tethyan and Higher Himalaya, which gradually decreases in Lesser Himalaya and Outer Himalaya. Waterfalls and rapids particularly in the Higher Himalaya and the inner Lesser Himalaya punctuate their longitudinal profiles. In pre-Quaternary time, the Lesser Himalaya converted into gently sloping to flat area with mature topography resembling peneplained surface due to sustained denudation. Due to rapid upliftment in Quaternary the mature topography was lifted up, therefore in lower parts of the transverse profile of valleys show youthful topography while upper part mature to old topography e.g. in Himalayan Tons river. These rivers change their course frequently in rainy season and shift several tens of meters within one rainy season. Hence, they create interstate and international disputes. The Yamuna River dried up the river Saraswati due to river capture. Luni River, which presently flows through the southeastern part of the Thar Desert, was once part of the Himalayan system. Drainage analysis has been carried out in many isolated areas, seventh is the highest order recorded for river Ganga. Earlier workers were of the opinion that relation between two major rivers at their confluence is "<- shapped" but this arrangement is oversimplification of the reality as other deflection directions are also present. Unusual flow directions or river diversions resulting from aggradations in upstream segment have been observed in many rivers 15% of the total discharge of river Ganga at Hardwar comes from glacial melt. The runoff in major river basins vary between 7.59- 24.16 percents.

Dokriani glacier contributed 6238000 cubic m water discharge and 15.1×10^{4} tons of sediments flux in ablation period between May- October 1994 and direct influence of rainfall in glacial suspended sediment is marginal (Thayyen, 2003).

In the Bakiya khola section of the Central Nepal Siwalik on the basis of stratigraphy and sedimentology, estimate for palaeochannel slope gradient varies between $2.14–5\times10^{-5}$ m/m to 1.05×10^{-3} m/m and the range of palaeovelocity was 0.22 m/s to 4.93 m/s and the slope of the upper part of the Siwalik must be steeper than 2.60×10^{-2} m/m (Ulak and Nakayama, 2003).

3.2.3 Evolution Of Himalayan Rivers

The Indus, Brahmaputra and Ganga river system are joined by the thousands of major and minor tributaries of different lengths and orders in the Himalayan region. Himalayan rivers are originating from the Outer, Lesser, Higher, Tethyan and Trans Himalayan or Tibetan regions. Source wise large number of medium and major rivers are originating in the Higher Himalaya followed by Lesser and Tethyan Himalaya. Rivers like Indus, Brahmaputra, Satluj, Gartang, Pareechu, Karnali, Sun Kosi, Arun, Amo, Subansiri are having their source in Tibetan part of the Himalaya. By and large major and medium rivers of the Lesser, Higher, Tethyan and Trans Himalayan terrain flow in the southerly direction. Rivers originating from inner regions pass through one or more tectonic belts; and in general the length of a river is proportional to its location of source in interior region and also to its age and complexities in its evolution. Himalayan Rivers show both simple and complex evolution patterns. In case of simple pattern of evolution, rivers evolved along the slope on newly uplifted surfaces e.g. Outer Himalayan Rivers and in complex pattern rivers evolved along the foot of mountains, cut the major structural trends, carved valleys along structural elements or follow combination of these factors.

The Outer Himalayan Rivers are relatively of short lengths and are developed on lately uplifted Neogene Siwalik rocks and younger alluviums. River valleys are mostly consequent in nature and developed on the northern and southern slopes of folded Siwalik or Outer Himalayan range. It seems that valleys grew in length by headward erosion towards crests of the gradually uplifting Outer Himalayan terrain. These consequent valleys are now slowly adjusting with the structures and lithology. The mild sinuosity and also little shifting from the direction of maximum slope indicate this. Thus they show feeble characters of subsequent valleys. Parallel, trellis, rectangular drainage patterns indicate influence of tectonic factors like transverse faults, joints and lithologic contacts etc. However, widely prevalent dendritic pattern indicates main role of the head ward development process and homogeneous lithology on simple sloping surfaces. As soon as rivers flowing in consequent valleys come down to valley floor level or in foothill areas, they start flowing parallel to mountain range. e.g. Suswa in Doon Valley till they join with major streams flowing in that area (Anantharaman, 1985). In cases of longer rivers perfect consequent nature is not discernible and rivers are deflected, thus fall in the category of subsequent.

The age and evolution of rivers, particularly their length, originating from interior regions is complex and is controlled by particularly topographical, structural and tectonic settings occurred in Quaternary period (Srivastava, 1982). Topographically, towards the end of Pliocene, Himalayan terrain was considerably worn down to stage of maturity to early old stages by persistent active denudational processes, as upheaval activities remained dormant for many million years. Huge thickness of Siwalik sediments (6-7 km) are derived from the northern source and were brought by many rivers, which were flowing transverse or oblique to foreland Siwalik basin. This rapidly deposited (6-7 mm/year) sediment indirectly suggests favorable climatic conditions of sediments formation such as warm tropical climate, relatively favorable slope and moderate to heavy rainfall. Transport of sediment in massive volume requires efficient and well-developed drainage system or network. In western Pakistan Himalaya evolution of drainage was controlled by antecedence, superposition, capture, ponding, avulsion and faulting. The ancestral river Indus probably came into existence at least 20 Ma ago i.e. in the middle of the tertiary along the axis of the island arc and established itself across the Ladakh arc due to growth of the isthmus between two plates (Shroder, 1989).

Quaternary tectonic movement rapidly uplifted the entire Himalayan terrain and many existing rivers had shown remarkable agility and developed their valleys with a pace to sustain their activities in this condition. Some of these rivers have achieved success in maintaining themselves and today we recognize these rivers as antecedent rivers e.g. Indus, Brahmaputra, Satluj, Arun etc. Most of the Himalayan Rivers fall in antecedent category, this is evident from the fact that out of very large number of rivers present in the Higher and Lesser Himalaya. These rivers are cutting the roughly east-west trending regional structures but rivers have not developed their valleys in these structures for long distances. Thus structural control in valley development is lacking. Only handful numbers of rivers are coming out in the plain after crossing the outer Himalayan or Siwalik range. In a study of Uttaranchal rivers, it is observed that the southerly flow of major rivers, except river Yamuna, is not controlled or influenced by regional east west trending major folds axes or faults planes or zones. Influence of trend of older, major structures is very limited at best for some kilometers only(Fig No. 3.1). However, influence of younger smaller structures capable of introducing weakness in the rocks like faults is more pronounced. The trends of mountains control the flow direction of larger segments of major rivers, which are oblique to folds or faults traces, even powerful rivers like Ganga and Yamuna needed the help of faults to cut through the Siwalik for entering in the plain (Fig. No.3.2). The rapid rise of Siwalik, a difficult obstacle, forced the realignment of drainage in the hinterland of the Himalaya to maintain their flows and only limited number of rivers were able to challenge this barrier successfully. Out of these, three (Ganga, Yamuna and Kali) originating from interior area and three, Ramganga and Gola from the southern Lesser and Outer Himalaya are entering into the southern plain.

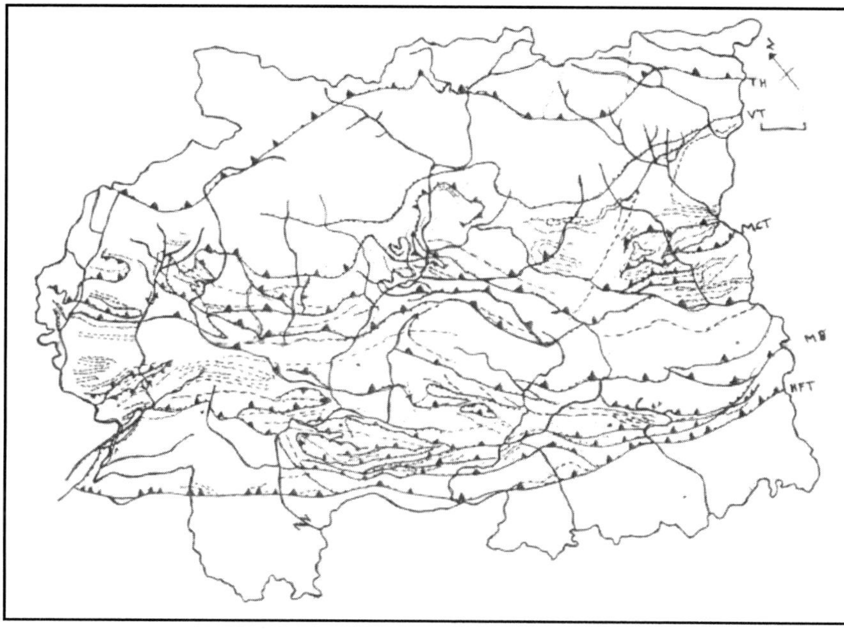

Figure 3.1. Drainage defying the structural control in Uttaranchal. Abbreviation : HFT - Himalayan Frontal Thrust MBT - Main Boundary Thrust, MCT - Main Contral Thrust, VT - Vaikrita Thrust and TH - Tethyan Thrust

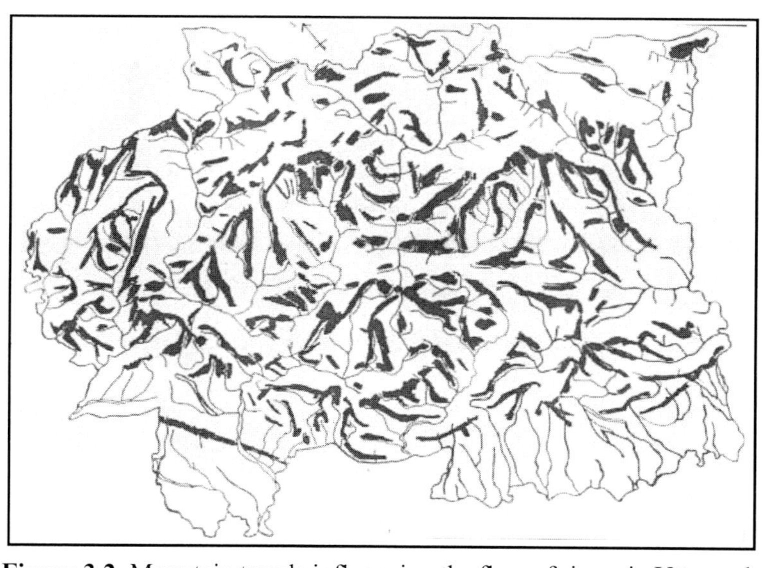

Figure 3.2. Mountain trends influencing the flow of rivers in Uttaranchal

In Quaternary, the Himalayan topography is extensively rejuvenated and in southern portion structures like folds and faults became active. Their activities have caused spectacular upliftment and increase in surface area of the Himalayan terrain. Thus the realignment of network was needed. Rivers lengthen their valleys mainly by headward erosion and in some cases by river capture. The general slope of the area and trend of mountain ranges dictated the new alignments. Topographical changes produced by the evolving nappes, tectonic and structural weakness rock types, water availability etc. Variable influence is exerted by the above-cited factors in overall development of the valleys (Brookfield, 1993). It is commonly observed that one factor may dominate over others in a particular segment of the valleys and other factor may dominate in other segments e.g. Ganga, Jhelum, Ravi, Beas, and Trishuli etc. The evolutionary history of individual rivers differs from the other adjoining rivers but controlling parameters for all rivers would remain same. Sometimes evolution of the river is by and large controlled by only one factor e.g. Yamuna River its entire 150 km long Himalayan stretch developed along strike-slip fault.

The prominent Trans Himalayan Rivers namely Indus, Brahmaputra and Satluj are evolved from Kailash range in Tibet. The Indus follows WNW and Brahmaputra follows ENE course along the suture zone. Tibetan course of then river is basically dictated by the headward erosion and time-to-time domain of there by glacial material. Their sudden southward turn around Nanga Parbat in case of Indus River and Namcha Barwa for Brahmaputra River show inability of river to cut across rapidly rising mountain and at the same weakness created in southern side through development of hairpin like eastern and western syntaxes. The Satluj River flows in ESE direction in Tibet and turns to SE as it touches the foot of Leo Pargil Peak in Spiti region. In Tibetan region its development is controlled by headward erosion along the tectonic weakness.

Rivers like Satluj, Beas, Ravi, Giri and Gambhar of Himachal Pradesh do not show any evidences of tectonic origin (Srikantia and Sharma, 1976).

3.3 Lakes

Himalaya is dotted with hundreds of lakes of large and small sizes. In general, lakes are mainly confined in the interior region. Tibetan lakes like Mansarovar (500 sq km), Rakshak (360 sq km), Yamdrli (870 sq km), Gunchu (100 sq km) and Trigu (130 sq km) cover large areas and many of them contain saline water due to restricted or no outlets. Many of these lakes are formed in graben structures produced by normal faulting or in tectonic depression or of glacial origins. Himalayan glacial lakes are formed by valley obstructed by morainic barrier, depression due to surface irregularities created by glacial drift, kettle holes left by melting moraines, valley obstructed by ice barrier, erosion, supraglacial, cirque or ice eroded hollows.

The important Himalayan lakes are Nainital, Dal, Wular, Kedar Tal, Saptrishi Kund, Badani Tal, Begnas, Renuka, Chhanga, Deotal, Dodital, Gohana Tal, Hemkund, Khajiar Tal, Manimahesh, Khurpa Tal, Phewa Tal, Physmda Tal, Rupkund, Shastra Tal, Tso Lhanio, Mahendra, Gosain Kund, and Spanggur Tso etc. These lakes have different origin for example Phewa and Begnas of Nepal and Renuka of Himachal Pradesh are ox-bow lake, depression for Nainital lakes are of faulted origin; Kedar Tal, Saptrishi Kund, Vasukital are of glacial origin, Gohana lake in Alaknanda valley is of landslide origin. Lakes formed by blocking of rivers by landslide or snow avalanche are temporary in nature. Moraine –dammed lakes are formed near glacier terminus. In Satluj basin 22, of 3.3 to 27.9 hectare areal extent, and in Chenab basin 31, 0.3-105.3 hectare areal extent, such lakes are present. Some of these lakes such as Geepang Lake have grown from 27 hectare in 1976 to 55 hectare in 2001.

Lakes are relatively younger in origin as they are formed in Late Pleistocene or in Holocene time. Bhimtal – Naukuchiatal Lake of Kumaun Lesser Himalaya and Goting Lake in Tethyan Himalaya were formed around 40000 years ago by reactivation of faults. Naini Lake was also formed around this time by faulting. These lakes are considered as dangerous as they may burst without notice and caused devastating flash flood in down stream areas. Indus River is dammed by landslide in December 1840 in between Haramosh and Nanga Parbat and produced 270 m deep lake. The lake blasted after 6 months and washed away entire Sikh army camping 350 km in down stream location on riverbank. Similar floods are also occurred in 1833 and 1858 by Indus, in river Shyok in 1932, in Alaknanda (1893), in Satluj (1819) in Chenab (1790) etc. a moraine-dammed lake of Mingbo valley, Dudh Kosi region, east Nepal bursted on September 3, 1977 and its impact in the form of flood was recorded at 70 km downstream region. Degeneration of lakes like Dal, Nainital, Phewa, Wular is taking place with alarming rate. Size and depth of lakes are decreasing and their water is getting polluted due to sedimentation, discharge of city waste in the lakes, encroachment of banks and increasing tourists activities. In Uttaranchal 127 lakes covering 2.49 sq km of glacial origins have been identified with the satellite imageries.

Salt and Brackish Water Lakes

There are many salt and brackish water lakes present in the northern interior parts of the Himalaya. The important lakes are Chang Chenmo, Lingzi Tang, Kyaghr Pangong Tso, Spanggur Tso, Tsokar and Tso Morari lakes of northwestern Ladakh. Many of these lakes freeze in winter months. Tso Morari Lake is situated at an elevation of about 4900 m and has length of 22 km and width varies between 5-7 km with maximum water depth 30 m. Tsokar Lake of 10 sq km area is situated in the Salt Lake valley, of 20 by maximum 7 km size, at 5000 m elevation

Palaeo Lakes

Presence of palaeo lakes of Quaternary age has been recorded through lacustrine sediments at many places in the Himalaya. These lakes were formed by the blocking of river valley by landslides or by uplift of the riverbeds along the faults running across the rivers. Many palaeo lakes like Lamayuru Lake (existed between 35000-1000 before present, BP), Tsofar Lake (3000-9000 BP) and Kenlung lake existed >35000 BP (Valdiya and Kotlia, 2001). Wadda palaeolake formed along the east to west oriented Almora Thrust in eastern Kumaun. Nearly 7-13 m thick sediment was deposited in the lake between 360000- 2000 years duration before present. The western Dhauli was ponded about 40000 before present (Valdiya, 2001). Hanse-Kashmir, Kioto, Karewas, Kathmandu etc. are other examples from the Himalaya

3.4 Glaciers

Nearly ninety seven percent of world's glaciers are present in the Antarctic and the Arctic regions, area wise out of 15 million sq km glaciated area, more than 14 million sq km area occur in these regions. A small part of Himalaya besides Alps, Rockies and New Zealand Alps are capped with valley and mountain (cirque, niche, crater, ice apron and compound basin) glaciers. Perhaps glaciers were covering much larger area in Pleistocene. The Siachen glacier, developed between 3800-7500 above mean sea level, of the Karakoram region of India and Pakistan is the largest non-polar glacier in the world.

An *Inventory of the Himalayan glaciers* has Himalayan glaciers are remnant of last ice age. Glaciers are confined in higher ranges of Himalaya. These glaciers, either valley or hanging glaciers, can be divided into transverse or longitudinal one. The transverse glaciers are of large size and terminate at higher elevation. The altitude of summer snow line varies between 4600 m in Arunachal Pradesh, 5000 m in Bhutan to 6000 m in western Himalaya (Raina, 2005). Rise in heights of snow line is attributed to scarcity of moisture in NW region. Estimate of Himalayan glaciers varies in a very wide limit that ranges between 5000-15000. In Indian territory 5067 glaciers are spreading over 38039 sq km area. The state wise number of glacier, occupied area and average length in km is as follows, Jammu and Kashmir (3136, 32000, 10.29 with maximum localization in the Karakoram range), Himachal (402, 1349, 3.35), Uttaranchal (917, 3530, 3.87), Sikkim (450, 197, 1.50) and Arunachal (162, 228, 1.41). Basin wise Indus Basin has 3538 glaciers while Ganga and the Brahmaputra have 1020 and 662 glaciers respectively. According to *Inventory of The Himalayan Glaciers* (1999) published by the Geological Survey of India, the Indian Himalayan region is divided into two namely the Ganga and the Indus basins, which are further divided up to fifth order basins. Total 5243 glaciers holding 142.88 cubic km ice volume covering 37959 sq km area are present within the boundaries of India. In 2004 figure reached to 7613 and may cross 9000 marks. Sah and Philip (2005) reported 1439 glaciers in Uttaranchal covering 4060.04 sq km area with ice reserve of 475.43 km^3. They have divided glaciers distribution in eleven catchments or basins. The catchment wise presence of glacial lakes, number of glaciers, area of glaciated area in sq km and ice volume in cubic km is: Tons (12, 102, 162.58, 17.42), Yamuna (8, 22, 10.4, 0.451), Bhagirathi (30, 374, 921.46, 129.928), Bhilangana (2, 19, 112.84,13.47), Mandakini (10, 40, 81.64, 5.98), Alaknanda (43,457,1434.61,170.369), Pindar (1, 43, 158.99, 15.01), Ramganga (-, 7, 6.74, 0.322), Goriganga (10, 128, 561.35, 69.18), (Dhauliganga (6,135, 373.19, 34.6) and Kutiyangti (4, 112, 236.24, 18.67). In Himachal Pradesh, 10 percent surface area is covered with permanent ice cover while 60 percent area comes under snow cover during snowfall and valley wise Chandra Bhaga contains 200 glaciers, 80 in the Beas-Parvati, 54 in Ravi, 159 in the Spiti, 68 in the Satluj and 40 in the Malunag. These glaciers are generally 2-4 km in length and occur above 3500 m altitude.

Among the largest glaciers are Siachen (74 km in length, second largest outside the polar world), Hispar (61), Biafo (60), Baltoro (62), Hispar (60), Zemu (26), Gangotri (26), Gasherbrum (26), Yalung (20), Ngo-Jubka (20), Kanchanjunga (16) and Kedarnath (14). The thickness of ice is 550 m in Fedchenko glacier, 250 m in cirque region of Gara glacier while in Zumu glacier it is 200 m. Glaciers show daily movement between 2-108 cm.

The 72 km long and 2 km wide Siachen glacier is situated on the north facing slopes of the Karakoram Mountain. The sidewalls of the glacier are steep to vertical and avalanches prone. This glacier feeds the river Shaksgam. The Gangotri glacier (109.03 sq km) of the Uttaranchal is 30.2 km long and 0.5-2.5 km wide. This valley glacier is situated between 4120 to 7000 m heights in Uttarkashi district. Raktavarna (15.90 km length and 45.34 sq km area), Chaturangi- Kalandini (22.45 km length and 72.91 sq km area) and Kirti (11.05 km length and 31.28 sq km area) along with 18 other minor glaciers are main tributaries glaciers of Gangotri glacier. Out of 258.56 sq km estimated glacierized catchment area has 200 sq km ice covered area of the Gangotri glacier holds nearly 20 cubic km ice. In and around glacierized area full range of erosional and depositional features have been developed. The erosional features are U-shaped valley, arêtes, horns, truncated spurs, cirques, crags and tails, waterfalls, rock basins and glacial lakes. Among important depositional features are varieties of moraines like terminal, lateral, medial, ground, surface, snow avalanche fans, snow-bridge, dead ice mounds, boulders and bogs and talus cones. Many supraglacial lakes are present in the middle part of the glaciers. Gaumukh is most famous snout of the glacier. Between 1936-96 glacier is receded by 1147 m giving an average rate 19m/yr and between May 1996 and October 1999 glacier is receded by 76 m (Naithani et al., 2001). The Barashigri glacier retreated by 170 m in seven years ending in 1963.

Parbat glacier of 48.44 sq km and 126 m thick has retreated 588m during 1990-2001 with a rate of 52m/yr and between 1960 to 1990 5991 m in length and 8.3 sq km in area probably because of location of the glacier (90%) in the lower altitude, 5200 m snow line, range and is expected to retreat 1461m between 2001 and 2022. The retreat rates of other glaciers are 2.6 m/yr for Bilare range in the Satluj valley, Shaune in Satluj 26.4 m, Meola in Dhauliganga 20m/yr, and Miyar in Chenab 17.2 m/yr (Kulkarni et al., 2005). In Garhwal region, the Gangotri Glacier advanced to 40.5 km to Jhala during the Bhagirathi Glacial Stage from the present position at Gaumukh, 19 km upstream of famous religious place the Gangotri in Uttarkashi district.

The important glaciers of Uttaranchal are Arwa Bank, Baspa Bamale, Bhagirathi Kharali, Khatling, Gangotri, Jagin, Khulia, Raj, Raktavarna, Arun, Bandarpunch, Chaturangi, Jamdar, Swargarohni, Lambar, Pindari, Ratiya, Mana, and Nanda Devi etc. Of Nepal are Annapurna, Dhaulagiri, Manaslu, Ganesh Himal, Langtang, Chhoba-Baware, Rolwaling, Khumbu, Kumbhkar, Kanjiroba and Mt. Everest.

Glaciological researches have revealed that glaciations were not synchronous everywhere in the Himalaya. Timing of last glacial maximum varies between 60-18 Ka in different regions of the Himalaya. The earliest glaciation in the Himalaya covered more areas while younger events progressively covered lesser and lesser areas. This is attributed to decrease of transport of moisture deep into the Himalaya due to upliftment of the mountain range, change in style of glaciation, earlier glaciers probably developed on flat land where as later glaciers were developed in the deeply cut valleys and on uneven land topography produced by the erosion of earlier event(s) of glaciations and lastly through change is climatic pattern which is controlled by the South Asian summer monsoon and the westerly jet streams. Multiple glacial events, at least three, have been identified throughout in the Himalayan and in Trans-Himalayan region while only one event is identified in the Kunlun Mountain situated to the north of Himalaya (Fort, 2005). In Swat region of Pakistan three stages of glaciations, the Laikot- Gabral- and the Kalam stage have been identified; in Karakoram Mountain three stages in the Pleistocene – the Shanoz, Yunz and the Borit Jheel and five glacial advances in the Holocene namely the Ghulkin I, Ghulkin II, Batura, Passu I and Passu II have been recognized. Similarly in the Lahaul area of Himachal Pradesh, Chandra-, Batal- (Batal I and Batal II) and Kulti stages have been identified. In Khumbu Himalaya of Nepal three glacial events one in the Pleistocene and two in the Holocene were identified (Owen, 1998). Glaciation in the Tibetan region is debated between an extensive glaciation i.e. ice sheet covering the entire Tibetan Plateau or in limited spread with multi glaciations in pockets confined to local mountains or uplands and in same places minor ice sheets developed (Kuhle, 1998, Rutter and Rakosh, 1998). At least two glacial advances during Quaternary are reported from Bara Lacha La and Shyok and Nubra valleys. In this region glacial processes were operated above 3000 m altitude. There are many filled up glacial lakes present in the area. This area exhibits wide development of glaciofluviatile and glacioeolean features (Pant et al., 2005)

Himalayan glaciers are described as warm or temperate i.e. temperature of the ice at pressure melting point. Himalayan glaciers show negative mass balance i.e. they are in receding phase (generally 5-10 m/yr and some 20-30 m/yr) and various types of moraines occur at different altitude level. The lowest level of terminal moraine is recorded at 2100 m level in Kashmir. Glaciers have extensively eroded the higher altitude and have produced all type of erosional and depositional features. Most common among there are U shape valleys, hanging valleys, Cirques, arêtes, ice scoured plain, depressions and mammillated topography, and among depositional features various types of moraines, esker, drumlins, glacial erratics, stone polygons developed where mean annual temperature lies between –30° to –40° C are present at higher altitude in all parts of the Himalaya. Glacial moraines at village Banli and Deoli in Himachal Pradesh are situated at altitude 1200 and 610 m respectively (Srikantia and Sharma, 1976). 3000 m was the lowest limit of glaciation in Bhutan.

There were 21 glacial cycles during the last ice age, alternating with equal number of interglacial warm periods. Not all glaciers in the Himalaya are in recession state, some glaciers like Hassanabad Glacier (9.5 km in two and half months in 1900 AD and Kutiah Glacier (113 m/day in 1953) have shown evidences of advancement. Batura Glacier destroyed a major bridge in 1974 because of abrupt change in position of the snout. The anomalous 'Little Ice Age' spanning between 1300-1850 AD (Pandey et al., 2003) was responsible of reduction of sea-surface temperature by 3-4° C and is probably caused by low solar activity and increased volcanism is characterized by advances of mountain glaciers in most part of the earth has been also recorded in Himalaya.

3.5 Solution Topography

Solution topography usually occurs in readily soluble rocks such as limestone, dolomite or sedimentary rocks having high amount of calcareous cementing material. Limestone caves are present in the Krol limestone in Sahastradhara area, Dehradun and in Nainital. These caves are generally few meters to few tens of meters in horizontal dimensions and several meters in height. In Sahastradhara, three to five levels of caves are present. Microcrystalline calcite grains in stalactites are of microbial (*Bacillus thuringiensis, Bacillus pumillis*) precipitation origin (Baskar et al., 2006). Repeated tectonic movements in the Himalayan terrain prohibit large-scale development of caves. Caves also occur in Doon gravel in Dehradun e.g. Tapkeshwar temple. In lower part of Doon gravel, very hard and strong conglomerate with occasional boulder length of more than one meter of calcareous cementing material and silty matrix is present. Removal of cement and matrix resulted into loosening of the boulders and formation of caves with good development of stalactites and stalagmites (~1.5 m length). Caves are also developed in massive sized loose boulders (~10 m long) in cavernous calcareous breccia on the Rishikesh-Neelkanth foot track. Small size sinkholes are present of the surface of these boulders. On the floor and roof, small size stalactite and stalagmites have developed. These caves have been developed as tourist sites or sites of temples. Hermits also reside in these caves.

3.6 Mass Wasting

Mass wasting refers to a group of process that involves the transfer of large volume of rock debris from higher to lower level under the influence of gravity. Processes like creep, solifluction, earth flow, mudflow, debris avalanche, landslides are included in it. Snow avalanche is excluded from strict definition of mass wasting but it is similar to debris-avalanche phenomenon.

In Himalaya, mass wasting is very common phenomenon because moderate to vertical slope of mountains are ideal for gravitative transfer of rocks, soils and glaciers. Almost all slopes are affected by slowest type of mass wasting process i.e. creep. It can be observed through tilted trees, electric poles, fence etc. During rainy season solifluction, earth flow and mudflow are common processes occurring more or less in most part of the Himalaya.

Landslide is most dangerous and destructive process occurring in all parts of the Himalaya. Every year more than 200 persons die because of it and properties like houses, road, bridge, agricultural field; power, communication and water lines of crores of rupees are destroyed. Road damage blocks traffic movement and supply of goods causing hardship in temporarily cut off part. Landslides also temporarily block the flow of the rivers, which at times burst without warning. In massive landslides millions of cubic meter rock mass get transferred over a large area e.g. landslide in the Langtang valley in Nepal. Scientific study of landslide is being carried out extensively in different parts of Himalaya and remedial measures like plantation, retaining walls, slope dressing, wire and jute netting etc. are successful to some extent in preventing it. However, complete check on landslide is not possible (Gupta et al., 1995).

Snow Avalanches- In winter or early summer months, snow avalanches occur generally in the interior region having high altitude. Its impact is severe on population and properties because of its confinement in remote as well as in populated area.

3.7 Volcanoes

There is no sign of present volcanic activity in the Himalaya. However, volcanic eruptions time to time take place in the Andaman Nicobar islands. Himalayan volcanic activities ceased in the Late Tertiary. Tectonic activities in the Himalayan arc are in progress, as before, therefore, complete cessation of volcanic or magmatic activities cannot be ruled out. It may occur in future. Presence of geysers or boiling water springs at many places suggests anomalous temperature condition in a few km depths. This anomalous situation could be caused presence of cooling of hot magmatic rocks or by small size magmatic pockets or by frictional heat generated by tectonic turmoil.

3.8 Cold Deserts

Deserts are defined as regions, which receive annually less than 25 cm rainfall. The deserts are classified into hot and cold deserts depending upon climatic condition. In Himalaya cold deserts area are Ladakh-Karakoram, Lahaul, Spiti, Pooh, part of Kinnaur district, Bharmour in Chamba district of Himachal Pradesh; upper Jadhganga valley of Uttarkashi, Mana and Niti tracts of Chamoli district and Malla Johar, Darma, Chaudan and Byans valley and Unta Dhura tracts of Pithoragarh district of Uttaranchal and barren parts of northern Sikkim in Lanchung and Lachen chu valleys. These areas receive very scanty rains as they are falling in the rain shadow regions of southwest monsoon or located at very high altitudes (Negi, 2002).

3.9 Marshes

The southern slope of Siwalik range was a marshy land (Terai) about half century ago. Marshes were also present in Duns and in northern slopes of Siwalik. These marshes are now converted into fertile agricultural land at lower altitude. Many high altitude glacial lakes due to degradation are converting into marshy area.

3.10 Weathering and Erosion

Himalayan terrain is subjected to all types of weathering processes. In the lower altitude chemical weathering plays dominant role but the process could not go in advanced stage i.e. transformation of solid rock into clayey material or large scale change in chemical composition because weathered product quickly brought down to valley floor level mainly by rain water. That is why thick weathering profiles are generally absent on sloping surfaces. Physical weathering by melting and freezing of water is characteristically occur in higher altitude, It's evidences are presence of vugs and cavities in rocks, patterned ground or stone rings, small hollows on flat surfaces, fresh angular rock debris of few centimeters to many meters length. Decay of dead animals and plants release organic chemicals and acids in the soil. These substances on interaction with rock increases fertility of soils but on the other hand set a slow reaction process of rock destruction. In Himalaya no specific study of organic weathering from geological angle is reported. In Karakoram region oily lustered deep – brown desert varnish is present.

Himalayan terrain experiences subtropical, temperate, severe cold, permafrost, snowy, and cold desert climatic conditions. The tropical and temperate zones are subjected to moderate to heavy rainfall. Thus huge run off is generated from the slopes of mountain. The sediment contributions of Himalayan Rivers are far in excess than the peninsular rivers. This runoff carries sediments in large quantity as less than 3 percent geographical area of world produces more than 10 percent global sediment output. The over all suspended mass transport by all rivers of the world is estimated to be 13.5 billion tons (Bt) with additional 1-2 Bt as bed load and flood and 2.13 Bt in dissolved form. The estimated rate of erosion in the Brahmaputra catchment is 953-ton/sq km/yr. Indian rivers are famous for their sediment loads 1.2 Bt/year thus they have deposited 6 million cubic km sediments in the Bengal Fan and Ganga-Brahmaputra delta in past 20 Ma.

Calculated average annual sediment load carried by the Brahmaputra, Ganga and the Indus river are 597, 329 and 100 million tones (Mt) from basinal area of 0.58, 0.75 and 0.97 million sq km respectively (Subramanian, 1993). In another estimate the mass sediment transport by the Ganga- Brahmaputra is calculated 1670 Mt/ yr. Sediment load of the Ganga River at Hardwar is 16 Mt/year that increases to nearly 800 million tones at Farrakha in West Bengal. For example this phenomenal increase in sediments load is due contribution of several mighty tributaries. Similarly Yamuna supplies 140 Mt sediments annually to Ganga River at Allahabad.

The long-term denudation rate of the Himalaya estimated from the volume of Indian Ocean sediment (8.5×10^6 cubic km) is 0.2 mm/yr. The present average denudation rate is estimated between 1.0- 1.8 mm/year. The average in the Higher Himalaya is 1.7 mm/yr, 1.9 mm in Arun Valley, 2.5 mm in Sunkoshi Valley, and 1.7 mm for Kumaun Himalaya annually. Erosion rate in Dabka watershed of Kumaun Himalaya varies between 2.21 to 3.57 mm/yr. The exhumation or denudation rate of Mandi Granite (Himachal Pradesh) varies between 833 to 875 m/Ma during 8.6 to 4 Ma ago and in last 4 Ma i.e. up to present rate of denudation decreased to 410 m/Ma. A relief of 1760 m is developed on Gangotri Granite in 0.9 Ma i.e. between 2.4 to 1.5 Ma duration. This gives 2-mm/year average rate of denudation. The denudation rate in Raikot glacier is 4 mm/yr. In other glacierized area it is estimated around 2 mm/yr. The average rate of erosion by rivers is four times less of glacial rate of erosion.

The high amount of chemical weathering in the Himalayan region is related with topography resulting from tectonism and climate. Chemical weathering not only produces stable residue (minerals) but also facilitates chemical dissolution of elements in the waters by a complex cycle. Rainwater absorbs atmospheric carbon dioxide and this water on reaction with silicate and carbonate rocks produces bicarbonate ions in waters. Higher global temperature and greater rainfall thus would facilitate more interaction between carbon dioxide and water and more weathering is likely to occur. This would have influence on global climate over a long period of time, fall in temperature, but not in short time range. The global carbon dioxide transport to ocean is estimated to be 0.28-0.30 giga tons/yr or 8.7×10^{12} mols/yr and 12.3×10^{12} mols/yr for silicate and carbonate weathering respectively. 25% of total load reaching the ocean today comes from the 5% of the earth surface i.e. catchment of the eight largest rivers- Ganga, Brahmaputra, Chang Jiang, Xijiang, Irrawady, Indus, Mekong and Huanghe- draining the Himalayan- Tibet region. The amount of carbon dioxide consumed in silicate and carbonate weathering by rivers draining Himalaya-Tibet region ranges from

54- 832×10^9 and 24- 997 $\times10^9$ mols/yr. For river Irrawady, silicate-weathering figure is 832×10^9 mols/yr while for river Ganga figure is 471×10^9 mols/yr. The total carbon dioxide consumption by rock weathering by six Himalayan rivers is 4×10^{12} mols/yr. This consumption is four times higher than those of the Amazon basin and account for nearly one-sixth of the global carbon dioxide i.e. 24×10^{12} mols/yr. The higher carbon dioxide consumption by the Himalayan rivers might influence global climate in long run (Sarin, 2001).

3.11 Thermal Springs and Geysers

Thermal spring catalogue prepared by the Geological Survey of India (1999 and 2001) records presence of nearly 340 thermal springs in India, out of which 123 present in the Indian Himalaya. Himalayan terrain with southern Tibet is designated as Himalayan Geothermal Belt (HGB). It is 3000 km in length and contains more than 500 thermal springs, a majority of them in the Tibet. The life of a thermal spring varies between 10000 to 100000 years without any change in temperature in the same duration. It is also observed that at one location the hot water from subsurface reservoir is coming out from several nearby places e.g. in Yamunotri at half a dozen thermal springs are present in 200 sq m area.

Thermal springs are classified on the basis of reservoir temperature as high- (>200°C), intermediate (200°-150° C) and low (150° C) temperature while in another classification temperature limit is placed at more than >150°C, 150°-80° C and <80° C. The surface temperature of hot water varies from 20°-25° C to 80°-90°C (boiling point). The reservoir temperature of thermal springs can be estimated by the hydrogeochemical, isotopic and direct measurement methods. Hydrogeochemistry of hot waters of selected thermal springs is given in the Table No. 10.1.

Region wise distribution of thermal springs in Himalaya is as follows: Jammu and Kashmir (37) Leh, Chumathong, and Barabati in Puga Valley. Shyok, Beas, in Nubra Valley and also at Changlung, Panamilk, Pulthang and Chumathong. The waters of these springs are mineralized in many cases and temperature varies between 65°-75°C. In Himachal Pradesh thermal springs are present at Jan, Kasal, Manikaran, Balargash, Khirganga, Kalath, Kharal Bihal, Vashisth, Tatapani. Temperature of water varies between 33°-95°C; Uttaranchal (37) springs are reported from 14 hydrothermal systems with a minimum proven potential of 130 MW electricity. Thermal springs at Wazri, Banas, Janakichatti, Yamunotri of Higher Yamuna Valley are most promising (Biyani, 1991), Tapovan, Badrinath, Aungi, Bhukki, Jhyan, Gangnani, Joti, Junja, Kali Manep, Moran, Rameri, Rapavan, Sera and in other locations. In Nepal (20) thermal springs occur along either MCT or MBT Tatapani, Bhatekosi, Tamakosi, Syabrubensi, Kali Gandaki, Chainpur, Mayang dogi, Sekaru, Seti khola Nayagaon, Chisapani, Thuman, Janakpur, Jamla and Roidukhola; in Arunachal Pradesh two thermal springs are reported, and 9 thermal springs locations are present in Bhutan. The important locations in Bhutan are Landrai, Raongri, Tinfari-Jiti, Gasa and Ratikhola (11) hot springs are reported from the Rigo, Maju and Chetu valleys discharging water of 97°-102°F in an atmospheric temperature condition 36°-48°F.

Geysers are reported from Yamuna valley, Alaknanda valley, Puga in Leh and other places. At Yamunotri in Yamuna Valley geyser is discharging almost non-stop, steam jet of nearly 1 m length. Geysers are associated with aquifer having more than 180°C reservoir temperature (Biyani, 1991 a and b, 2002 c and f)

Plate 3.1. **A**- Meandering Tons River near Kishau, Uttaranchal. **B**- Dipping thickly bedded sandstone, Nahan, Himachal Pradesh. **C**- Kaliasaur landslide near Srinagar, Uttaranchal (Courtsey - PC Bahu Khandi). **D**- Meandering Alaknanda River at Srinagar, Uttaranchal. **E**- Torrents in minor seasonal river, near Haridwar, Uttaranchal. **F**- Panoramic view of lesser Himalaya in background Manikot Hill, Uttaranchal.

Plate 3.2. **A**- Tensional cracks in Debri cover over lesser Himalayan rocks in Himachal Pradesh. **B**- Photograph showing Kharsali terrace on left hand side and fault scarp towards right, Yamuna valley, Uttaranchal. **C**- Slope elements (Waxing, free face, waning) in Tons river. **D**- A view of Chaukhamba peak in background, in foreground Alaknanda river is flowing through V-shaped valley.

Plate 3.3. **A**- Steep gorge of Bhagirathi river, in Uttarkashi, Uttaranchal. **B**- Tilted trees showing creep in Siwalik rocks of Sirmour district of Himachal Pradesh. **C**- Kaliasaur landslide near Srinagar, Uttaranchal, (courtesy – P.C. Bahukhandi). **D**- Triangular landslide and debris cone complementing each other in Tangri landslide, Rudraprayag, Uttaranchal (courtesy – P.C. Bahukhandi). **E**- Hanging bench due to ground subsidence Neelkanth, Uttaranchal. **F**-Removal of fractured rock material due to heavy rainfall, Uttaranchal.

3.12 Tectonic Landforms

Landforms are function of structures, processes and stages. However, in recently tectonized areas many landforms show prominent imprints of tectonic activities. These landforms are called tectonic landforms. In some parts of the Himalaya, landforms owe their origin to actively progressing subsurface tectonic movements. HFF forms the 60-90 m high scarps in Ramnagar, uplifted terraces, ponds, lakes pressure ridges, triangular facets, bulge between Mohand and Chhutmalpur, soil development from the north to south, tilting of Dun Gravel 10-15 degree in Ramnagar Jim Corbett park, canyons in the Siwalik (Nakata et al., 1984). Ground subsidence, slope failure and ground swelling in Ghagar River near Chandigarh. Thrusting of the middle Siwalik of the Soan Valley on the older terraces in Hoshiyarpur, in Tons valley three to seven levels of depositional terraces have been recorded.

DIMENSIONS OF HIMALAYAN GEOLOGY
A. K. Biyani (2006) : pp.39-87
Satish Serial Publishing House, New Delhi
E-mail: ssp@ndf.vsnl.net.in, hkjain1975@yahoo.com; website: www.satishserial.com

Adhibhutam ksaro bhavah purusascadaivatam (Geeta, 8.4)
The physical nature is known to be endlessly mutable.

CHAPTER - 4

STRUCTURE AND TECTONICS

On set of Tertiary era initiated spectacular changes at global level. On global level one of the notable outcome was integration of formative Alpine orogenic belt with yet to evolve orogenic belt of the east and encircle a large linear portion of the earth with mountain. Deformation resulting from the collision of continents first obliterated the Tethyan Ocean and subsequently evolved a thoroughly deformed landmass in the form of Himalayan Mountain (Fuchs, 1982) but as such its physiographic appearance started acquiring shape since Miocene time. High relief of the Himalayan terrain facilitates easy study of the structural and tectonic elements in vertical dimension.

Structures in the Himalaya are divided into two categories pre-Himalayan and Himalayan. Early Proterozoic or possibly Late Archaean rocks are also presents in the Himalaya. Such antiquity of rocks suggests, based on geological records of elsewhere, that in these ancient rocks structural and tectonic architecture of pre-Himalayan deformational (pre-Tertiary) event(s) should be present. Pre-Himalayan deformation is believed to be either of Proterozoic or Cambrian or Permian or Late Mesozoic/Early Tertiary age. The powerful Tertiary orogenic movement has obliterated the evidences of earlier deformation(s) too great extent; hence pre-Himalayan structures are either very feeble or are matter of speculation (Joshi and Tiwari, 2004). However, Some dominant Himalayan structures believed to be of pre-Himalayan origin and age are reactivated during Himalayan movement. Sharma (1977) has reported evidences of Caledonian/Hercynian orogeny in the Kulu-Rampur belt of the Himachal Pradesh. The main deformation, the strongest one, has taken place in the Miocene period i.e. during main Himalayan orogenic phase.

The vast dimension of this latest orogenic innovation is bound to show a spectrum of structures (e.g. faults or folds) suggesting non-uniform orogenic environment in the orocline. An example of this occurrence of Alpine type of folds in Lesser Himalaya and Higher Himalaya, and Jura type folds in Outer Himalaya and in the Tethyan Himalaya tectonic belts.

This chapter opens with a narration of regional faults of fundamental significance followed by structural architecture of tectonic belts covering various features of deformational origin.

4.1 Tethys Ocean

Mythological, Tethys was a sister and consort of Okeanos – the Greek god of oceans. In geological literature, Suess introduced this term in 1893 for a shallow epicontinental sea existed between Late Proterozoic to Early Tertiary. This ocean was positioned between southern Gondwanaland and northern Eurasia on a supercontinent named Pangaea. In its long geological history, ocean more or less continuously recieved sediments for a mountain to be evolved in later period of time, thus in a classical sense it was viewed as a geosyncline. Tethys was obliterated by the Tertiary tectonic movement and gave rise to birth of Alpine-Himalayan mountain range. This simple model was subjected to scrutiny by the later workers in the light now development in tectonics. Deitz and Holden (1970) suggested that the Tibetan block constituted northern part of Indian continent before fragmentation of the Pangaea. Following this observation, Sengor (1984) suggested that due to Late Palaeozoic rifting, Tethys was bifurcated into two oceans those he named as the Palaeo- and Neo-Tethys. This Permian rifting had impact on the northern margin of the Gondwanaland and caused separation of the Tibetan landmass from Indian continent as well as creation of other small sized continental blocks like Cimmeride continent.

The Neo Tethys was situated between India and Cimmeride continents. Due to northward drift of Cimmeride continent, plate movement, the NeoTethys grew in size and simultaneously Palaeo Tethys started shrinking and closed finally. The Triassic Cimmeride orogenic belt of was evolved from obliteration of the PalaeoTethys. The NeoTethys basin shows complete Wilson cycle i.e. rifting, spreading, closure and upliftment. This basin in its history was filled with 45 km thick clastic dominated sediment. On the passive Tethyan Indian margin 4.5 km thick shallow water carbonate and siliciclastic sediments were deposited in Jurassic and early cretaceous. Indus-Tsangpo Suture Zone represents the remnant of NeoTethys. To the north of suture zone in NW Himalaya, Eurasian plate margin or Karakoram plate comprising of the Pangong Metamorphites, Karakoram Batholith and Karakoram Tethys and to south of Suture zone, the Indian plate margin is constituted by Tso-Morari Crystallines, Suru Crystallines and Tethyan sedimentary sequences of Kashmir and Zanskar.

Splitting of Tethyan Ocean into Palaeo- and Neo-Tethys and thereby placing large portions of Mangolia and Central Asia with in the Tethys eased the problem plate reconstruction. However, Bhat (1998) rejected the splitting of Tethys on stratigraphical and tectonic ground and suggested great antiquity of Tethyan Ocean. According to him, this ocean was existed even in Late Archaean time and entire sedimentary sequences of the Tethys and Lesser Himalaya were deposited in it. Palaeontological input reveals regression of the Tethyan Sea was started ~ 50.6 Ma ago (Late Ypresian) in the Indus Suture Zone, ~ 49 (end of Ypresian) and ~46 (Early Lutetian) Ma ago in the Zanskar Tethyan Zone and Himalayan foothill belt respectively in the Western Himalaya (Mathur and Juyal, 2000).

4.2 Alpine Fold Belt

The mountainous area between Gibraltar to Middle East is designated as Alpine Fold Belt. Its tectonic evolution began with the fragmentation of the Pangaea that occurred in Early Jurassic. The fragmentation of Pangaea produced besides major plates, several minor/

micro plate like Adriatic, Messina, Aegean, Ionian, Turkish Laventine Caspian, Iran etc. and among major plates are African and Arabian plates and of late Indian plate in south and Eurasian plate in the north. Evolution of fold belt is result of subduction of major plates and micro plates through a temporary, now vanished network of mid oceanic ridges, continental margins, islands arcs, back arc basins, transform faults and subduction zones. Peak deformation in this fold belt is reached in the Late Mesozoic. Within the Alpine Fold Belt six episodes basaltic gneiss formations, seven of ophiolites, three of ophiolitic subductions, eleven deformations and seven of high temperature pressure metamorphisms are recorded. Besides this, many a times continental margins have been created and destroyed. Ultimately the region is evolved as peripheral mobile belt encircling the supercontinent and subsequently transformed into an intra continental structure through collision. Now the mobility in this belt is believed to be in waning phase. The belt of Alpine mobility passes eastward into Asian Himalaya via the Caucasian Range of south western Russia, Anatolian peninsula, Cyprus, Elburz and Zagros of Iran and the desert hills of Afghanistan (Read and Watson, 1978 and Windley, 1984). Fig No. 4.1 to 4.4 show western and northern surrounding regions of Himalaya.

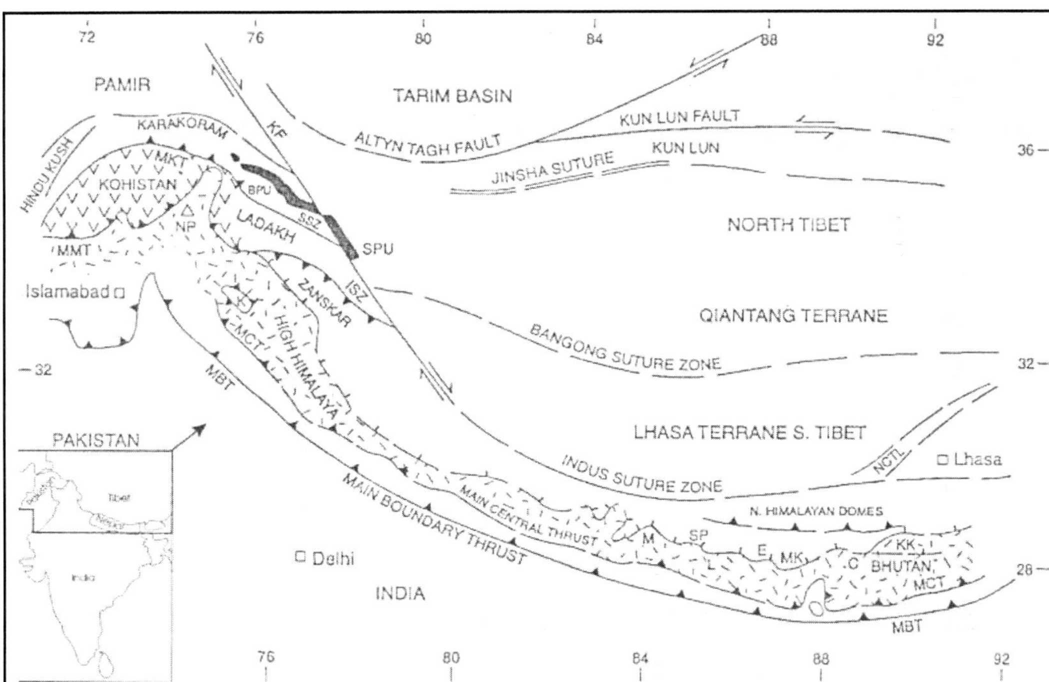

Figure 4.1. Major tectonic elements of Central and Western Asia (after Searle et al.,1999).

4.3 Himalayan Arc

Attention is immediately attracted to enigmatic regional arcuate structures. Tough square block is formed by three gigantic arcuate trends in the South Asia. Superb arcuate disposition of the Himalaya is its readily striking feature. To its south both on the east and west sides Myanmar and Sulaiman arcs are present.

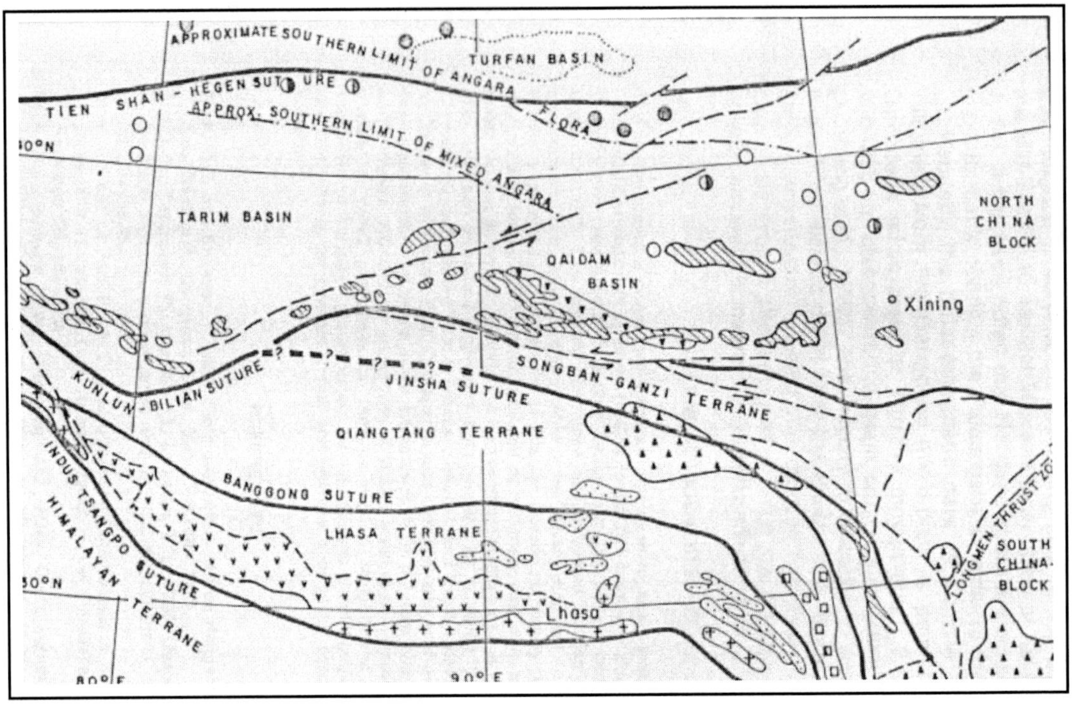

Figure 4.2. Terrane map of Tibetan and adjoining areas (after Sinha, 1997)

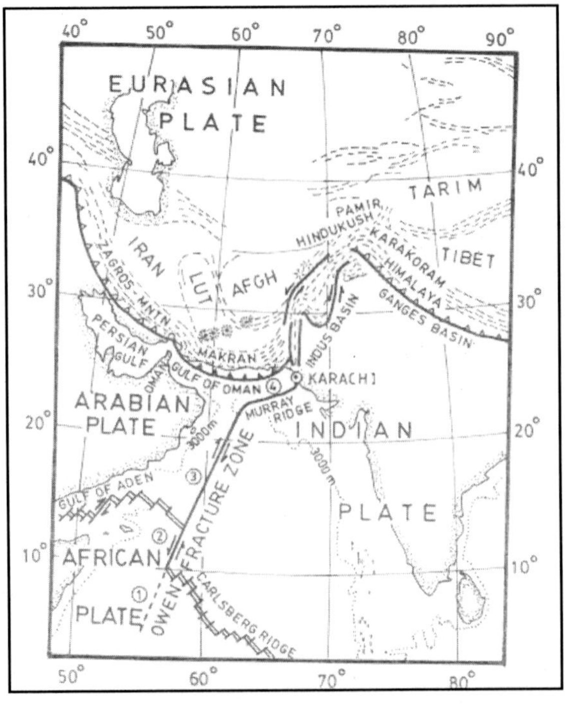

Figure 4.3. Tectonic map of Central west Asia (after Verma, 1991)

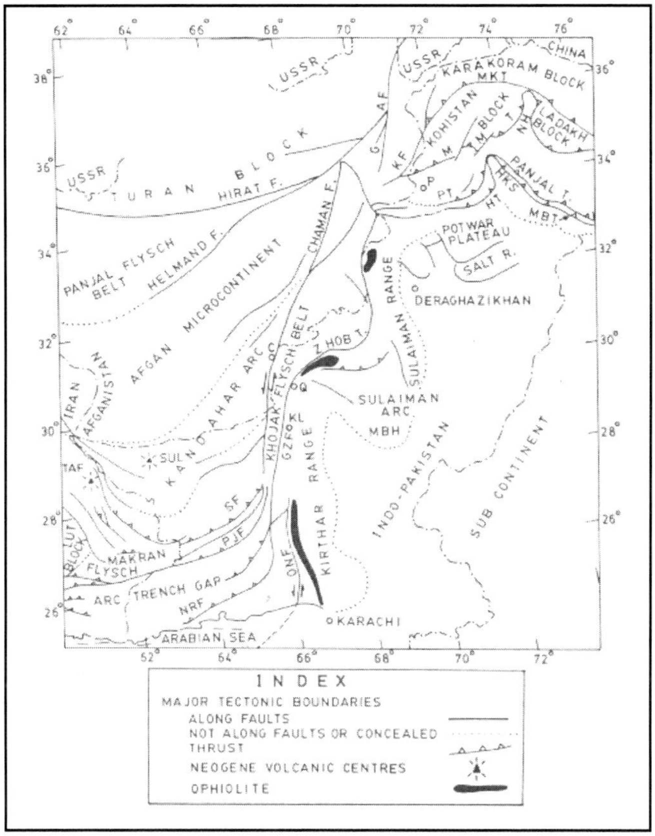

Figure 4.4. Tectonic map of south and south west of Himalaya (after Verma, 1991).

The Himalayan arc is convex towards south or towards peninsular India and has a length of 2200 km measured along the chord and 2400 km if measured along the length of the arc. The center of arc lies near the Lop Nor Lake or in the great Turfan depression in the Sinkiang province of the South China.

The arc is considered to be of primary or secondary origin. Wadia (1937) described it as a result of resistance exerted by the Deccan Graben to post-Tertiary tangential force from the north. Dubey and Bhat (1986) also suggested that arc is a primary feature produced by shear effect of Ninety Degree East Ridge. The shear exerted by the ridge has modified the geometry of the Himalayan basin. Crawford (1982) considered that arc is a part of periphery of a circle whose center is located in Turfan depression in China. It is produced as a result of earth expansion in the post-Paleozoic time.

Supporters of plate tectonics describe the secondary origin of the Himalayan arc. According to this theory arc is result of subduction or underthrusting of the Indian plate. However, a major difficulty in this view in that convex side of arc is facing on the wrong side. It is also proposed that arc is effect of Ninety Degree East Ridge and oroclinal rotation of the Himalaya (Klootwijk et al., 1985). It is also explained with indentation tectonics, northerly drifting rigid Indian landmass forcefully indented the Asian or Tibetan landmass.

4.4 Major Longitudinal Faults of Himalaya

Several major faults, mostly thrust, run parallel to the strike trends (NW-SE) of Himalaya from west to east (Kumar et al., 1989). Some earlier workers also called these faults as boundary faults implying they limit the depositional boundary of a particular mega stratigraphic unit. However, this concept has lost relevance now due to new data and observations gathered during detailed geological fieldwork and mapping. The important longitudinal faults of the Himalaya are-

4.4.1 Himalayan Frontal Fault

It is also known as Himalayan Frontal Thrust (HFT) or Main Frontal Fault/Thrust (MFF/MFT). This fault for the first time was reported by Nakata (1972) as a superficial reverse fault boundary between the Indo Gangetic Plain and the Outer Himalaya on the basis of photogeological and field evidences in Kumaun and Nepal. The Himalayan Frontal Fault is intermittently exposed as a set of echelon discontinuous or intermittent fault, which cut the Siwaliks obliquely. It is a blind fault in many segments. This is the youngest regional fault came into existence 1.6 Ma ago, in the Early Pleistocene, when the foreland basin was in the intense compressional phase. Its nature changes between reverse to thrust faults in different exposed sections of the Himalaya. In Nahan area emergent MFT has produced 5-10 m wide intensely crushed cataclasite-gouge zone in the Middle Siwalik. Rocks lying to the north of the MFT are always anticlinally folded and are considered as ramp anticlines produced by the fault propagation, fault bend folds. HFF is seismically and tectonically highly active but at some places it appears as inactive fault. HFF is viewed as youngest south-vergent, in-sequence thrust. HFF shows numerous geological and geomorphological evidences of its active nature such as it is causing folding in alluvial terraces, tilting the older terraces more steeply, longitudinal folding, shearing and back thrusting in the Siwaliks of Nepal, as well as of other rocks, Siwaliks are thrust over the IGP near Nepalganj, Surahi khola, Chhatru and Karnali gorge area in Nepal. A set of four faults in Kalabagh area near Nainital produced a depression along HFF, which was subsequently filled up with sediments. HFF forms escarpment as high as 60 m in front of IGP, fault scarplet on alluvial fans, upwarping and northward tilting of river terraces and ponds and lakelets by blocking the streams. The Main Doon Thrust in Nepal brings Lower Siwalik in contact with the Upper Siwalik. In Mohand area near Dehradun, HFF is known as Mohand Thrust, thrusting of the Siwalik rocks over piedmont fan and famous Kangra earthquake of 1905 elevated the Mohand Anticline, situated adjoining to HFF, by 10.3 cm. In Kosi Valley of upliftment in three phases, totaling 160 m, is observed along HFF in Holocene time. Around south of Doon Valley the upliftment and slip rates in the HFF are 6.9±1.8 and 13.8 ± 3.6 mm/yr respectively. Siwaliks and southern alluvium are subducting in the Himalaya with 20±4 mm/yr rate (Valdiya, 2003). HFF decreases the reflectance power of peat/lignite to 0.25 to 0.45 from 0.75 observed near the MBT. In western Himalaya Dutta and Samanta (2005) identified genetically related structures in HFF viz. ramp anticline of Janauri type, emergent thrust with eroded folded structure or Mohand type and structures with emerging back thrust type or Hardwar type. In HFF folds are ramp anticline originating through mechanism of fault bend fold or fault propagation folding. The detachment plane, ramp, form about 1000 m depth within pre-Tertiary metasediments. The ramp angle varies between 20°-30°.

4.4.2 Main Boundary Thrust (MBT)

It is also regionally known as Main Boundary Fault (MBF) and was first described by Middlemiss as a tectonic contact between the Outer and Lesser Himalaya and stratigraphically separates the younger Tertiary from older rocks. The highly active MBT is a set of three to four parallel running faults and region wise various names have been assigned to it like Muree Thrust in Jammu, Krol Thrust in Himachal and Uttaranchal, Mahabharat Thrust in Nepal and Gaur Thrust in eastern Arunachal Pradesh. The fault marks the northern boundary of Siwalik molasse. The north-dipping fault makes a 5-200m wide zone and shows very steep dip (50°-60°) to vertical dip on the surface but flattens in the depth (20°-25°). At a few locations (e.g. Ambala district in Haryana) the fault is dipping in south and apparently shows characters of normal fault (Dutt, 1982). The MBT is a zone of crushed carbonaceous clayey rock and quartzite. Many secondary faults and thrusts branch off from MBT, these faults/ thrusts diverge in westward direction and merge with the main fault in the east.

The fault brings in varieties of lithologies in juxtaposition such as Mesozoic-Cenozoic shelf carbonate succession over Tertiary sediments in the Kohat region of north Pakistan, older Tertiary and younger Tertiary in Muree against Siwalik in Jammu, Lesser Himalayan sedimentaries and Siwalik through Krol Thrust in Himachal Pradesh and Uttaranchal, metamorphics and Siwalik in Nepal. In eastern Himalaya Gondwana abut against MBT. The fault is also responsible for metrication of Lesser Himalaya in south and in many cases granites (e.g. Amritpur Granite in Kumaun) of interior regions have traveled near to MBT. In Kohat region thrust faults imbricate and duplex and tight folds formed by the out-of–the-syncline faulting.

MBT and Main Central Thrust are described as intracrustal boundary faults (Valdiya, 1980). MBT is proposed as upthrust created by recent tectonic movement along syn–sedimentary basin or as listric fault while another view says it was originally south dipping fault. Srikantia (1989) totally rejected MBT as boundary fault and suggests that is not a thrust plane but makes a tectonic limit on the surface due to overlapping thrust sheet along different thrust planes at multistoried levels.

Krol Thrust

Auden (1934) has recognized this 400 km long regional thrust fault in the Uttaranchal and Himachal Pradesh. He visualized that nearly 4000 m thick largely sedimentary sequence and mildly metamorphosed basal part of Middle Proterozoic to Early Paleozoic rocks are thrust over the mainly Siwalik and partly on Subathu, Dagshai and Kasauli formations. Krol Thrust conceals the MBT in major parts of its length, however, both expose separately in the Tons valley. Loosely Krol Thrust is taken as synonym to MBT in Uttaranchal and Himachal Pradesh. The idea is disputed by many workers who proposed that it is a reverse fault originally developed as penecontemporaneous down to basin fault, which became an upthrust in post-Siwalik period. The Krol sequence is also proposed as autochthonous rooted to its foundation or paraautochthonous (slightly mobilized).

4.4.3 Main Central Thrust

Heim and Gansser (1939) observed a thrust contact between sedimentary and low-grade metamorphics of the Lesser Himalaya and crystalline rocks of the Higher Himalaya and named this contact as Main Central Thrust (MCT). MCT is also responsible for abrupt change in height of Himalaya i.e. sudden increase in topographic height of the Higher Himalaya terrain in comparison to the Lesser Himalaya. MCT is clearly recognized as ductile intracontinental thrust in preference to brittle thrust and is not a single discontinuity but a 5-20 km wide ductile shear zone is designated as MCT Zone. This thrust is also designated as Shontargali thrust in Bara in Pakistan, Panjal Thrust in Kashmir, Jutogh Thrust in Himachal Pradesh, Munsiari, Budakedar, Tayali or Panwali Thrust in Uttaranchal, Crystalline Thrust in Nepal (Pecher, 1975, 1977) and Darjeeling Thrust in Sikkim, Paro Thrust in Bhutan and Bomdila Thrust in Arunachal Pradesh. MCT dips with 25°-35° in east in NW Pakistan and in general in N and NE directions with 5°-67° and is believed to behave as listric thrust in deeper level of the crust. Its steep dip on the surface outcrops is due to post-thrust deformation. MCT is an active thrust showing vertical and horizontal displacement of more than 20 and 100 km (Makalu-Everest region in Nepal) respectively. On the surface, MCT has developed extensive zone of mylonitisation (~15 km wide in Nepal), cataclasis and produced strong stretching lineation and other shear zone structures, and also show evidences of mineralizations (Bouchez and Pecher, 1981;Biyani 2004c and 2005 a). Spencer (1995) is of opininon that in the west of Kashmir identification of MCT becomes complex and real identification is not easy, therefore, beyond Kashmir MCT term should not be used. Some workers have applied modern thrust terminology (Boyer and Elliott, 1982) and identified a set of thrusts instead of single thrust plane and designated the thrusts as MCT I, MCT II, MCT III and so on and described these as floor thrust, branch thrust and roof thrust with local names as prefix (Bahuguna and Saklani, 1988; Bhatt and Saklani, 1990).

Being mainly responsible for origin and evolution of the Himalaya, this thrust has been studied and discussed thoroughly in all sectors of the Himalaya and of late it is associated with many controversies. MCT is evolved about 28 Ma ago (Manaslu Granite in Nepal) in Oligocene (Edward, 1995). It is also proposed that this rupture is premetamorphic in age and movement continued in Miocene to Pliocene time. MCT in Nepal is considered as pre-Himalayan in age.

Few workers also considered MCT as the boundary of Indian plate and believed that evidences of subduction are concealed under the crystalline rocks. The MCT Zone shows very high concentration of pervasive, flattening type of strain with bi-directional strain profile. In Yamuna Valley, palaeostress calculation suggests that 7 km wide ductile zone associated with MCT is formed in moderate stress regime (20-200 MPa) by various creep processes (Biyani, 2004). In field, MCT displays various type of contacts like sharp (e.g. In Yamuna Valley), gradational or occurs as imbricate or schuppean zone (Bist and Sinha, 1990). MCT is poorly defined in the Sikkim and Bhutan, no marked change in grade of metamorphism is observed in this contiguous region

Attempts have been made to redefine MCT by modifying original criteria of Heim and Gansser, for example, Valdiya (1976) prefer break in metamorphic isograd (index mineral garnet is followed by kyanite and thus omission of staurolite index mineral in metamorphic rocks) and thus shifted the boundary original MCT to further north and envisaged earlier recognized Vaikrita Thrust of Kumaun Himalaya as new MCT, while MCT of Heim and Gansser is designated by him as Munsiari Thrust.

4.4.4 Tethyan Thrust

The contact between crystallines of the Higher Himalaya and low grade metamorphic of Haimanta Formation and other sedimentary are rocks are also described as Tethyan Thrust or Dar Martoli Fault, Malari Thrust, Zanskar Thrust/ Fault, Zanskar Shear Zone, Trans Himadri Shear Zone, Trans Himadri Thrust, Annapurna Detachment Fault, Chomolonga Thrust, GontoLa Detachment etc. This contact is low to steeply dipping (10°-70°) while in depth (> 30 km) it flattens to 2-3°. Fault forms a 2.2-6.7 km wide ductile shear zone characterized by 250 m wide zone of intense mylonitisation with minimum offset of 35 km along detachment plane and vertical displacement of 5 km (Valdiya, 2005). Some workers interpreted this contact as normal fault contact and designated it as Jhala Normal Fault in Garhwal Himalaya. The normal fault contact between these two tectonic units are also interpreted in Himachal Pradesh, and in Kashmir and is visualized due to NE-SE extension leading to gravity collapse of Tethyan zone rocks. However Thakur (1992) observed that normal fault contact is not a primary feature and cited several evidences of occurrence of Tethyan rocks in the Higher Himalaya and further argued that the Paleozoic sequence has traveled for several tens of km in south direction. The apparent normal faulting post date the south directed overthrust thus suggests change in mode of deformation. Zanskar Thrust makes the basement cover fault or contact and is top down-to-northeast normal fault that regionally has been described as Zanskar Shear Zone (ZSZ). ZSZ is probably resulting from gravitational spreading of Tethyan cover sediment as a wedge of the basement rocks, the Higher Himalayan Crystalline, rapidly uplifted. It was also suggested that ZSZ have an earlier phase of thrusting and later phase of normal faulting (23-18 Ma). Three phases of deformation namely the mesoscopic to megascopic southwesterly verging isoclinal fold of the first phase followed by open to tight northeasterly verging fold along with axial plane cleavage in second phase and in the third phase folds were associated with acidic intrusives are reported from the Annapurna region (Brown and Nazarchuk, 1993). In different regions this fault was activated in different duration and limit of movement can be constrained during 23.5 –11 Ma ago i.e. in the post Early Miocene to Middle Miocene.

4.4.5 Indus Suture Thrust

The Indus Suture Thrust also known as Counter Thrust or the Zanskar Thrust draws the boundary between Himalaya and Trans-Himalaya. It is dipping in the south. The Trans-Himalayan terrain is comprised of the formations of Indus Suture, Shyok Suture, Karakoram Block and of Kohistan region. To the further north of the Indus Suture Thrust in the Ladakh and Pakistan the Main Mantle Thrust or Northern Thrust that dips in the north is present.

Based on the structural, stratigraphic, magnetic, and geochemical evidences Dubey (2004) considers these thrust faults as listric faults and proposed that during pre- Himalayan time these were normal faults and in compressive Himalayan orogenic phase they inverted into listric thrust faults.

4.4.6 Karakoram Fault

The oblique Karakoram Fault trending in NW-SE direction is a regional strike-slip fault of more than 1000 km length, outcropped between North-West Tibet cutting across Shyok Suture and to the south Kailash thrust system and to further southeast. The fault exhibits dextral displacement of ~120 km. The fault frequently branches out at many places, for example, in the Pamir region its splays are described as Rangkul, Murghab and Karasu faults. To the SE of Khalsar, the fault branches out which are initially run parallel and the northern branch near Rutog turns into thrust.

The fault is perceived as very active one and has formed triangular facets as high as 2 km and scarps of 50 m height in the Indus valley region of the Ladakh. Offsets of several hundreds of meters with extremely high slip rate (3 cm/yr) in the Holocene fan sediment are common. The ^{40}Ar-^{39}Ar isotopic age determination suggests that the movement in the Karakoram Fault was initiated ~14 Ma ago (Upadhyay, 2002, Bhutani et al., 2003).

4.4.7 Main Himalayan Thrust

A lowly northward dipping subsurface detachment surface or zone has been identified on the basis of focal depth of the recently struck earthquakes in the Himalaya. This detachment surface is described as Main Himalayan Thrust or Basement Thrust Fault (BFT) or Himalayan Detachment Fault. Over HFT and in the Outer Himalaya this detachment zone is present between 3-8 km depth, 10-18 km depth in the Lesser Himalaya of the Uttaranchal and 15-18 km depth in the MCT zone. This plane of detachment also represents a plane of decollement. The detachment occurs between the top of northward moving and converging Indian plate and southward propagating wedges of the Himalayan rocks. One view suggests that all major thrusts of the Himalaya are branch out from this detachment zone. The detachment lying below the Outer and Lesser Himalaya slips during the great earthquakes and under Higher and Tethyan Himalaya the detachment slips aseismically.

4.4.8 Salients and Reentrants

In a salient or virgation faults or folds form an arcuate belt convex towards the foreland and in a reentrant the belt is concave towards the foreland. Foreland is the area in front of a thrust sheet towards which the thrust sheet moved and the area behind the thrust sheet is called hinterland. Kangra and western end of Doon valley are examples of reentrants and east and west syntaxial bends are magnificent examples of salient.

4.5 Transverse Faults and Lineaments

Himalayan terrain is cut by many transverse faults, many of these are identified as of strikes slip nature and few of these are of unrecognized nature. These faults are either perpendicular or oblique to Himalayan strike. Most of these faults are tens km to a few hundred km in length, and are confined with in the domain of Himalaya but some are extended across the Himalaya i.e. they are traceable from south of Himalaya to Tibet (Qureshi et al., 1988; Raiverman, 1992, Thakur, 1988, Virdi, 1979). These faults show variable displacement of terrain, in most case only few hundred meters and do not disturb the geological settings on large scale; however, they play prominent role in the development of drainage. In most cases they are of neotectonic origin and they represent adjustment of stresses either due to rotation of landmass or impact of subsurface highs of peninsular India landmass or caused by further push in northward direction of Indian plate or release of residual stress. Many of these faults are active in nature and movement in them is continuous or discontinuous in nature.

A brief description of the important transverse faults or lineaments of Himalaya is as follows:

1. *Ravi Lineament* - It is a dextral strike slip fault trending in NE-SW direction. It is responsible for development of Ravi Syntaxial (reentrants) and has considerable influence on the course of Ravi River.
2. *Rupar-Manali Lineament*- It is also described as Mandi Shear, having dextral nature trending in NNE-SSW direction. Satluj River flows along this lineament for some distance.
3. *Chandigarh-Bilaspur Lineament*- It is very long lineament trending in NE-SW direction and finally joins with Rupar-Manali Lineament.
4. *Kalka-Rampur Lineament*- It is a very long lineament.
5. *Ambala-Kotkhai Lineament* - This dextral slip lineament controls the evolution of Gitel River.
6. *Yamuna Lineament* -This left lateral ENE-WSW, lineament originally identified in the Outer Himalaya but now extended to entire Himalayan course of Yamuna river i.e. till the source of Yamuna River at Yamunotri. In the Higher Himalaya it displaces the terrain by 100-200 m and is associated with couple of second order strike-slip faults and with other lineaments (Biyani 1999; Biyani, 2003). Raiverman extended this lineament further in the north in Tibet.
7. *Bijnor-Kalsi Lineament*- Dextral lineament having NNW-SSE trend.
8. *Ganga Lineament* -This right lateral lineaments run along the Ganga Fault, and is located between Hardwar-Rishikesh Narendra Nagar. In the interior region it controls the development of Huini River while between the Hardwar and Rishikesh lineament controls holy Ganga River. It shows displacement of 1200 m and is active in nature.
9. *Mithawali Lineament* - It is sympathetic to the Ganga lineament and shows N-S trend between Rishikesh and Lalkhand before merging with Ganga Lineament at Rishikesh. This dextral fault limits the eastern boundary of famous Doon Valley.
10. *Kotdwar-Devprayag Lineament* - It shows N-S trend, controls the Kot River and also, limits the eastern margin of Chandpur and Nagthat formations, in Mussoorie Syncline.

11. *Najibabad-Chamoli Lineament* - This lineament is of left lateral nature, trending in NE-SW direction, draws the straight course of Nayar River and parallel to this Doggada Fault is present.
12. *Ramnagar-Kosi Lineament* - NNE-SSW running wrench fault controls the part of the Kosi river in Kumaun.
13. *Haldwani-Karanprayag Lineament* - This is interpreted as deep-seated dextral lineament trending in NNW-SSE direction.
14. *Tanakpur-Gonai Lineament* - This dextral wrench fault shows NNW-SSE trend and controls the Sarju River.
15. *Tanakpur Lineament* - This lineament shows change in strike from NW to NE in the Outer Himalaya and then follows the course of Sarda River in Nepal.
16. *Arun Fault* - It is present in Nepal and is a vertical fault and cut across Tumlingator window.
17. *Dangarjong Fault Zone* - This wide fault zone lineament directed in N-S direction and makes a graben having a throw of 2700 m.
18. *Chimung Ri Fault* - It is present in Bhutan and shows ENE-WSW trending fault.
19. *Thangrung Fault* - This fault flattens the Ku-Ri anticline.

Other transverse faults in Bhutan are Pangkhar Fault, Damkar – Kharungla Fault, Yongbari Fault, Dechhiling Fault, U-Ri Fault and Thinypar Fault. In Arunachal Pradesh important transverse faults are Bame Fault (trending in NNE-SSW direction), Roing Fault (trends in NW-SE) and the Tidding Suture is a narrow tectonic lineament marks the boundary of Indian and Central Burmese Plate.

4.6 Extensional Faults

Normal faults are common in many orogenic belts but in the Himalaya due to dominance of thrust faults of compressive origin, they had been overlooked in the past. Normal faults and associated structures are reported from many region, however, the size of such features are comparatively small.

In Yamuna Valley, several normal faults of regional dimension are present e.g. Hanumanchatti Normal Fault and Banas Normal Faults are making a conjugate pair and another pair dipping towards each other is constituted by Kharsali and Untagad Normal Faults and a small depression resulting into formation of a graben called Kharsali. In Lahaul-Spiti region a series of high angle (42°-80°) normal faults originated in Miocene time are identified. The minimum horizontal extension across these faults is 26% and vertical displacement is reported to be of 1.24 km in a 7 km wide zone. In 2-3 km wide Zanskar Shear Zone, normal faults are reported to formed by earlier phase of non-coaxial ductile shearing. In Chandra Valley of Himachal Pradesh, horst and graben structure as gravitational normal fold present only in upper crustal level are also reported. Small-scale extensional structures are also reported from Zanskar, Spiti, Satluj (Kulu-Rampur and Larji window zones) and Tons, Bhagirathi Valley showing NE-SW extensional phenomenon. Grabens are reported from Upper Satluj Valley.

In western Nepal, Surkhet-Garahi Fault is nearly 90 km long. Annapurna Detachment is another major normal fault (Brown and Nazarchuk, 1993). Thhakhola-Mustang area of Nepal, Thhakhola Normal Fault shows down to east vertical component having displacement of 3 km that brings in contact mid crustal footwall rocks of the Higher Himalaya with the sedimentaries of Tethyan Himalaya.

The Himalayan extensional features are comparatively small size structure confined in upper level of crust and are produced in later part of deformation in the orogen. Their origin is explained by different mechanisms such as gravitational collapse of rising orogen, inclined stress axes, as relief feature to main compressive strain, upwelling of magma, slumping of over thickened crust during under thrusting of Indian plate, exhumation of hot remobilized Higher Himalayan Crystallines in ductile regime, as post-orogenic features and cessation of slip on the MCT, formed at the base of orogenic wedge and related with new deeper thrust faults and a new base to deforming Himalaya. Extensional tectonics is explained by a group of above-mentioned processes.

4.7 Structure of Tectonic Belts

4.7.1 Outer Himalaya

The Outer Himalaya comprises of 7 km thick, normally sequenced autochthonous rocks of Siwalik Group or Churia Supergroup in Nepal and few pre-Siwalik Tertiary groups or formations. These rocks show in general broad open anticlines and synclines of Early to Middle Pleistocene age, with sub-horizontal fold axis trending parallel to Himalayan strike. In Jammu region, Upper Siwalik shows broad open fold while Lower Siwalik depicts open upright folding. In Ramnagar area of Udhampur District two sets of folds are recognized. The macro scale NW-SE trending F_1 folds are broad, open, and nearly symmetrical with their axes parallel to regional strike and are associated with reverse faults and the F_2 folds, trends in NNE-SSW direction, are open and plunging in NNE direction. In Nahan (Himachal Pradesh) folds are tight while in Garhwal and Kumaun area folds are asymmetrical and occur in the conjugate fashion. In Bhutan, Siwalik and Gondwana rocks are gently folded in N-S direction while in Arunachal Pradesh doubly plunging anticlines and synclines showing NE-SW and ENE-WSW trends, however Dafla Formation is tightly folded and here Siwaliks are terminating against Roing Fault.

In eastern and central sectors simple homoclinal structure is disturbed by faults and thrusts. In western region, Jura type folds and intrabasinal thrust hading in SW direction. Large Outer Himalayan synclines are partially filled up with recent gravel and developed plain areas. Folding and faulting in Outer Himalaya is preceded by overthrusting of Lesser Himalaya along the MBT for at least 25 km. The MBT at many places in contact with Upper Siwalik while Lower Siwalik is thrust over Quaternary gravel. The Outer Himalaya is subjected to simple but intense deformation. Raiverman et al. (1990) have shown regional size 54 thrusts, 28 anticlinally axes and, 29 synclinal axes and 17 faults in their regional geological map of western Outer Himalaya between rivers Ravi and Yamuna. In interior regions of the Siwaliks duplex, imbricate fan system, pop-up structures, snake head anticlines and snake head duplexes, fault propagation folds antiformal stakes, over steps thrust systems and other structures have been observed. In Doon Valley Siwalik rocks shows two phase of deformation under NE-SW and NW-SE directed compressive stress regimes (Joshi, 2003).

In foothills of the western Himalaya, shortening is estimated by the balanced cross section methods. In Hoshiyarpur- Janauri-MBT transect reported shortening is 22% or of 23.4 km and in other transects the shortening in terms of percentage and km are 71.3% or 60.4 km in Bharatpur-Cheri- Darla- MBT transect, 61.1% or of 55.9 km in Subathu Depression, 26% or 10.6 km in Mohand-Dehradun transect and 51.1% or 11.2 km in Tanakpur-MBT transect. Such a large amount of shortening is also doubted on inadequate subsurface geological inputs like variations in three dimensional geometry of basin because of subsurface faults and thrusting of the Lesser Himalayan rocks and estimations in few sections or transects with simple restoration of the deformed sections.shortening of 20 km or 40% has been reported from the Bagmati section of the Siwalik hills of Nepal and the thrusting along the MBT caused minimum shortening of 25 km (Schelling, 1992).

Salt Tectonics Presence of salt tectonics is reported from outer fold-thrust belt of North West Frontier Province in Pakistan. Decollement condition was initiated by sediment load and sliding caused by compressive forces forced the salt beds to behave as viscous fluids under stress. Structures like piercement, turtle and non-piercement normal faults are formed.

4.7.2 Lesser Himalaya

The Lesser Himalaya is nearly 15 km thick tectonic unit of autochthonous, paraautochthonous and allochthonous rocks of Middle Proterozoic to Early Tertiary age situated between the MBT in south and the MCT in north. This belt of fluctuating width is considerably wide in Uttaranchal and Himachal Pradesh. However, it is extremely narrow in eastern Sikkim and in Nepal and Bhutan because of overthrusting of crystallines of Higher Himalaya, but again widens in the Arunachal Pradesh. This belt is mainly made up sedimentary rocks but a part of it is metamorphosed in very low-to-low grade conditions and have medium to high-grade crystalline metamorphic rocks transported from northern regions as nappe and klippe. The belt also shows fair occurrences of intrusive and extrusive igneous rocks having autochthonous and allochthonous nature (Gairola, 1982). Gondwana rocks occur along the southern boundary of Lesser Himalaya in Nepal and further in east and also few outcrops of Gondwana with in the Lesser Himalaya in Nepal. Several thrusts have descended to the level of MBT e.g. Krol Thrust in Uttaranchal and Himachal Pradesh and Baxa Thrust in Bhutan. Aravalli strike (NNE-SSW) has also been recorded from several places in Uttaranchal suggesting that the entire belt is an extension of peninsular India.

The belt shows highly complex structural and tectonic settings. Besides polyphase deformation area has been extensively invaded by large number of thrusts or nappe sheets and closely spaced fault. Subsequent vigorous erosion at many places reduced the nappes and overthrusts sheets into small klippen and deep erosion also exposed the underlying autochthonous rocks in large number of windows.

In Kumaun Lesser Himalaya four lithotectonic units namely Shali-Deoban paraautochthon, Shimla Slate-Damta Unit Krol Nappe and Crystalline Klippen Unit are recognized by Thakur (1992) while six units by Valdiya (1980).

In Kumaun Lesser Himalaya, Valdiya (1980) has shown 6 major or primary thrusts namely the Almora, Munsiari, Vaikrita, Ramgarh, Berinag and Krol; 5 minor or secondary thrusts viz. Chandpur, Deoban, Ghes, Jogira and Nahan; 7 strike faults (Aglar, Bhimtal, Duram, Dunda, Alaknanda and Kalapani) and 24 transverse faults in the regional geological map. Four schuppean zones namely Chamoli Schuppean Zone (6 minor thrust sheets) Purola Schuppean Zone (5 minor thrust sheets or scales), Pipalkoti Schuppean Zone (3 thrust sheets) and Askot-Baijnath Klippen schuppean zone are recognized in the region.

In Rudraprayag area of Chamoli district of Uttaranchal, 3 phases of folding are identified F_1 is most intense event leading into formation of tight isoclinal fold e.g. Kharpatia Syncline. F_2 event is feeble leading into formation of open anticline and syncline plunging to north or south at low angle along with development of crenulations. F_3 folds are broad open folds plunging in NE or SW direction at moderate angle e.g. Mandakini anticline. In Murbhem area of Garhwal four phases of folding are recognized; asymmetrical to isoclinal, open recumbent, shear and chevron type folds are formed due N-S, NE-SW, NW-SE and E-W movement involving flexure and shear mechanisms. In Dudatoli area five phases of deformation were recognized, the earliest one is of pre-Himalayan in age and the remaining phases were produced during the Himalayan orogeny. Anticlines and synclines dominate in Bageshwar- Kapkot area (Bhattacharya, 1979).

The Chamba Nappe of western Himachal extends into Kashmir and there it is designated as Kashmir nappe having Early Precambrian to Upper Mesozoic fossiliferous strata.

In western Nepal, three lithotectonic units namely Chail Nappe, Rumkum Nappe and Paraautochthonous Krol are present in Lesser Himalaya. In Central Nepal, two zones namely northern Midland zone and southern Mahabharat zones are recognized. The Midland zone is made up of northern Fore Anticlinorium I comprising of minor Ramsango, Tumlingator and Tamur anticlines and southern Fore Anticlinorium II comprising of Sunkoshi, Okhladunga, Dhankatia regional anticlines. In between these two anticlinoriums, Kathmandu synclinorium is developed which is doubly plunging in nature. Mahabharat zone is tectonically made up of tight synclinorium in eastern Nepal while in the south of Kathmandu forms monoclinal structure and in the western Nepal open folds.

In Bhutan folds trend in WNW-ESE in western region, east-west in the central area and east eastnorth - west westsouth in eastern area. In Siang district Arunachal Pradesh, three deformational events are recognized. F_1 folds are tight, isoclinal to reclined in nature showing E-W to WNW-ESE axial trend and northerly dipping axial plane. The F_2 folds are tight isoclinal to close fold and F_3 is open type folds.

4.7.3 Higher Himalaya

The Higher Himalaya is more than 20 km thick thrust slab comprises of metamorphosed sedimentary and igneous rock, called Higher Himalayan Crystallines (HHC), of Proterozoic age. Crystallines were deformed by Tertiary collisional process and are overthrust at least about 80-100 km in SSW or SW along the MCT. This 30-50 km wide polyphase deformed and metamorphosed region is considered as geoanticline ridge separating southern Himalayan geosyncline and northern Tethyan syncline (Pecher and Le Fort, 1986). Another view describes it as Main Axial Zone bounded by deep- seated fault on either side. Still another view considers it as a continuation of northern peninsular mass. These rocks show many cratonic characters and deformed during Tertiary orogeny, therefore, described as remobilized cratonic mass (Biyani, 2004).

Three tectonic units namely Chail Thrust, Jutogh Thrust and Vaikrita Thrust units respectively constituted by low, medium and high (upper amphibolite facies) grade metamorphic rocks in Himachal Pradesh and Uttaranchal and also in eastern part are identified in the Higher Himalaya. In the Kashmir and further western regions distinction between the Lesser Himalaya and Higher Himalaya crystallines is not as sharp as it is to the southeast of Kashmir.

HHC shows remarkable uniformity in their structural characters. MCT is now recognized as MCT zone as along this shear 5-15 km wide ductile shear zone is characteristically developed in which rocks are mylonitic and are displaced in the south direction in general (Biyani, 1994). Fold development in MCT zone occurred in ductile regime that operated in a relatively narrow but intensely deformed belt (Jain and Anand, 1988). While in the inner areas, homogenous flattening follows buckling, parallel to layering (flexure slip) under brittle regime operating under moderate to low intensity condition of deformation. This implies that folds formed in MCT area in deep-seated condition while in Lesser Himalaya and other areas in shallow depth. This is further supported by estimation of flattening strain which is 4 percents is the Outer Himalaya, more than 45 percents in sedimentary and crystalline nappes of Lesser Himalaya and in the MCT zone flattening strain is more than 90% (Bhattacharya, 1996).

HHC shows variable phases of deformation in different regions. One to six deformational phases are identified in the HHC, for examples, Dubey and Bhat (1986) reported through experimental study one phase deformation associated with the Himalayan orogeny. Two phases of deformation are reported from Yamuna Valley (Biyani, 2004), Garhwal area (Gairola, 1975) and Sikkim (Gangopadhyay, 1979). Three phases of deformation are reported from large number of locations of Kashmir, Himachal Pradesh, Uttaranchal, Nepal, Bhutan and Arunachal Pradesh in which first and third phases are considerable weak in nature while second one is most strong. In the four phases deformational schemes (e.g. Zanskar in Kashmir, Annapurna area in Nepal, Garhwal-Kumaun of Uttaranchal) first phase (F_1) folds are isoclinal to recumbent nature and are responsible for development of axial plane schistosity. They generally plunge towards NW-SE, however reclined folds plunging in NE-SW direction are also recorded. In second phase folds (F_2), tight style is observed, overturning towards SW. They are generally coaxial to F_1 and plunging either in NW or SE. F_4 folds are locally developed are chevron to kink in nature with steeply dipping axial plane and showing shear sense top to the south or south west. Each phase of deformation is characteristically associated with either ductile or brittle or transitional type of deformation in the Garhwal Himalaya

(Jangpangi, 1982). In the Kulu- Rampur and Larji windows area of Himachal Pradesh six phases of deformations are identified by Pande et al. (2003). Initial deformation, D_1, before the initiation of MCT shortened the entire terrain by developing F_1 folds between Eocene to Late Oligocene-Early Miocene, in D_2 deformation exhumed Lesser Himalayan crystalline nappes along Jutogh and Chail thrusts of the MCT zone produced structural fabric, D_3 deformation occurred prior to MBT and fault-bend folds, developed in autochthonous or paraautochthonous units, D_4 deformation developed cross folds in D_5 deformation Kulu-Rampur antiform with extensional features developed in the Jakhri-Sungri and Banjar thrusts and D_6 deformation was solely related with extensional tectonics in the Pleistocene,

Mesoscopic ductile shear zones of the Bhagirathi Valley reveals NNE-SSW maximum horizontal compression synchronous to movement of MCT (Tripathy and Srivastava, 2005). In Garhwal Higher Himalaya, crustal shortening of 193-260 km is calculated in Higher Himalayan Crystallines, the crystallines were evolved through break-back thrust mechanism (Srivastava and Mitra, 1994).

Higher Himalaya: A Remobilized Craton

Craton (from the Greek 'Kratos'- hard) term was introduced in the geological literature by Stille for a region comprised of shield surrounded with Precambrian platform. The term shield was coined by Tyrell for a block of continental dimension, which has not undergone tectonic disturbances, apart from gentle vertical movements, since early geologic time. Though shield was subjected to deformation, metamorphism and granitization in early geological history. An Archaean nucleus, not always, in shield is surrounded with a number of Precambrian tectonic provinces. Platform is described as the stable part of the Precambrian continental crust having undeformed overlying sediments of 1-3 km thickness of little relief (Pichamuthu, 1985). Wetherill et al. (1962) in Read and Watson (1975) suggested that cratons were formed between 3800-3000 Ma duration, however, Muratov (1977) suggested that craton is the region of >2400 Ma age having a nucleus, which should have nucleus forming processes were limited between 3800-1500 Ma. Cratonic regions possess several distinguishing characters and many of such characters are present in the Higher Himalaya. For examples, signature of the pre Himalayan deformations have been recorded from many places, this region is solely comprising of Proterozoic rocks may be of Archaean affinity and several workers consider this region is an extension of peninsular shield of India, displays several episodes of acidic and basic magmatisms and which are well comparable with world wide magmatic episodes, metamorphic rocks show poly phase metamorphism and deformational histories, Higher Himalaya acted as basement for deposition of the sediments of the Lesser and Tethyan Himalaya with history of reactivation of the northern basement in Cambrian time, shows negative Bouguer anomaly of -200 to –300 mgals, showing fairly wide spread signature of base metal mineralizations etc. Moreover signatures of reactivation or remobilization have been identified in other cratons also, for examples, Wyoming and Montana craton (2700 ma) have reached near to coastal area (Windley, 1984) or Dharwar craton of peninsular India is rising with 3 mm/yr rate and crustal thickness of craton in Tamil Nadu and Kerala is 4-5 km more than adjoining Dharwar region (Valdiya, 2001). Pichamuthu

(1985; pp. 79) observation is significant "... belief that cratons once stabilized are vulnerable by younger tectonic events...".

4.7.4 Tethyan Himalaya

A more than 10 km thick sequence of clastic and carbonate rocks deposited in the long, narrow epicontinental platform of Higher Himalayan crystalline during Late Proterozoic to the end of Mesozoic or even in Early Tertiary. Due to subsequent deformation of Himalayan orogeny, the continuity of rock is broken and variably displaced. The exposures of the Tethyan rocks occur in Karakoram and in Kashmir Valley and also in Himachal Pradesh, Uttaranchal, Nepal, Bhutan and Sikkim and also in the Tibet. At many places, rocks show original north verging synclinorium structures evolved from Jura type (broad, open) folding. These rocks exhibit open folding on the southern margin of Tethyan zone, domal upward to upright folding in the middle part and further in the north southerly isoclinal-overturned to recumbent folds and also towards the suture zone situated to the north. Regional fold axis runs in NW-SE direction in older rocks while in younger Mesozoic sequence changes to N-S in the Martoli Formation. Tethyan rocks show cross folding plunging in NW-SE direction while the Kuti Shales shows disharmonic folding. At some places two phases of deformations are recorded the earlier (D_1) is diffused isoclinal to reclined in nature while D_2 is comparatively stronger.

Large scale faulting and thrusting in general are absent in Tethyan Himalaya. Two sets of faults, one-strike faults trending NW-SE to NNW-SSE and other transverse faults trending NE-SW to E-W are reported from Himachal Pradesh.

4.8 Nappes in Himalaya

Arnold Escher vonder Linth described first time, in the nineteenth century, large-scale movement of rocks in the orogenic Glarus area of Switzerland. Since then long distance displacement of rocks (allochthon) have been recognized from all parts of the world. The term nappe is now used for those rocks that have been displaced at least 10 km from their roots zone. Nappes are divided into fold nappe (displaced large size recumbent or isoclinal folded regions) and thrust nappe (displacement of rocks due to thrusting) and also as thin skin nappe i.e. involving displacement of comparatively less thick sequence of sedimentary rocks and thick skin nappe showing movement of thick sequence of crystalline sequence.

Himalaya is a young fold-thrust belt of 2500 km length hence thrusts and nappes have been recognized practically everywhere. Pilgrim and West (1928), Wadia (1928), Auden (1934, 1937) and Heim and Gansser (1939) are among pioneers who initially recognized nappes in western Himalaya. Some nappes have traveled to spectacular distance of the order 100-150 km. Majority of the nappes have traveled in the south direction due to compressive pressure of northward subducting Indian plate. A few nappes have traveled in the north direction due to listric nature of faulting.

Himalaya has been investigated by large number of geologists and often they worked in isolated small domains, so, large number of nappes have been identified in the entire territory and some of these later become controversial as far as their true nature are concerned e.g.

Krol nappe. Nappes have been identified and described by the workers in the regions of their study (a part of the state or state or countries of their work) and only very late regional correlation is attempted between different nappes.

Nappes in Pakistan

Between Swat to Kaghan Valley in Kohistan part of northern Pakistan, Besham, Swat, Hazara, Banna, Lower Kaghan and Upper Kaghan have been recognized as major nappes. Precambrian granitic and biotite rich orthogneisses, metasedimentary gneisses and schist, marble and amphibolite dominate Besham nappe. In the west it is overlain by the Swat nappe. The base of Swat nappe is marked by ductile shear zone called Alpurai Thrust. Precambrian schist and quartzites of Tanwal Formation constitute the Hazara nappe. Banna nappe comprises of chlorite bearing slate interbedded with limestone of Banna Formation. All these rocks show four episodes of deformation and peak metamorphic condition was experienced by these rocks around 40 Ma ago. Metamorphic with volcanic intercalation Bibai and Gogai nappes are present in the northeast of Quetta.

Nappes in India

A state wise description of nappes is as follows:

Jammu and Kashmir

In Jammu and Kashmir, Wadia (1928) identified allochthonous rocks in Kashmir Valley and designated it as Kashmir Nappe. This is also known as Salkhala Nappe. Synclinal Precambrian low grade Salkhala (= Chail) and Dogra slates and overlying fossiliferous Palaeozoic and Mesozoic rocks of Tethyan affinity are transported along the Panjal Thrust over Lesser Himalayan rocks. These rocks make southern contact with foreland autochthonous rocks while root of nappe lay in the north of Kashmir basin in the Higher Himalaya. The Panjal Thrust is believed to be NW extension of Suketi Thrust or Chail Thrust. At some places in Jammu, the contact between autochthon and allochthon is gradational in nature.

The Kashmir Nappe also continues in the western Himachal Pradesh representing a basement thrust and is folded into antiform and synform. Recumbent folding along the sole of nappe suggest translational pattern of the thrusting, which otherwise is upright. In Chamba area, Himachal Pradesh and eastern Kashmir, Fuchs identified two nappes namely Chail Nappe (low grade metamorphic rock) and Crystalline Nappe (Jutogh Nappe or medium grade metamorphic rocks) having root in Zanskar Range.

In the Ladakh, Shyok and Indus Suture rocks show thrusting in the south and north direction respectively. The average thrusting is 20-30 km in range and exceptionally it may be of 80 km length.

Himachal Pradesh and Uttaranchal

In Himachal Pradesh and Uttaranchal, following crystalline and sedimentary nappe of regional dimension are recognized.

Himachal Pradesh	Uttaranchal
A-Crystalline	
Vaikrita	Vaikrita
Jutogh	Almora
Chail	Garhwal
B-Sedimentary	
-	Berinag
Jaunsar	Jaunsar
Krol	Krol

Vaikrita Nappe

High-grade (upper amphibolite facies) metasedimentary rocks of Early Proterozoic age (2300-2400 Ma old) constitute this nappe. It is 5-7 km thick thrust slab in which movement has taken along Vaikrita Thrust or MCT I (19.2 Ma ago). The rocks are intruded with Tertiary and older granites and basic sills, dykes and flows. It is equivalent to Upper Crystalline Nappe of Fuchs. Different opinions have been expressed about displacement in Vaikrita Nappe, one views says that there is not much displacement in the nappe as it is situated near root zone, other says it a superficial nappe or paraautochthonous in nature and the third views says displacement in this nappe may be as much as of 100 km.

The equivalent of Vaikrita Nappe in other regions are Tibetan Slab (Lombard) or Upper Crystalline Nappe (Fuchs) or Khumbu Nappe (Haden) in Nepal, Thimpu in Bhutan, Darjeeling series in Darjeeling in West Bengal and Chasilakha Takhtsang Gneiss or Sela, Group in Arunachal Pradesh. Klippen of Vaikrita Nappe occur extensively in Kashmir, Chamba, Nepal, Sikkim and Bhutan.

Jutogh Nappe

Its equivalent in Uttaranchal is called Almora Nappe, which is one of the largest nappes in the Lesser Himalaya. Its southern boundary in Himachal Pradesh is defined by Jutogh Thrust or MCT II; while in Uttaranchal by South Almora Thrust (SAT) and in north its limit is defined by North Almora Thrust (NAT). It is made up of medium grade metamorphic rocks. The inverted sequence in the Jutogh Nappe is attributed to recumbent folding. Some worker suggested that the stratigraphic sequence is in normal order challenged this view. Its root is identified in Manal ridge, Wangtu area of north of Rampur window in Himachal Pradesh and in the Badrinath Nanda Devi region in Uttaranchal. In the root zone (10-12 km wide) deep-seated steep to vertical dipping rocks are present. It is equivalent to Lower Crystalline Nappe (French) or Scale Zone or Kathmandu Unit of Central Nepal and Paro and Bomdila in the Eastern Himalaya. In Uttaranchal region, nappe shows on either side thrust bound synformal structure and is divided into Ramgarh and Almora groups. Its unmylonitized part shows polyphase (F_1 and F_2) deformation of Proterozoic time. Several klippen like Amri, Banali, Askot, Thal, Chankari, Dharanghar, Baijnath and Nandprayag represents extension of Almora Nappe in Garhwal Kumaun region of Uttaranchal.

Chail Nappe

Heim and Gansser described the southern boundary of the Chail Nappe as Main Central Thrust. It is equivalent to Salkhala Nappe in Kashmir, Bajaura Nappe in Himachal Pradesh, Ramgarh Nappe is equivalent to Bijni Thrust sheet in Uttaranchal and Daling Nappe in Sikkim and Bhutan and Arunachal and Samchu and Sumar Nappes. The Chail nappe and its subsidiary units, having displacement up to 100 km in south, have been recognized all along the Himalaya.

In Uttaranchal, the Chail Nappe is made up low to medium grade metamorphic rocks (=Munsiari Formation) of Archaean age (2600-2850 Ma Old). These rocks have traveled along Munsiari Thrust (10 Ma) = MCT III or MCT of Heim and Gansser. The Chail rocks are intruded by granites and show 8-15 km wide zone intense mylonitisation in frontal region. The Chail acts as the basement for the Lesser Himalayan rocks and its root is not exposed as is overlapped by Jutogh Nappe.

Sedimentary Nappe

Jaunsar Nappe

It is mainly developed in Himachal Pradesh and partly in the western Garhwal, Uttaranchal and is believed to represent a superficial nappe. The rocks of Jaunsar Group comprising of Mandhali, Chandpur and Nagthat formations constituting the basal sequence of Inner Krol Belt and are moved over the foundation of Deoban Group. Jaunsar is correlated with Shimla Group rocks of Himachal Pradesh.

Garhwal Nappe

Auden (1937) first time reported this nappe in the Garhwal region of Uttaranchal. It is made up of two thrust sheets namely the Bijni Nappe (less traveled) and the Amri Nappe (more traveled) of crystalline nature containing schistose phyllite, quartzite, gneisses, migmatite and gneissic granite (Kumar and Dhaundiyal, 1979; Rawat, 1979)). The crystalline rocks are equivalent to the Chandpur and Nagthat formations and are thrust over the Krol Belt. The root of this nappe lies in Higher Himalaya along the MCT. Banali and Satengal klippen are extension of Garhwal Nappe. The Garhwal Nappe has 80 km length in south in Garhwal and 95 km in western Kumaun and 115 km in eastern Kumaun. Garhwal Nappe in general forms a broad open synform. Shanker and Ganesan (1973) divided Garhwal Nappe into Amri, Lower Bijni and Upper Bijni units. Garhwal Thrust has brought down these rocks. This nappe is considered equivalent to Ramgarh Nappe or Chail Nappe.

Fuchs instead of Garhwal Nappe used the term Chail Nappes due to presence of several subsidiary units. In between Ramgarh Thrust and South Almora Thrust, Bijni Nappe unit is recognized. At many places Ramgarh Thrust brought rocks of Chail or Ramgarh in contact with Nagthat Formation of Krol Nappe. Saklani (1993) identified, like Fuchs, it as Chail Nappe and divided into Chail Nappe I, Chail Nappe II and Chail Nappe III on the basis petrological and structural characters.

Berinag Nappe

It is made up of coarse grained massive to bouldery sericitic quartz arenite. Berinag Thrust and Munsiari (Chail) Thrust demarcate its lower (southern) and upper (northern) boundaries respectively. This nappe is developed in the inner Lesser Himalaya.

Krol Nappe

Krol Nappe or Krol belt is nearly 400 km long having more than 6 km thick mainly sedimentary sequence is situated between Kunihar (Himachal Pradesh) in the northwest and Nainital (Uttaranchal) in the southeast and is further extended into farwestern Nepal. The sequence comprises of sedimentaries like carbonates, conglomerate, shales and sandstones and very low-grade metamorphics like phyllite, quartzite and slates. Subsequent deformation of this Belt has resulted into formations of several synclines named as Pachmunda, Krol, Raigarh, Khanog, Saindhar, Nigali, Korgai, Mussoorie, Garhwal and Nainital. Auden (1934) recognized this belt as nappe with Krol Thrust at the base in the eastern and Giri Thrust area in the western Himachal Pradesh. However, the allochthon nature or Krol Belt sequence is strongly refuted by Rao (1968). Bhargava (1976) and Valdiya (1980) suggested parautouchthon nature of the Krol Belt.

Bhargava (1976) divided this belt into Outer Krol Belt (a superficial nappe) and Inner Krol Belt. The sediments of the Krol Belt are deposited on the foundation of Shali/Deoban Formations. The Outer Krol Belt, developed between Kunihar to Kill, is comprised of the Shimla Group followed by the Blaini, Infra-Krol, Krol and Subathu formations and is structurally bounded by the Krol Thrust to the west and the Giri Thrust in the west. The Inner Krol Belt extends from Halog to Nainital, as a whole thrust bound synform along the northerly dipping Chail Thrust and southerly dipping Tons Thrust and is comprised of the Mandhali, Chandpur, Nagthat, Blaini, Infra-Krol, Krol, Tal and Subathu formations and is moved over foundation as Shimla Group as superficial nappe.

Arunachal Pradesh

Tectonic setting of Bhutan Himalaya continues in the Arunachal Pradesh. The low-grade rocks are overridden by high-grade nappe, klippe of Gondwana rocks are also reported from outer Himalaya. Thrust sheets or nappes propagated as in piggyback fashion, Siang, Siyam and Rikar are the three thrust sheets developed in the Arunachal Pradesh. They are present in descending order and towards the core Abor Volcanics are exposed along the Yingkiong and Dalbuing Faults (Santra, 1979). However, Acharyya (1978) divided nappes into two categories namely the lower nappe system consists of sub nappes of Permian Gondwana, Daling Baxa System and Late Precambrian to Middle Late Palaeozoic, and the Upper Nappe system made of gneissic rocks of mainly of the Paro Group.

Nappes in Nepal

Hagen initially explored the geology of Nepal in 1950's and 60's. He identified multitude i.e. a pack of nappes in Nepal and divided these into two categories large nappes and secondary nappes. In Lower Nepal Himalaya two units namely the Nawakot-Piuthan unit (sedimentary or low grade metamorphics) and the Kathmandu Unit (Crystalline) at the base of Lesser Himalaya were primarily recognized. The former unit is equivalent to Krol Nappe and latter is equivalent to Garhwal Nappe/ Almora Nappe of Kumaun. The Nawakot-Piuthan Unit comprises of four nappes namely the Lowest Nawakot (I), Paraautochthonous Piuthan Nappe (II), Bajang Nappe (III) and paraautochthonous Dadeldhura Crystallines (IV). Around Kathmandu and in the east Hagen identified five nappes with in Kathmandu Nappe. In further east, Khumbu Nappe is equated with Kathmandu Nappe Hagen divided Khumbu Nappe in three nappes and the Upper most Khumbu III nappe is further subdivided in three subnappes. Lumbu, Sumba and Upper Kangchindzonga areas in Higher Himalaya are root zones of crystalline nappes.

Recently, Upreti (1995) presented nappe geology of the central and far western Nepal and identified following crystalline nappes Kathmandu Nappe, Jajarkot Nappe, Bajhang Nappe, Dadeldhura Nappe and Karnali Nappe or Chakhure-Mabucrustallina Klippe. In Kathmandu Nappe and Jajarkot Nappe the metamorphic sequences are capped by lower Palaeozoic rocks without tectonic break and in the Karnali Nappe Tethyan rocks are thrust over the crystalline rocks in a conformable sequence. The Makalu Nappe in the eastern Nepal is thrust over the Barun Nappe. However in Barun-Meladu section of Arun Valley, Kalvoda identified following nappe in descending order Chamolangma partial nappe, Makalu partial nappe, Barun Nappe and below the MCT Tinjur Nappe and Buranse Nappe, he further advocated that few of these are relief nappes (Jaros, 1982).

Bhutan

In Bhutan, Jangpangi (1974) identified lower synclinorium Sumar nappe, overlying Baxa Formation, occupying nearly 800 sq km area and is comprising of low grade rocks such as phyllite, quartzite, slate, phyllitic schist etc. The upper Thimpu Nappe comprising of biotite schist, biotitic gneisses and garnet-kyanite gneiss. It is equivalent to adjoining Darjeeling Gneiss or Vaikrita Formation. It shows nearly 70 km displacement and underlying Sumar Formation is exposed due to erosion.

Nappe of Tethyan Rocks

The miogeosynclinal Tethyan rocks show SW or NE verging folds. Allochthons of the Tethyans are present in the Kashmir and Chamba in India and in the Phulchowki area of Kathmandu in Nepal; otherwise Tethyan is essentially of autochthonous nature.

Suture Zone Nappe

The eugeosynclinal sequence of suture zone occurs as 'exotic' blocks on Tethyan sequence in Pakistan, Jammu &Kashmir, Himachal Pradesh, Uttaranchal and Nepal. Since some part of these sequences had moved as superficial thrust sheet over a distance of 30-80 km in south. These transported rocks include volcanics, flysch, and radiolarians, chert mixed with Palaeozoic-Mesozoic exotics of basement rocks. These transported rocks occur at Amlang la, Jungbwa, Ship Cha, Kiogar, Karzak Fu, Kiogar, Dargai and Spontang. In general, it is observed that in various klippen typically imbricate platform sediments have been overthrust by sheet of ophiolitic mélange, which is overlain by slices of ultramafics compositions observed in various klippen. It is believed that in root zones these suture zone rocks must have suffered intense buckling and uplifting

Jungbwa Ultramafics Nappe

This nappe covers an area of 3500 sq km and has 500 m thickness. It is made up of predominantly harzburgite that is subsequently serpentinized (antigorite). Its composition indicates that high P-T rock represents to upper mantle environment, brought up on the surface by obduction process and finally this month slab was thrust in south direction.

Origin of Nappe

Origin of Himalayan nappes is explained by two mechanisms; one view says that nappes are formed at the expense of sequeezing from one or more root zones and the other view advocates that the nappes are evolved due to decrease in width of original synsedimentary structure. However, Ashgirei explains that root zone of nappes lies in Main Axial Thrust, located at the boundary of Higher Himalaya and Tethyan Himalaya and nappe formed in two periods first in the Early Riphean and second after Middle Eocene. Root zones could have been in Riphean and Phanerozoic geosynclinal basins that were subsequently compressed and overlying nappe conceals actual root area. Gansser suggested that nappe formation might be due to decrease of initial width of synsedimentary structures. There are different roots for different nappe. Nappe formed in two stages; pre-Eocene nappes are related collision along Indus Suture and Miocene nappe are related with development of MCT. Shovalman recognized three stages of nappe evolution i.e. pre-Eocene, Middle Miocene and Pliocene.

4.9 Himalayan Suture Zones

The region of collision where one plate underthrust beneath the other plate is called subduction zone or suture zone as in this zone two plates of contrasting nature are joined or stitched together. In case of the Himalaya subducting plate was Indo-Australian plate and stationary plate was Eurasian plate. The main development of suture zone has taken place in the Tibet and is called the Indus Suture but in Ladakh and North Pakistan the Indus Suture is bifurcated or forked into two sutures, the 100 Ma old northern suture is called Shyok Suture in India and Pakistan and is Bangana Suture in Tibet and 55-60 Ma old southern one is called Indus Suture. Fig No. 4.5 shows a geological cross-section running across these cross-sections. In Ladakh, Ladakh Batholith separates these sutures by different types rocks. The Tethys Sea existed before the evolution of Shyok Suture is called Palaeo-Tethys while at the time younger Indus Suture is called neo-Tethys. However, the sea was one continuous body but because of change in tectonic condition, this distinction is made for convenience.

Indus Suture represents intra-oceanic island arc setting i.e. Mariana type and Shyok Suture exhibits ocean-continent island arc setting i.e. Andean type (Ahmad et al., 2005).

Many workers do not believe in Indus Tsangpo Suture Zone as boundary of Indian plate, they consider actual boundary lies in further north in Tien-Shan mountain, hence according to them Himalaya is intra continental mountain range.

4.9.1 Indus Suture

Indus Tsangpo Suture Zone, abbreviated as Indus Suture, is narrow tectonic belt of more than 2500 km length uniting the Indian and Eurasian plates together into one landmass and is mainly developed in Tibet. It runs along the Indus River in the west and along the Tsangpo on Brahmaputra River in the east, however, a small and tectonically highly complex portion of the suture occurs in the Ladakh of Kashmir and in the Pakistan. Further north of Indus Suture in Tibet, another suture is developed and the land block between these two sutures is known as Lhasa Block. Several continental size strikes-slip faults are present to the north of Indus Suture and they have displaced the terrain to considerable distance and have also assisted in consumption of Indian plate.

The Indus Suture Zone is tectonic unit lying between Tethyan Thrust in south and Indus Thrust in north and in Pakistan it is called Main Mantle Thrust (MMT). It is composed of 5 km thick ophiolites, few hundred metres wide narrow belt of ophiolitic mélange. Cretaceous-Eocene flysch and variety of igneous rocks formed in diverse tectonic setting, representing obducted fragments of Neo-Tethys Sea and is also called Indus Tectonic Belt. Ladakh oceanic magmatic arc (Ladakh Batholith) separating the Indus Suture and Shyok Suture is related subduction of Indian plate.

The Indus Suture displays block-in-matrix fabric i.e. diverse blocks are imbricately arranged within the disrupted turbiditic matrix. In Spiti-Zanskar region Indus Suture is mainly represents volcano- sedimentary rocks of tensional origin. Blue schist rocks occur as slices within the ophiolitic mélange. The blue schist is derived from volcanic, pyroclastic and pelitic parental materials and glaucophane and lawsonite are characteristically present in these rocks. Mafic inclusions in blue schist suggest alkaline affinity. Ophiolitic mélange is not a true ophiolitic sequence but a mixture of ophiolitic material and tectonically emplaced litho units of diverse origin called olistolith and olistostromes. Ophiolite mélange zone is composed of disorganized ophiolitic breccias, mudstone olistolith of mafic volcanic, shallow water and pelagic limestone, disrupted turbidite, red, green, gray and white chert, magnetite, harzburgite, blue schist, amphibolite, scaly cleaved serpentinite, pyroxinites, fine grained dolerites and basalt. Matrix is silt to clay size particles of sandstone, siltstone and serpentinite. Among sedimentaries, flysch is deposited over pelagic sediments and is thrust over Indus Formation. Molasse were deposited in two separate fore-arc intermontane basins. Molassic sediment is dominated by conglomerate; it is 400 m thick in Kailash area and 6-7000 m thick in Shigastic belt. These belong to Upper Cretaceous to Paleocene in age. Chert, radiolarian mud stone, radiolarites are siliceous deposits present in Indus Suture and show equatorial condition of deposition and are obducted around 45 Ma ago.

Indus Suture Zone (ISZ) represents a steeply dipping zone. The seismic profile of this area indicates a vertical fault zone and Moho is faulted and shows a throw of 20 km. In ISZ principal thrust dips in SW and major folds have N vergence while in all regions reverse relationship is present. The ISZ has been strongly squeezed due to tectonic activities of later age and its klippen are normally transported to 20-30 km distance in south direction and exceptionally up to 80 km, (for example in Maltu area). Due to deformation and metamorphism, ophiolites are deformed, foliated and serpentinized. Two prominent thrusts are recognized in the zone, one between ophiolite mélange and Indus molasse, and the other one between Khardung Volcanics and Shyok Ophiolite Melange. ISZ represents to an accreted trench developed in convergent trend setting. Dras Volcanics of this area is yielded 170-150 Ma age i.e. the age oceanic crust.

Structurally, folds in suture zone are overturned in style, trending NW-SE to WNW-ESE direction and show large amplitudes.

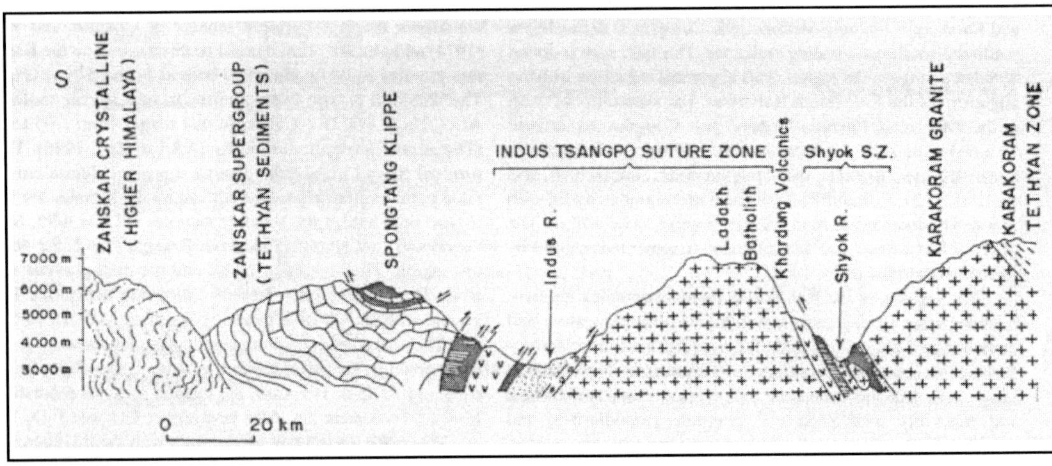

Figure 4.5. Geological cross-section across Indus and Shyok suture zones (after Thakur, 1981 b)

Evolution of ISZ

Presence of Late Mesozoic pelagic clay, radiolarian chert and orbitolina limestone, and calc-alkaline lava throughout in the Indus Tsangpo Suture Zone suggests presence of close to Middle Cretaceous a deep, narrow, elongated basin representing oceanic trench and island arc pair. This pair was developed at the northern margin of Tethyan basin adjacent to Tibet-Karakoram continent. The estimated 300-500 km gap between India and Tibet was gradually closed due to subduction and of Tethyan Sea was completely closed probably in Middle Eocene. The squeezed up trench is represented by the Dras Volcanics. Ophiolite and ophiolitic mélange indicate obduction as Indian plate pushed further in the Eurasian plate. Due to consumption of Indian plate in Late-Cretaceous Paleocene, granitic bodies were formed in Ladakh and Shigaste areas. At the same time proto-Himalaya was evolved.

4.9.2 Shyok Suture

The Shyok Suture also described as Northern Suture in Pakistan and is developed between Kohistan arc and Karakoram micro plate of Pakistan while in Ladakh it is situated north of Ladakh Batholith. Karakoram block representing the southern margin of Eurasian plate and is situated north of Shyok Suture. In Pakistan Main Karakoram Thrust (MKT) draws southern boundary of Shyok Suture. The time span of development of Shyok Suture is constrained between 100-70 Ma. The Karakoram Batholith is a major composite body of granitoid rocks intruded into sedimentary and metamorphic rocks of Paleozoic-Mesozoic age. Granodiorite, quartz-monzonite, tonalite; S-type, syn-collision, two mica granites are present in the composite body. In the further eastern side of Karakoram, pre-collision I-type granite of Jurassic to Early Cretaceous age are present. In Pakistan between Shyok and Indus sutures i.e. between Main Karakoram Thrust and Main Mantle Thrust (northern boundary of Indus Suture) following sequence is present in ascending order Jizal complex, Komila Amphibolite, Chilas Complex, Ladakh-Kohistan Batholith, Rakaposhi Volcanic complex, Yasin Group/other isolated sedimentaries and Chalt Melange. In Ladakh, pre- (100-83 ± 9 Ma), syn- (63-43 Ma) and post (15-9 Ma) magmatic bodies stitch the Karakoram area with Eurasian and Indian plates. Kohistan island arc belt is situated between Indian and Karakoram plate. This arc comprises of ultramafics, gabbroic complex, metavolcanics, mid oceanic ridge type tholeiitic basalt and calc alkaline andesite. The Shyok Suture Zone is essentially an ophiolitic mélange zone comprising of basic volcanics, flyschoidal, sediments, pillow lava, andesite, dacite, pyroclastics, serpentinite, gabbro and diorite and is intruded by granitoids by both on southern and northern margins.

Rocks of Shyok ophiolitic mélange and Karakoram area have thrust to the south. In Shyok Suture, three generations of folds have been recognized, the F_1 folds are mainly reclined in nature, plunging with 30°-50° in N20°E-N50°E directions and this movement is also responsible for formation of a major antiform and synform couples. F_2 are moderate to tight upright minor folds trending in NW or SE directions. F_3 are minor upright, chevron and kink folds trending in NE or SW direction.

4.10 Tectonic Windows

Tectonic windows are identified in Lesser, Higher and Trans-Himalayan regions. In the Lesser Himalaya, Bhargava (1982) identified 19 windows and divided the Lesser Himalaya for the purpose of windows into three zones namely the Outer, Central and Inner Zone. In the Outer Zone, three thrust sheets namely the Outer Krol Belt, Inner Krol Belt and Daling Thrust sheet are identified and are resting on the Tertiary or Gondwana sediments. In the Outer Krol Belt are Solan and Anji (Subathu rocks are surrounded with Infra Krol and Kewala Khal (Sanjauli Formation surrounded with the Blaini Formation) windows are present. In the Inner Krol Belt, Bidhalana and Pharat windows (Shimla Group rocks are surrounded with Subathu, Chandpur and Blaini-Krol rocks) are situated near Dehradun and in the Sikkim, Rangit window (Rangit Pebble of Baxa and Gondwana, rocks an surrounded with Daling Group rocks) are present. These windows are complete and thrust sheets are described as superficial nappe. In Central Zone, partial windows are exposed. Shali Window, (with in Chail rocks, Shali rocks are exposed; Pilgrim and West, 1928), Deoban Window (Jaunsar rocks exposed in Deoban group rocks), Nawakot (Kathmandu) Window (quartzite, phyllite of Sanguri Formation and stromatolitic limestone of Nawakot Group are exposed in Kathmandu Thrust sheet) and Daling Window (Daling rocks in Darjeeling Crystalline) are present. In the inner window zone, Kishtwar Window (Rampur quartzite and phyllite in crystalline Salkhala Thrust sheet), Rampur (Rampur Group in Khamrada, Salkhala Thrust sheet, Larji (Larji rocks in Rampur Thrust Sheet), Tejam – Pithoragarh (Tejam rocks in Almora-Jaunsar Thrust sheet) Galwa Outer, Galwa Inner, Tamba Kosi Outer, Tamba Kosi Inner, Tumlingator and Okhladunga – Tabling windows are present. Window in window is reported from Kulu-

Rampur-Larji area where Proterozoic rocks of Kulu-Rampur-Larji window tectonically lie over early to middle Riphean rocks of Larji window.

In the Siang Window, the Palaeogene sediments are exposed mainly in the Abor volcanics. In Subarnsiri district, Arunachal Pradesh, sedimentary rocks of the Menga Formation, Sipi Formation and Raje Quartzites are exposed in the crystalline Daporijo Gneisses and Mara Schist in Sipi-Mara window.

In Higher Himalayan, Lesser Himalayas low-grade metamorphites are exposed within crystalline sheet as window. In Ghuttu Window in Bhilangana Valley of the Uttaranchal, Chandpur Formation of the Lesser Himalaya is exposed in the Ghuttu Member of the Higher Himalaya (Islam and Thakur, 1988). Along Yamuna River, three eyelid windows at Banas – Phulchatti, Janakichatti and Yamunotri are identified by Biyani (2004 c). In these areas, Paligad Quartzite of the Lesser Himalaya is surrounded with crystallines of theHigher Himalaya.The Lesser Himalayan rocks are exposed due to deep erosion along the normal faults. Okhladunga, Arun, Tamur and Darjeeling Windows show Lesser Himalayan formations surrounded with the Higher Himalaya rocks (Meier and Hiltner, 1993).

4.11 Unconformities in Himalaya

Frequently interrupted sequences of Late Archaean (Munsiari Formation of Kumaun; 2600-2850 Ma old) to Recent sediments constitute the stratigraphical records of the Himalaya. Each interruption or break indicates presence of unconformities and in some unconformities time gap is of many hundred million years while in some it is just few tens of thousands years. Ravishanker et al. (1989) meticulously compiled the stratigraphical column of the Himalaya and identified total fifteen super sequences in it. According to them a super sequence is a grouping of rocks formed in different parts of Himalaya during a particular geological time and is separated by underlying and overlying unconformities. In short, supersequence is a block of homotaxial lithounits demarcated by unconformity. A brief description of these supersequence are as follows.

Supersequence I

The oldest rocks of Late Archaean or Early Proterozoic age like Munsiari Formation, Suru Crystalline, Himal Group, Thimpu, Sela Group, Kanchanjunga etc. constitute this supersequence.

Supersequence II

Constituted by Palaeoproterozoic Bafliaz, Rampur-Shali-Sundernagar Darla, Garhwal Group, Dharagad-Deoban, Dailekh, Pokhra, Lakharpata, Chunthang, Daling, Baxa, Paro, Bomdila and other formations.

Supersequence III

This Mesoproterozoic supersequence comprises of Gamir-Bailu Ramban, Salkhala, Kulu, Shimla, Jutogh, Vaikrita Mandhali – Chandpur – Nagthat, Basal Martoli, Bhimphedi, Jaljala, Shumar, Dirang and Lumla groups and formations.

Supersequence IV

Represents the duration between Upper Neoproterozoic to Middle Cambrian. Blaini-Krol-Tal, Ramsu Machhal, Batal, Milam, Kunzumla, Tistuing are the important groups of this sequence.

Supersequence V

The time range of this sequence is Ordovician to L Carboniferous. In this duration mainly Tethyan rocks were deposited e.g. Ralam, Garbyang, Muth, Phulchowki and other formations.

Supersequence VI

Lower Permian Agglomeratic Slate, Panjal Volcanics, Ganmachidam–Gechang, Aulis Basalt, Gondwana, Bhareli–Abor Volcanic are important formations/groups of this sequence.

Supersequence VII

Upper Permian and partly of Triassic formations such as Zewan, Kuling–Chocolate Limestone–Niti –Kalapani Yamune formation are forming this sequence.

Supersequence VIII

This partly Triassic and partly of Jurassic age supersequence is recognized mainly in Tethyan Himalaya e.g. Kioto–Lapthal formations.

Supersequence IX

Representing the Upper Jurassic and Lower Cretaceous rocks of Tethys e.g. Wuyan, Lilang – Tandi, Quartzite series.

Supersequence X

Upper Cretaceous–Lower Palaeocene Manikot Shell Limestone, Sangcha Malla Formation, Balchadhura Volcanics and other formation are represented by this supersequence.

Supersequence XI

Upper Paleocene to Middle of Middle Eocene Subathu, Kanji, Bhainskoti, Dalbuing, and Geku Formations are incorporated in this sequence.

Supersequence XII

Muru, Dharmsala, Kasauli-Dagshai, Dumri/Suntar formations of Middle Eocene to Oligocene age are included in this sequence.

Supersequence XIII

Widely developed Siwalik Group of Miocene to Middle Pleistocene is included in this supersequence.

Supersequence XIV

Middle and Upper Pleistocene, Older Alluvium, Shyampur Formation, Upper Karewa, Diklai Boulder Bed and Hapoli formations are included in this.

Supersequence XV

Newer Alluvium terraces of Holocene age are included in this super sequence.

4.12 Syntaxial Bends in Himalaya

The Himalayan strike takes sharp turn towards south or southwest at its eastern and western ends. Such regions of sudden strike change are described as syntaxial bends, here the rocks show hairpin or loop likes appearance and along the axis of loop matching geology occur on either sides. Two syntaxial bends namely the eastern and western are recognized in the Arunachal Pradesh and Pakistan respectively.

4.12.1 Eastern Syntaxial Bend

Proterozoic to Cenozoic rocks occur in thrust bound lithotectonic Siang, Siyam and Rikar units and paraautochthonous zone comprising of Abor Volcanics, Yanking Formation and Dalbuing Formation in the Siang gorge of Arunachal Pradesh. The strike deflects NW-SE to NE-SW. The lithostratigraphic succession of the Eastern syntaxial Bend area is as follows:

Siwalik Group		
Paraautochthonous		Dalbuing Formation
Abor Volcanics		
Gondwana Group		
Nikte Quartzite		
	Nellenge Thrust	
Rikor Thrust Sheet		Rikor Group
	Siyam Thrust	
Siyam Thrust Sheet		Siyam Group
	Siki Nala Thrust	
Siang Thrust Sheet		Siang Group
	Trans Himalaya	
Trans Himalaya		Tuting Volcanics
Tuting Granite		

Structure and Tectonics

Syntaxial bend shows three phases of deformation and metamorphism Proterozoic age is assigned to M_1 and D_1 event while D_2 is post-Eocene characterized by thrusting movement and M_2 is pre-thrusting. The Pleistocene age D_3 event is responsible for origin of syntaxial bend. Major antiformal structure in syntaxial bend comprising of several minor antiforms and synforms namely Menchukha-Monigong Synform, Siyam Antiform and Synform, Igo Synform, Yang Sang chu Antiform and Synform, Simans Antiform and Synform and Roing

Synform. Several regional faults such as Tuting Igo (N-S orientation and nearly 120 km length) Siman Simen Fault (N-S, 100 km), Yamune Fault (NW-SE, 75 km), Bame-Dumro Fault (NE-SW) and Nikte Riew Fault are present in the Bend.

Origin of the Eastern Syntaxial Bend

E-W compressive force generated from the eastern direction by moving Burmese platform east in Pleistocene is responsible for the formation of ESB. However, Wadia in 1931 proposed that ESB is caused by bending of Himalayan orogen against northeast projecting foreland of Shillong Plateau, and some workers have also suggested it is due to joining of two units of different nature along a tectonic contact (Singh, 1995).

4.12.2 Western Syntaxial Bend

The Western Syntaxial Bend (WSB) is now situated is NW Pakistan. The NE-SW strike of the Himalaya appears to terminate at Nanga Parbat (8119 m) just near the deep gorge of Indus River. However, instead of termination, strike suddenly changes the direction to the south and then to the southwest. All the geological formations here take sharp U shape or hairpin like bend and the geology across the bend axis is like a mirror image. Proterozoic granite and gneisses, Salkhala Group and Dogra slates and Palaeozoic, Mesozoic and Cenozoic rocks are present in the bend. In syntaxial bend, continuity of stratigraphy, structure and rock types on the Kashmir flanks of the bend pass over into Hazara regions of Pakistan without any tectonic discordance, even the individual folds are traceable from one side of the loop to other side (Wadia, 1930, 1937).

4.13 Upheaval Phases of Deformation in Himalaya

The Himalayan tectonic mountain shows complex structural and tectonic set up. Two to six phases of deformation are proposed for the evolution of entire range and are correlated with major deformational activities having impact over very wide region in the Himalaya. In a smaller region not necessarily all the phases of deformation are present. At macro level usually two to four events of deformational phases are present.

Wadia (1975) recognized three important phases of deformation the first was the post-Nummulitic (Eocene) age and this phase was culminated in the Oligocene, second was Middle Miocene (most powerful) and third was post-Pliocene to Middle Pleistocene in age, He also proposed that slow movements were started from Permian time and upheaval were continuous during Tertiary. Raina (1982) recognized three phases and called them Ladakh Phase (Late Cretaceous–Early Eocene), Dharmsala Phase (Middle Miocene) and Siwalik Phase (Plio-Pleistocene).

Krishnan (1982) identified five phases of deformation in the Himalaya. The first movement took place in Upper Cretaceous, second in Upper Eocene, third in Middle Miocene (most powerful), fourth in the end of Pliocene and final phase in Early Pleistocene. Valdiya (1984 a) also recognized five phases and designated them as Karakoram Phase (Cretaceous–Palaeocene), Malla Johar Phase (Late Eocene to Oligocene), Sirmurian Phase (Middle Miocene), Siwalik Phase (Late Pliocene-Middle Pleistocene) and Holocene Phase (<1.8 Ma).

Kumar (1977) described six deformational phases for Himalayan rocks of Arunachal Pradesh. Each phase has produced own style of folds. Early two of these phases belong to Proterozoic orogenic movement (Wutaian-Luliangian and Zhongyuean – Sibaoan). He also recorded evidences Pan-African (also by Valdiya, 1993), Tanzanian and Insignia orogenic movements, beside these four other phases were related with Himalayan orogenic movement.

4.14 Suspect Terrane

A terrane is defined as an area characterized by internal continuity of geology (stratigraphy, fauna, structure, metamorphism, igneous activities, metallogeny, geophysical characters etc.) that is bounded by faults or mélanges representing trench complex or cryptic suture across which neighboring terrains may have similar geological record (proximal terrane) that may only be distinguished by present terrains boundary representing telescoped oceanic lithosphere while suspect or displaced terrain are allochthonous block accreted to the continent over considerable period of time. These allochthonous traveled, before accretion, great distance. A suspect terrain is a geologically mappable unit characterized distinct geological history from adjacent block. The process of amalgamation of terrain different nature is called mosaic tectonics or accretionary collage. The terrane model has been applied on regional level for example entire Himalayan Alpine region (Fig No. 4.6)

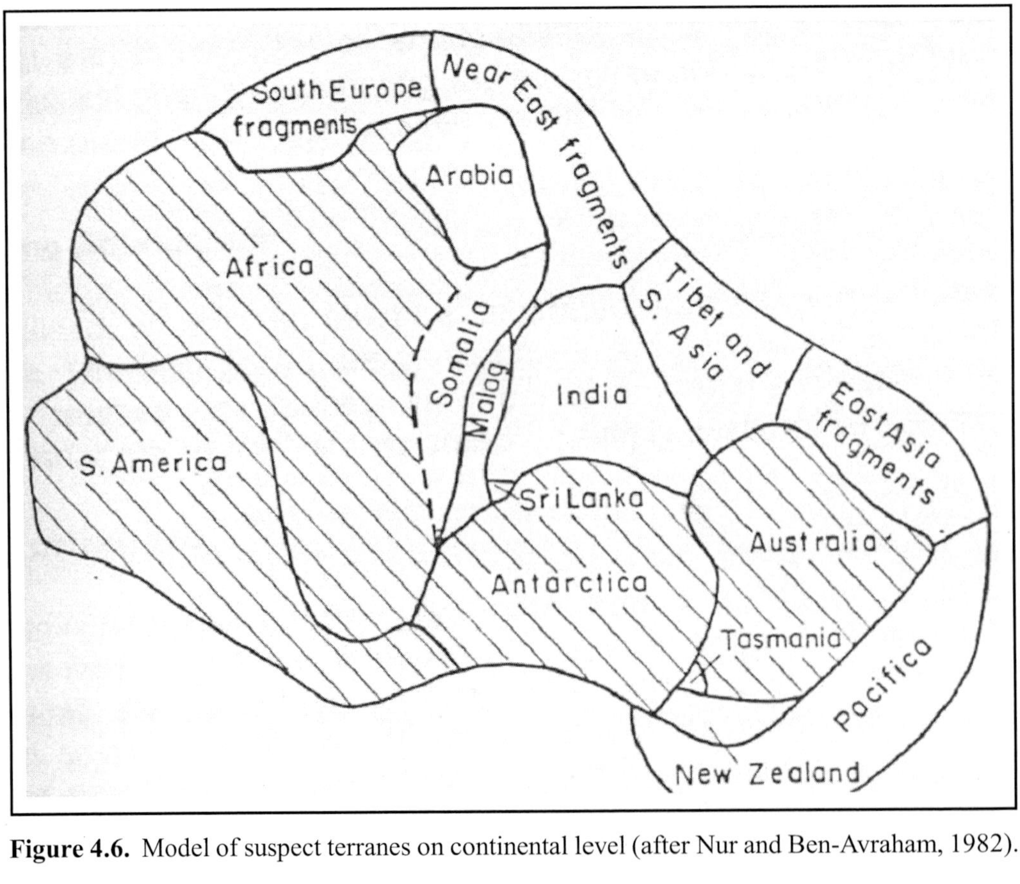

Figure 4.6. Model of suspect terranes on continental level (after Nur and Ben-Avraham, 1982).

In Himalaya, first time this idea was applied by Radhakrishna (1989) who proposed that longitudinal units such as the Outer Himalaya, Lesser Himalaya, Higher Himalaya etc. are representing suspect terrane. These terrenes are bounded with thrust and normal faults (Fig No. 4.7).

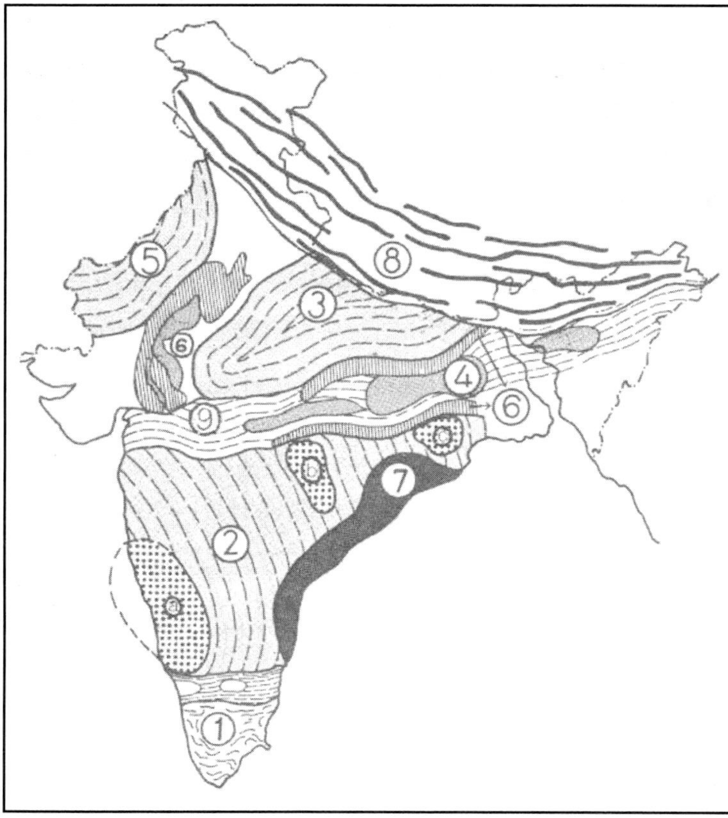

Figure 4.7. Suspects terranes in India (after Radhakrishna); (1) Gneiss-granulite Terrane of Tamil Nadu-Kerala, (2) Early Proterozoic magmatic Terrane-the southern Peninsular block with Archaean nuclei of Karnataka, Bastar, Singbhum as oval block, (3) Early Proterozoic magmatic terrane-the northern Bundelkhand block, (4) Chottanagpur terrane, (5) Marwar Terrane, (6) Early Proterozoic metamorphic fold belts of central and western India and Aravalli, (7) Mid-Proterozoic Eastern Ghat terrane, (8) Phanerozoic accreted terranes of the Himalaya, (9) Central Indian Tectonic Zone.

4.15 Neotectonics

Neotectonics is the study of deformation and movement, which have taken place since the beginning of Quaternary (Pleistocene and Recent) period. However, some have extended this time boundary to Middle Miocene. Quaternary begun about 1.8 Ma ago, it means primitive men have also witnessed deformation and movements. Deformative or tectonic movements started with the collision of plates (60-55 Ma ago) and they have shown episodic, waxing and waning, behavior. Indian plate is even today actively subducting beneath the Eurasian plate; therefore, deformational processes are operational in the Himalaya since the onset of the Tertiary. The neotectonic and active tectonic (very recent) movements are latest part of main Himalayan movements. The impact of neotectonic movements is far more intense and visible as well as raises the curiosity of society in general. Thus the aim of neotectonic study in the Himalaya is to identify and analyze the impact of deformations occurred in post-Tertiary duration. Neotectonics study helps us in recognizing the types of deformational movement (compressive or tension, or continuous or discontinuous), their relative severity and intensity (rapid or slow), impact on surface (change in elevation, i.e. uplift or subsidence and horizontal shifting or combination of these) and their effect on landforms. Neotectonics movements not only create tectonic features such as folds, faults, warping of land with change in elevations but also generate earthquakes. Neotectonics also produces a set of readily recognizable landforms, for example, river terraces, steeply sloping scarps, relatively straight course of rivers, gorges, chain of hot and cold water springs, waterfalls, entrenched meanders, abandoned river channels, two or multi stage topography etc (Biyani 2002 a).

Neotectonic study can be classified into quantitative and qualitative. In quantitative study rate of plate movement, magnitude and orientation of stress and strain, horizontal and vertical displacement along fault planes (total and annual average), magnitude and annual average of upliftment or subsidence of a large area, age of particular tectonic activity or even age of geomorphic features is measured and interpreted (Krishna, 1986). In qualitative approach relative assessment of movement is made through the study of landforms.

In the Himalayan region, evidences neotectonic activity are practically present everywhere because Indian plate is still actively underthrusting beneath the Eurasian and Burma plates is the north and east respectively. A brief assessment of neotectonic activity is as follows:

Plate Movement

The Indian plate is drifting in north direction with an average rate of 5 cm/year in Quaternary period. It means in last 1.8 Ma, nearly 90 km wide stretch of Indian plate is consumed beneath Eurasian plate. In fact, consumption is not a smooth affair along the suture zone but is distributed among different tectonic elements within the Himalaya and also to the north of the Himalaya. Fig No.4.8 indicates rate of deformtion and plate movement in Himalaya and adjoining areas.

Major Tectonic Features

The Himalayan Frontal Fault, NE Syntaxial Bend in Arunachal Pradesh, Tidding Suture are results of neotectonic activities. Various 'duns' (synclinal valleys) like Dehradun Valley, Kathmandu Valley, 5 km rise of Tibetan plateau and folding in Siwaliks are the result of neotectonic deformations in the Himalaya.

Earthquakes

Himalaya is a highly earthquake prone area; due to northward drift of Indian plate, strain is continuously accumulating in the tectonic belt and is being released time to time in the form of earthquakes in different places not only in the Himalaya but also in the Indo-

Gangetic Plain and in the peninsular India in the form of back strain. Earthquakes are discussed in the chapter 10.

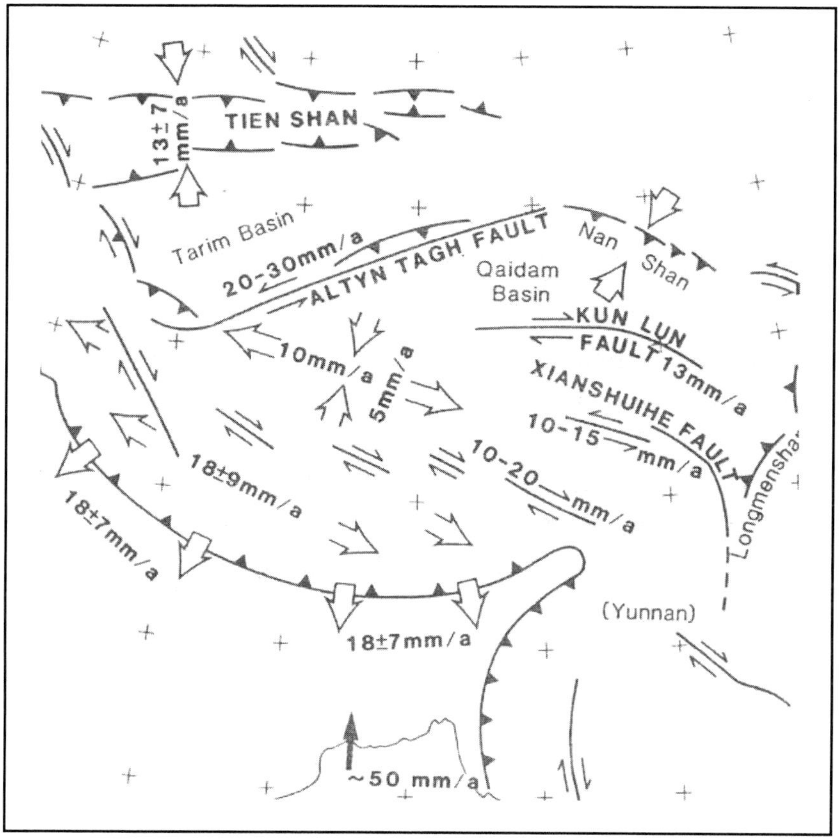

Figure 4.8. Rate of deformation and movement of tectonic blocks (after Molnar and Lyon-Caen, 1989)

Upliftment of Himalayan Terrain

Upliftment in the Himalaya has been quantitatively calculated in many locations by dating, triangulation, Global Positioning System (GPS) and by fossil record techniques. The Himalaya is rising with 5 mm/year and upliftment of 3-5 km has taken place at many places in the Quaternary period. This rate is five times faster than measured in the Alps. Three to five or six phases of rapid upliftment have been recorded in Quaternary. Most spectacular example of upliftment is Tibetan plateau (average height 5100 m), which was just few hundred meters above mean sea level before Quaternary. The Pir Panjal range is rising with 3.5-10 mm/year rates in Kashmir to become barrier between Karewa of the north and Siwalik in the south while Great Himalaya is rising there with 0.7 to 1.1 mm/year. Karewa sediments of Kashmir were deposited in a lake, which was formed due to upwarping in the footwall of MCT, originally few hundred meter above msl, now situated between 1700-1800 above msl. The Nanga Parbat has shown upliftment of 6 km in last 1.3 Ma with a rate of 10 mm/year and of 10 km in the past 10 Ma. Kargil molasse (Liyam Formation) contains fossils like fan palm, Trachycarpus, rhino skull, charaphytes gyronites (Livistone major). The present-day analogues of these fossils occur between 1600-2100 m altitude while in Kargil they are recorded at 5400 m level. This indicates more than 3300 m upliftment of molasse since Upper Oligocene. Karakoram ranges rose up with 2 mm/year rates. Doon Valley shows interesting effect of recent tectonics. Kangra Earthquake of 1905 rose its level 13.4 cm while adjoining Mohand anticline rose by 10.3 cm in one go and between 1926 to 1975 duration average uplift rate was 1 mm/year and after 1975 measurements show evidence of sinking. Older neotectonic activity not only tilted the Quaternary Doon Gravel by 6°-15° but also lifted them up by 60 to 90 m above the plain of valley and between MBT and HFF Siwalik is annually rising with 0.8 to 1.0 mm rate. An upliftment of Siwalik between 30-45 m relative to Lesser Himalaya is recorded in many sections. Nainital rose to 2600 m along the MBT in Quaternary. Kathmandu Valley has more than 500 m thick lake deposits at 1360-1550 m elevation at present. The fossil records show that area was having the same elevation before Quaternary. The present upliftment rate in the Himalaya appears to be very fast. Had this been the case since beginning then height of the Himalaya would have been several times of the present (Shroder, 1989). Raiverman (2006) observed that in first 50 million years of the Cenozoic rate uplift was very gentle in the central Himalayan ranges this was followed by rapid upliftment till 3 million years ago and severely accelerated upliftment in last three million years. However, the rate of upliftment inferred from rate of exhumation of plutonic igneous and metamorphic rocks of different sectors of the Himalaya are not uniform. The difference is as high as of fifty percents. Karakoram metamorphic belt is fastest exhuming region (0.9-0.7 mm/yr) in the Himalayan terrain.

Subsidence in Himalaya

Geodetic survey pointed out subsidence in several parts of the Himalaya viz. in Pala-Maneri Hydroelectricity Scheme area near Uttarkashi has shown subsidence 0.7 to 8.8 cm during 1973-74 and 1977-78 and 1.8-20.0 cm during 1973-74 and 1985-86, on the east bank of Ganga River. Doon Valley has subsided by 6.0 cm during the period 1974-77 to 1991 and with respect to Dehradun the Mussoorie Hills have subsided by 7.0 cm after Uttarkashi earthquake of 1991, across Ganga Tear near Hardwar 0.6 to 3.9 cm between 1974-75 to 1978-79, in Shanan hydel scheme area of Himachal Pradesh 1-5 cm subsidence has been recorded during 1975-80 and Ghaggar–Shimla line has shown subsidence of 2-4 cm in the period of 60 years (Issar, 2000).

Stress Direction

The direction of maximum stress axis (SH max.) in Himalaya (Fig no. 4.9) varies from NNW-SSE in the Quetta region of Afghanistan to N-S and NNE-SSW along the Himalayan frontal arc, NE-SW in the eastern syntaxes and to the E-W in the Burmese trend (Rajendran et al., 1992).

Formation of Structures

Several structural features like graben (Kharsali in Yamuna Valley and also in Nepal and Bhutan), horst (in Nepal), warping of sediment (in the northern region); folds (in Siwalik), overthrusts and nappes (in Siwalik) are formed by neotectonic deformation.

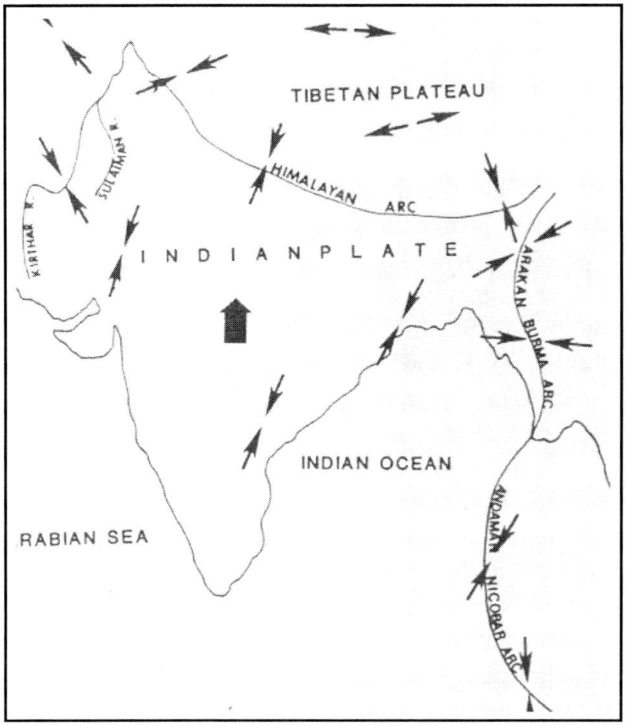

Figure 4.9. Maximum Compressive stress direction in Himalaya and the adjoining areas. Thick arrow indicates direction of absolute plate motion. (after Rajendran et. al., 1992).

Impact of Neotectonics on Topography

Several geomorphic features related with tectonic have been mentioned earlier. It is commonly observed that in transverse profile lower part of relief shows youthful topography while upper part of the relief shows mature or old topography. Three to six levels of terraces are recognized in majority of the Himalayan valleys (viz. Yamuna valley) and also due to tectonic activities several hundred meter wide valleys turn into narrow vertical gorges in Dehradun and Nainital area. Due to neotectonic movements in Doon Valley river terraces (in the Rispana, Song, Bandal rivers) and sub-Recent fans have been displaced. In Raipur (Dehradun), rocks of Chandpur Formation overrides the Doon Gravel indicating tectonic movements along the MBT. Seismites are reported from the Tons and Kathmandu Valleys. Pressure ridges are reported from Siwaliks of Western Nepal and in Doon Valley (Philip and Sah, 1999). Srinagar Thrust (North Almora Thrust) not only differentially uplifted the terraces but also shifted the palaeochannel in Alaknanda Valley. Fault depression, sag ponds, fault saddles, and diversion and off set of streams and pressure ridges are reported from the outer Himalaya of Nepal. Pull-apart antecedent Satluj Valley has carved out a more than 1000 m deep canyon due to Late Quaternary upliftment. Similarly Brahmaputra, Kosi, Karnali, Bhatekosi, Trishuli, Ganga and Indus are antecedent rivers; these rivers enter into Lesser Himalaya through gorges. The Lesser Himalaya is uplifted with a rate of 0.7 to 1.1 mm/year, which is controlled by intracrustal boundary thrust and other faults. Dhauladhar range during uplift blocked the Ravi River and a lake of 20 by 3 km size was developed due to blocking. Intermontane lakes like Nainital lake, Bhimtal lake, Nachiketa Tal (lake) etc. are formed due upliftment of Nainital area along Main Boundary Thrust and also by faulting in the interior region (Valdiya, 1993). Lakes existed in the Gilgit, Hunza, Skardu, Indus and Zanskar rivers in Ladakh area, which are vanished now. Neotectonic Lamayuru Lake was formed 35000 year ago and was persisted 1000-500 years before present. Tsokar Lake (4572 m above msl) of 125 sq km size was formed in a graben when palaeo-Tsokar River got blocked due to faulting. Kelung Lake (159 sq km) formed by movement in Chandra Fault. Neotectonic movements also form many other lakes in this region. Fault scarplets, entrenched meanders, displaced fan surfaces, back tilted area, paired and unpaired terraces are commonly present neotectonic features in Himalaya (Narula et al., 1989).

4.16 Active Tectonics

Active tectonics is a new branch of geology, which deals with those tectonic movements that are expected to occur or have occurred with in a time span concern to society. This field of geology is mainly concerns with catastrophic activities of recent past and with those likely to occur in near future. Mostly it pertains to earthquake or rapid rise or fall of land level, which are causing harm to man and his surroundings. It is a multidisciplinary field involving study of neotectonics, seismology, palaeoseismology, tectonics, structural geology, geomorphology, geochronology, Quaternary geology and archeology. In active tectonics, emphasis is generally placed on palaeoseismology. Earthquake induced movements mostly occur on already faulted surface. These faulted surfaces in majority of cases preserve earthquake history of the area and with the help of seismites or seismogenic structures a fairly accurate assessment of earthquake frequency, magnitude and repeat intervals can be worked out (Nakata and Kizaki, 1986). Data generated in this way can be used for not only prediction of future events but also can be employed in planning for minimizing the impact of forthcoming disasters. Palaeoseismological investigations have been carried out in the Garhwal, Kumaun, and Himachal Pradesh, Doon Valley, Assam and in many parts of Nepal. It is reported that repeat time of an earthquake of 8.4 magnitude is 200 to 270 years and the entire Himalaya ruptures once in duration of 180 to 240 years. In Assam three palaeoseismic events occurred around 500±120, 1100±200 and 1500±250 years before present. Evidences like knick points in longitudinal profiles of rivers, abrupt change in channel gradient, tilting of fans and high gradients with in the fan, change in drainage patterns, tilting of terraces, dislocation of landforms and structures resulting from active tectonic deformations have been reported from the Kangra and Dhauli Ganga valleys.

Active Faults in Himalaya

Nakata was first to identify active faults in the Himalaya. Present day displacement in several Himalayan faults have has been identified by the photogeological, remote sensing field, topographic and other techniques. Direction and rate of movement in them are quantitatively measured by precision leveling or by GPS techniques in many faults of the Himalaya. In far northwestern Himalayan region, Transcurrent Karakoram Fault, Tso Morari Fault, Tsokar More Plain Fault, Chandri Faults, Sarchu Fault, Spiti Fault and Lingtu Faults are active faults. Raikot Fault in Pakistan shows upliftment of 1.0-1.5 cm/year causing lateral offset of Indus River in the east by 1.5 km and has captured the Soan River due to tilting and also tilted Pleistocene sediments by 400. Chaman Fault is also active strike-slip fault. Yamuna Sinistral Tear is of neotectonic origin. Displacement of 100-200 m is recorded across the tear plane in Higher Himalaya along with the development of secondary tears (Paligad and Ujargad strike-slip faults). In Outer Himalaya, Yamuna Tear has shown strike displacement of 30 cm in 4 years and between 1965 to 76 duration recorded vertical displacement in order of 0.7 and 5.7 cm. In the adjoining Ganga Tear the recorded respective horizontal and vertical displacement are 1.0 to 5.7 cm and 1.6 to 3.8 cm in a period of 11 years between 1974-85 (Arur and Hasija, 1986; Rajal et al., 1986). Precision leveling and geodetic triangulation survey recorded horizontal movement of 3-15 cm and vertical movements of 0.7 to 8.8 cm during four years interval. Krol Thrust, Nahan Thrust and Shali Thrust are active faults present in the Uttaranchal and Himachal Pradesh. Black gouge entrapped in the Krol Thrust yielded 20660 ± 450 years are suggesting reactivation of fault that time while in the Yamuna Valley same fault has shown movement of 4 mm/year. The transverse Sile Fault shows net horizontal slip 200 to 1700 m. In Marhan Fault and Doiyonwala Fault the horizontal slip is 150 m and 1600 m respectively in the latter fault carbonaceous gouge suggested reactivation of fault around 28100± 1850 years ago. The MBT has lifted Doon Gravel by 200 m in Late Pleistocene and Recent. To the west of Doon Valley, a high angle reverse fault with oblique slip within sub Himalayan hanging wall Trans Yamuna Active Fault System comprising of Circuital, Shaman and Bali faults is a left stepping en echelon group of faults of 14-19 km length, this system is sub parallel to and crosses the Krol Thrust (Oatney et al., 2001). In Kala Amb of the Himachal Pradesh HFF is displaced by the Black Mango Tear Fault and four low-angle thrust faults are recognized in artificially excavated trench-wall surface. Analysis of trench wall stratigraphy revealed occurrence of two large earthquakes in AD 1294 and AD 1423 and possibility of another earthquake in AD 260. Kuria Fault in Kumaun, has produced an offset of 100 m length in MBT. The 18 km long Sanghrkhola dextral strike slip fault (NW-SE) has uplifted southern block by 10 m relative to northern block.

In Nepal, MCT, MBT and HFF are active faults. These faults are dissected and displaced by many oblique to transverse strike slip faults, The Darma, Talphi, Tibrikat and, Dhaulagiri are high angle strike slip faults and the Barigad (showing crustal shortening of 5 mm/year), Rosikhala and Kalphu khola Faults occur around vicinity of MCT. Along MBT, Rangun khola (80 km) Surket-Garahi (20 km), Arun khola, Hafaura, Udaipur, Sapt Koshi and Mechi faults are recorded as active faults. HFF is recognized for 20 km length between the Karnali and Babai rivers. The strike slip faults show average displacement rate of 2 mm/year (Kizaki, 1995; Saijo et al., 1995). Danuwargaun active normal fault drained the Kathmandu palaeolake (Sakai, 2001). In Darjeeling the average rate of movement in faults is 3-4 mm/year. In Bhutan area several active faults are present.

In Arunachal Pradesh, older tectonic lineaments like Main Central Thrust, Main Boundary Thrust, and Himalayan Frontal Fault show reactivation in localized area. Mishmash, Lapland and Tiding thrusts and Lohitpur, Tellu Nala, Roing, Hunli- Ryalli and Angolin-Kadali faults are recently reactivated and are deflecting the rivers, abrupt rise in the mountain front fault scarps without development of gullies, landslides, gorges, blockage of rivers, upliftment of river terraces and damage to civil structures like bridges.

Several geomorphic indices like transverse topographic symmetry factor, hypsometric curve and integral, stream length gradient index, mountain front sinuosity index and ratio of valley floor width to valley height have been used to interpret active tectonics status in Doon valley. The valley is divided into six blocks showing broad geologic and geomorphic variability and is separated from each other by major lineaments. Tectonically these blocks fall in the very active, active and moderately active areas (Joshi, 2003). Fractal approach, a mathematical analysis in which study area is divided into small boxes, is also employed for study of the neotectonics and earthquakes in the Himalaya. In Karighat area of the Kumaun fractal analysis revealed that in geologically homogeneous region degree of ruggedness of the terrain resulting from the neotectonic movements correspondence well with the increase in fractal dimensionality (Joshi and Rai, 2003). Positive relationship between neotectonically produced kink beds and joints and valley trend has been reported from the Tethyan Chamba Nappe of the Himachal Pradesh. Sunmonu et al. (2001) observed that 78 earthquakes recorded during 1885-1995 of M7 or of greater magnitude shows positive response in multi fractal analysis and reveals heterogeneous fractal structure of the Himalaya implying that time wise earthquakes exhibit preferential distribution.

4.17 Experimental Tectonics

Successful attempts have been made in laboratories to produce common structural and tectonic features encountered in the outcrops and also to replicate tectonic features of a particular orogen or tectonic setup of a region on modeling clay, sands, papers, glass, gelatin, wax of different grades, cloths, metal and plaster of paris on the principles of scale modeling. Initial exercises were conducted on simple single layer homogeneous material and latter experiments were extended to multilayer materials of heterogeneous composition and of different textural characters. In a simple apparatus called Biaxial Press, experiments were designed in various variable conditions, controlled externally, like homogeneous and heterogeneous layer compositions confined and unconfined mediums, changing load pressure, layers of varying thickness and viscosity contrast, smooth or rough contacts between layers, pre-determined and variable stress application and controlled strain conditions, rheology etc. Thus simulated structures provided valuable insight in mechanism(s) of evolution of structure and at time results offer totally new interpretations of what were obtained from field study.

In Himalaya, experimental tectonic study is undertaken by Dubey (1996) whereby he proposed that many strike slip faults observed in the Himalaya e.g. Sairku strike-slip fault, which displaced the MBT in north of Doon Valley by 20 km; is in fact an oblique thrust ramp. This thrust ramps are formed due to change in mode of deformation conditions. In another study Dubey and Bhat (1986) discussed the tectonic evolution of Himalaya and drawn the conclusion that in initial stage listric normal faults produced by rifting played very important role. Figs. No. 4.10 and 4.11 show experimental evolution of oblique ramps and their field examples.

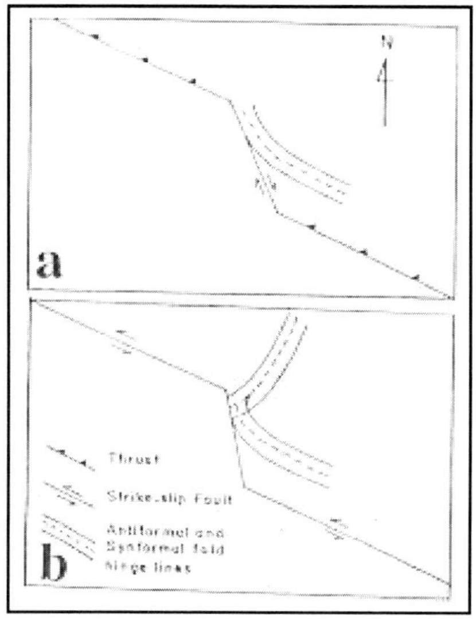

Figure 4.10. a, b. Experimental evolution of oblique ramp from fold after Dubey, 1999).

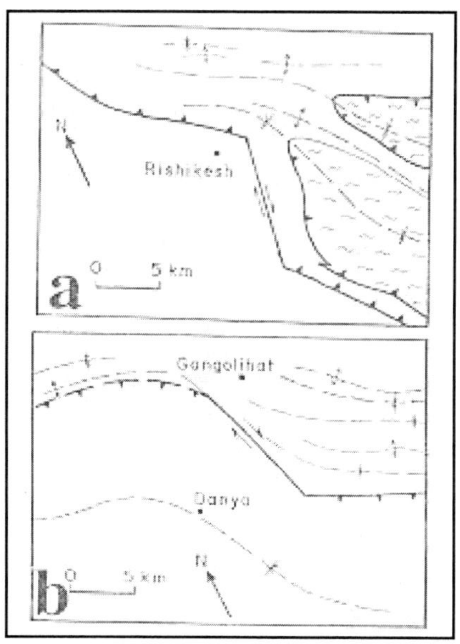

Figure 4.11 a, b. Verification of evolved ramp in Dehradun-Rishikesh and in Gangolihat- Danya sections (after Dubey, 1999).

4.18 Burmese Arc: An Interface with Indian Plate

In the eastern region, Himalaya abuts against the N-S directed Burmese Arc along the Tidding Suture in Arunachal Pradesh. The Burmese Arc is slightly convex in west direction and has a length of nearly 1100 km. To the south of Burmese Andaman arcs are linked with Sunda Arc of the south. In this region, Indian plate underthrust resulted into Arakan-Yoma, Chin and Naga Hills. In the Oligocene age, Indo Burmese Range is formed during Alpine

Orogeny; the Alpine type thrusts are present along the western margin of Burmese low land (Acharyya, 1974)

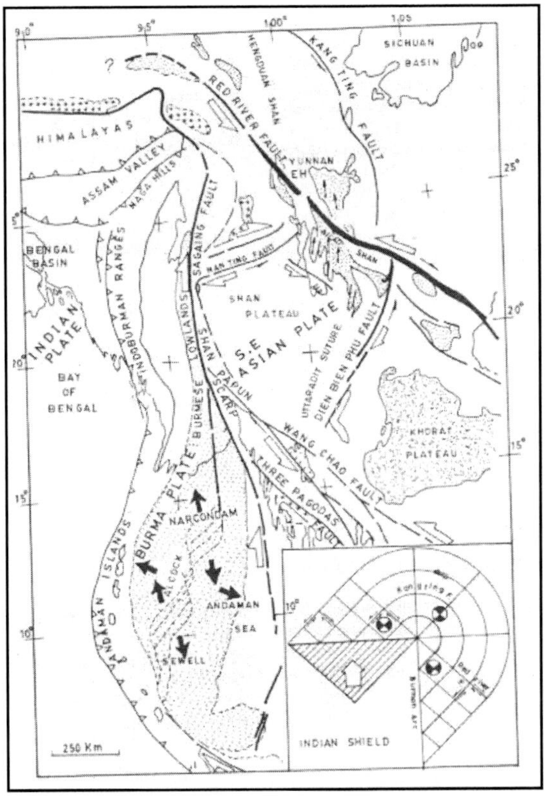

Figure 4.12. Regional map of Eastern Himalayan neighbors and Burmese and Andaman arc. Inset shows plain strain slip line fields created by indentation of Eastern Indian Shield (after Mukhopadhyay, 1992).

The boundary of Indian plate is situated in the west of Andaman Nicobar Ridge (ANR), along the trench axis. The Andaman microplate is situated between coastal Burma and N Sumatra shows fore and back arc nature in the north and south of the Andaman Sea (Fig No.4.12). The Burmese Arc is evolved since late Mesozoic in consequence to eastward subduction of Indian plate at the margin South Asian plate, a mosaic of microplates. It is one of the few regions in Asia and Europe where intermediate depth earthquake occurs in land environment.

Burma is characterized by the transitional tectonic set up between truly collisional phase (Himalaya) and typical consumptive margin (Sunda) and is divided into western fold belt also called Indo Burmese Range, (IBR) and Burmese lowland. The IBR is central 150 km wide region has north-south fold axis that formed due to east-west compressive forces

The important tectonic elements of Andaman microplates are Andaman Trenches, Andaman Nicobar Ridge, Nicobar Deep, West Andaman Fault, Andaman Sea and Mergui

Terrace, Seamount Alcodi is present is Andaman Sea. Above water volcanic arc is a part of Andaman Sea (Ray, 1981).

The Burmese plate has been created since Middle Miocene as consequence to opening of Andaman Sea by 460 km. Tapponnier suggested that active spreading under the Andaman Sea and lateral motion along the Saingang Fault in Myanmar are the result of indentation tectonics exerted by the Indian plate on Asian plate. Still subducting Indian plate below Burmese Arc provides evidences east dipping Benioff Zone (45°) till a depth of 180 km. The Benioff zone is extending below Myanmar lowland. Earthquake activities are controlled by the Benioff Zone and the fore arc and back arc zones. The focal mechanism solution yields double seismic zone model. Low angle thrust fault events occurring along the upper edge of Benioff Zone and shallow to intermediate depth earthquake point out normal fault solution (Mukhopadhyay, 1992). The Myanmar plate evolution is related with back arc spreading below the Andaman Sea since Neogene. A crustal seismic zone situated 60-80 km east of the Benioff zone is correlatable with back arc activity. The ANR seismic zone is 90-160 km wide and earthquakes are occurring at 5-230 km depth and Benioff zone dips with 55°. The rate of subduction of Indian plate is 8 mm/yr in north, 35 mm/yr in middle and 50 mm/yr in the south of Andaman Trench (Das Gupta et al., 1990).

4.19 Structural and Tectonic Interpretations in the Himalaya

Growing knowledge has certainly widen the chasm between differing interpretations about same set up and objects are redefined or reinterpreted in such ways that one has to prefer a particular view as to have a base for subsequent discussion. For example, an area that was described as allochthonous by one set of workers may be autochthonous or paraautochthonous for another set e.g. Krol Nappe, controversial position of MCT and others. However, despite large number of controversies, knowledge about Himalaya has grown both vertically and horizontally. Cross interpretations should be evaluated in the above context.

Structural and tectonic information is basic requirement for any detailed geological study. This information is obtained either by routine methods like geological mapping and interpretation of data collected from the field or of a region or by the aerial photographs and by satellite imagery; and, the other methods involve applications of various geophysical techniques (all of these having their own limitations) or by well loggings followed by the drilling. The later methods are very expensive and are used only when required for commercial purposes or in thoroughly funded scientific investigation. In case of Himalaya, most of interpretations are done with the routine methods though gathering of geophysical data is gradually increasing. Routine methods despite being quick and cheap are subjected to bias. These bias are involved due to personal preference for one particular view or on relying narrow lines of evidences or due to casual approach to problems. In net result, variations in approach result into different interpretations of a tectonic set up of the area. For example, in the Higher Himalayan Yamuna Valley, one geological cross-section depicts structures based on modern thrust tectonic and balanced cross-section (Saklani, 1993) while other one reports very simple set up (Biyani, 1998). In case of Shimla area, geological cross sections and tectonic interpretation given by West (1939). Ray and Naha (1971) and Dubey and Bhat (1991) differs considerably still another example is with three cross sections proposed from the Siwalik to Mount Everest in the Nepal by Hagen (1959) Lambard (1958) and by Bordet (1961) given in Vol. 9 of Atlas of Mineral Resources of the Escape Region. It is the case with many other regions also.

Main Central Thrust is another example of conflicting interpretations. Initially it was proposed as single plane of tectonic discordance, however, later workers visualized it as a pack of thrusts and designated as MCT I, MCT II or even MCT III with sub-thrusts in each category and few have visualized MCT as a broad zone of considerable width. Contact between two stratigraphical units even in the same locations are proposed as tectonic, unconformable or even gradational.

Cases have also been observed in the Himalaya where researchers have revised their opinions in the light of their experience about the structural set up of areas where they have worked after an interval of time. For example, Valdiya in 1962 described a huge regional inverted sequence of Saryu Valley Quartzite-Gangolihat Limestone - Sor Slate and Berinag Quartzite of Krol Belt in Pithoragarh area of the Kumaun Lesser Himalaya. Schists and gneisses of the Garhwal Nappe overlie this sequence. However, in 1980 he modified his view and separated the Berinag Quartzite from the underlying Gangolihat Carbonate sequence with a thrust. Further, in 1962, he considered these formations as part of Krol Nappe but in 1980, these appeared to him as autochthonous. Lately in 1995, he again modified his view considerably on the basis of new fossils finds and has correlated the carbonate sequences of the Outer (Krol) and the Inner (Deoban) Lesser Himalaya which in 1980 he was thinking to be different in hundreds of millions of years in age.

This is not the case only with Himalayan but also in many regions where tectonic setting is not simple. Such different interpretations require careful evolution of each view before coming to final conclusion.

4.20 Abnormal Thickness of the Crust in the Himalaya

The crustal thickness in Himalaya is double or more than double (70-75 km) than of average thickness (33 km). This is not the special case with the Himalaya only but also a characteristic feature of the other younger tectonic orogens like Andes. Different views have been expressed about the possible reasons for this abnormal crustal thickness in the Himalaya. It is accounted by folding followed by thrust stacking in the upper part (Ray and Naha, 1971); simultaneous folding and thrusting in the upper crust and magmatic addition in the lower crust during lithospheric rifting (Dubey and Bhat, 1991); under thrusting of Indian plate at steeper angle (Malinconico, 1989), almost horizontal underthrusting of the Indian plate (Bird, 1978), underthrusting at moderate angle with decoupling of the lower crust (Hirn et al., 1984), thickening as a result of compressional tectonics, along with underthrusting of Indian plate, transfer of mantle material into lower crust and vertical accretion (Mahadevan, 1994), by block faulting, by addition of basaltic material (Qureshi, 1969), by upwelling of hot mantle material and also due to deformation of less rigid Eurasia (Patton, 1978 in Miyashiro et al., 1982). The post-middle Eocene collision and continued convergence of the Indian plate under the Eurasian plate produced crustal shortening and as a result of this imbrications of Indian plate has taken place along the major intracontinental thrusts like MCT and MBT, metamorphic inversion and formation of Miocene leucogranites in the Higher Himalaya and intracontinental thrust like MCT and MBT.

4.21 Gneissic Domal Structures

Gneissic domes of several tens of km length comprising of core of granitic migmatites or gneisses with a mantle of low to high-grade metamorphic rocks are common feature in the orogenic belts. Various explanations are given for their origins namely the diapirism of low density middle crustal rocks or low density magma, extensional exhumation along low-angle footwall of normal faults of middle crustal rocks, having fault contact between the core and cover or exhumation of middle crustal rocks along thrust faults either by thrust duplexes or horizontal compression of gneissic rocks overlain by metamorphic rocks or by superposed or cross folding or combinations of these mechanisms.

Domal structures are reported from several places in the Himalaya. They occur in the Higher or in the middle of Tethyan Himalaya e.g. Tso Morari-Zanskar, Mabja, Kangmar domes are few examples of domes investigated in detailed. In Mabja Dome located north of the Kanchanjunga Mountain in the southern Tibet, migmatites orthogneisses are surrounded by Palaeozoic orthogneisses and Triassic Jurassic metamorphic, sillimanite grade at the base, and sedimentary rocks, on the top (Lee et al., 2004). In northeastern Pakistan several regional size domes have been identified e.g. Basha dome of 150 sq km area extending between Arendu to Niesolo comprised of orthogneisses, 20 km long and 5 km wide Bukpun dome and Arelter dome are other examples.

4.22 Sedimentary Basins in Himalaya

Sedimentary basin is defined as structurally controlled morphotectonic depression capable of receiving and preserving sediments with a geological history not withstanding erosion and non-deposition from time to time. Basin represents a unit of geological structure, which received sedimentary succession unique to it during a given span of time. A sedimentary basin characterized by distinctive sedimentary fill, single or multiple depositional cycle and environments, distinctive tectonic frame work and basin architecture, single or several phases of tectonic or thermogenic subsidences, one or more tectono-sedimentation episodes defining single or polyhistory basins and stratigraphic sequences related to tectonic episodes. Sedimentary basins are classified on the basis of tectonics, size, morphology and other parameters.

Lithologically, Himalaya is dominated by the sedimentary rocks of at least Early Proterozoic to sediments of Holocene age, which were deposited in number of basins of extremely large to very small size lakes with equally variable life spans. The oldest basin was existed in the Early Proterozoic in that sediments for Higher Himalayan metamorphics were deposited. This basin could have been extension of the peninsular Aravalli, Singhbhum, Bastar Iron Ore Group and Shillong Group of Meghalaya; the trend of Aravalli strike is present in the crystallines of Higher Himalaya in Uttaranchal. However, Higher Himalayan metamorphics differs from the Aravalli in several respects such as thickness (presently former is 20 km thick while latter is 3-4 km thick), absence of conglomerates and ultramafic rocks and mineralizations.

Next in importance were older Lesser Himalayan and younger Tethyan basins. The Lesser Himalayan basin is again considered extension of the peninsular Vindhyan basin. This mainly Middle to Late Proterozoic basin also extended into Cambrian and has nearly 15 km thick sediments showing low-grade metamorphism in between. The main component of this basin is further divided in to inner and outer Lesser Himalaya in Kumaun. In eastern Nepal and in the further east fresh water Gondwana rocks cropped out at several places. This fact further reinforces consanguinity of the Lesser Himalaya with peninsular India. Salt range of Pakistan though situated considerably in south of the Himalaya, 80 km, but also considered extension of Lesser Himalayan basin at present. Wadia (1975) stated that in structural, stratigraphical and physiographic features Salt Range are in contrast to Himalaya of the north but these two units are allied in their geological system and in their orogenic phases (pp. 132). Attock Cherat, Kala Chitta and Naushera regions also have rocks of the Lesser Himalayan affinities. This vast basin exhibits different tectonic characters in different regions and also provides proof of several cycles of sedimentations. In post-Cambrian period several minor basins developed over it because of incursions of sea. Younger deposits of marine incursions are designated as Binj, Kakara, Subathu formations in the Uttaranchal and Himachal Pradesh, Palaeozoics of Kashmir, Surkhet Group of Nepal and Lower Tertiary Yinkiong Formation of Arunachal Pradesh. The Tethyan basin of Late Proterozoic to Eocene age again has nearly 10 km thick mainly marine fossiliferous sediments and has repeated pulsating history of short durations and a good part of this would have been consumed by plate process. Towards the end phase (Cretaceous to Eocene) its water was receded in the south. The receding phase of the Tethyan Sea is testimonied by deposition of marine transgressive present over the Lesser Himalaya. Tethyan basin was eliminated when collision occurred between India and Eurasia. Sedimentary rocks deposited in these basins are present from one end to other end i.e. along the length of the present day Himalaya and this length has been derived by the fragmentation of Gondwanaland, which is geologically 105 Ma old event. Hence the size of the original basins would have been much larger.

The subducting Indian plate produced narrow but highly elongated basin in the suture zone. There were two suture zone basins in the west namely the Shyok and Indus and in the east of Ladakh only Indus suture basin runs along the plates boundary. The time duration of the sediments in the Shyok and Indus suture basins are Permian to Lower Eocene probably extending up to Miocene and Triassic-Jurassic to Oligocene respectively. Due to intense deformation formations are present in highly disturbed form.

To the north of Shyok Suture Palaeozoic-Mesozoic sediments deposited in the Karakoram basin are present. Next to develop was foreland basin in the south of the lesser Himalaya in which tertiary-quaternary sediments of initial marine to fresh water conditions nature were deposited. This foreland basin has been discussed in the next section.

The size of the basins also should not be evaluated on the present width of the respective divisions of the Himalaya but be evaluated after making correction of pre-folding and folding contractions and volume loss caused by the thrusting.

In the late Pliocene, Pleistocene and Holocene several minor lakes or basins of few hundred kilometers to several thousands of kilometers size such as Karewa in Kashmir, Pinjor dun, Doon in Uttaranchal, Kathmandu in Nepal, Diklai Boulder Bed in eastern region were developed in the southern and central part of the Himalaya. At present Himalayan lakes occupy few tens of meters to tens of square kilometers area. These lakes contain valuable sediments record of Late Pleistocene and Holocene.

4.23 Himalayan Foreland Basin

A peripheral foreland basin is formed as the elastic lithosphere of the approaching continent flexes under dynamic load of the frontal thrust belt of the overriding continent (Biswas, 2004). Foreland basin forms in collisional setup by continent-continent, arc-arc, and arc-continent collision. Evolution of Sirmur basin is shown in Fig No. 4.13.

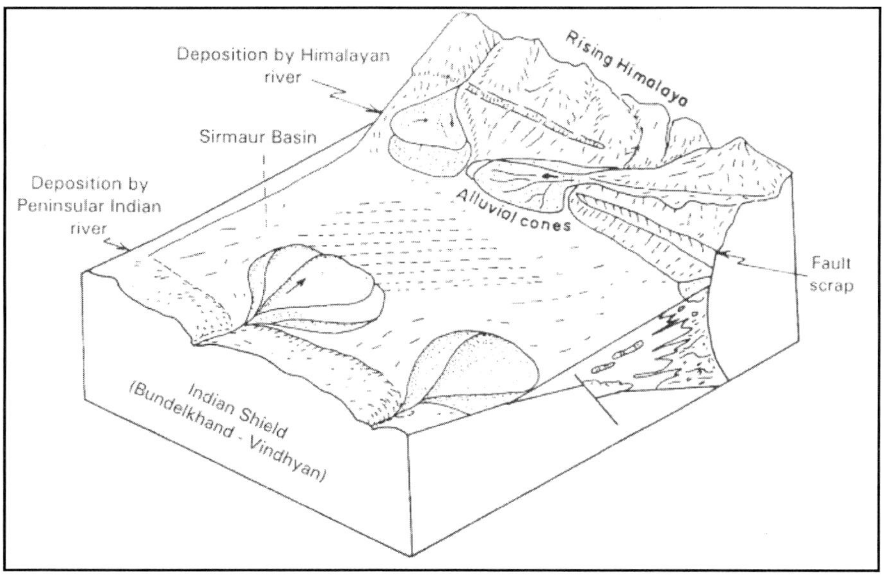

Figure 4.13. Development of Sirmur basin by sagging of the Indian crust (after Valdiya, 1998).

In between the peninsular shield of the India and to the south of the Lesser Himalaya a foreland basin, in front of an active orogen, came into existence due to powerful deformation in the then Himalayan terrain in the Middle Miocene. Initially the foreland basin term was employed for the Siwalik basin but now scope of the basin also extended to Indo-Gangetic alluvial basin, it is the largest basin in the world's formed by inter-continents collision related processes. 2400 km length of the basin swings its trend from NE-SW in the east to NW-SE in the west with mild convexity towards south. Shorter eastern (600 km) and the longer western (1200 km) arms of the basin are developed in the south. The basin is nearly 150–250 km wide in the central region.

This basin shows all major components of a foreland system namely an orogen i.e. Himalaya, deformed foreland basin deposits adjacent to the orogen i.e. Siwalik, a depositional basin i.e. Ganga Plain and peripheral cratonic bulge i.e. Bundelkhand granite.

The small foreland basin with minor subsidence probably originated in the Early Miocene and by the Middle Miocene basin completely established itself after lithospheric flexuring and subsidence. Topographic load on the hanging wall and sediments load within the basin are believed to be responsible for the lithospheric subsidence. During the Middle Miocene to Middle Pleistocene the northern part of basin or the Ganga plain was uplifted and thrust southward in response to movements and subsequent loading by the Himalayan thrusts.

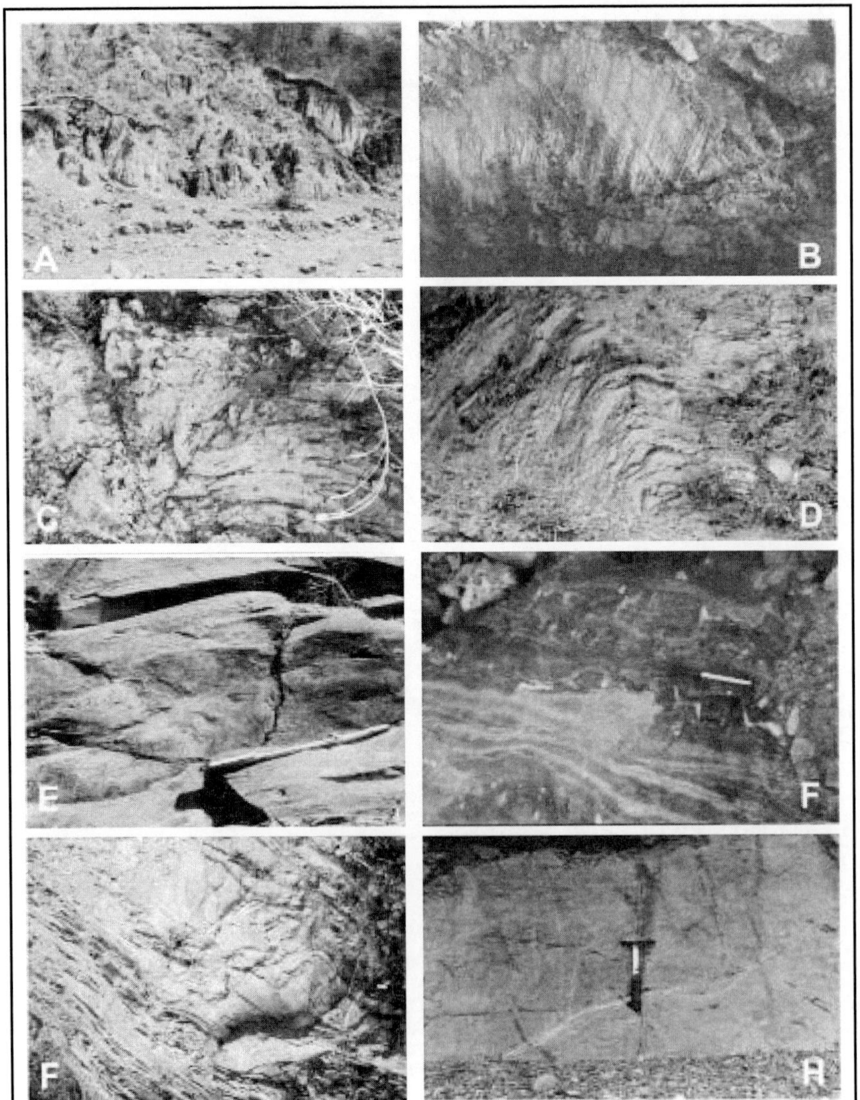

Plate 4.1. A- Crushed and burnt fault zone of Tal valley, Uttaranchal. **B**- Two sets of slicken lines in mylonitic quartzite near Hanumanchatti, Yamuna valley, Uttaranchal. **C**- Fold in Tal quartzite produced by fault propagation mechanism in Tal valley, Uttaranchal. **D**- Fold in lesser Himalaya, Himachal Pradesh. **E**- Expansion of joint due to weathering, near Neelkanth, Uttaranchal. **F**- Trident shaped cirque near Lari, Ganesh Himal, Nepal. **G**- Mild folding in Blaini shales near Dehradun. **H**- Flute casts in Tal sandstone near Mussoorie, Uttaranchal.

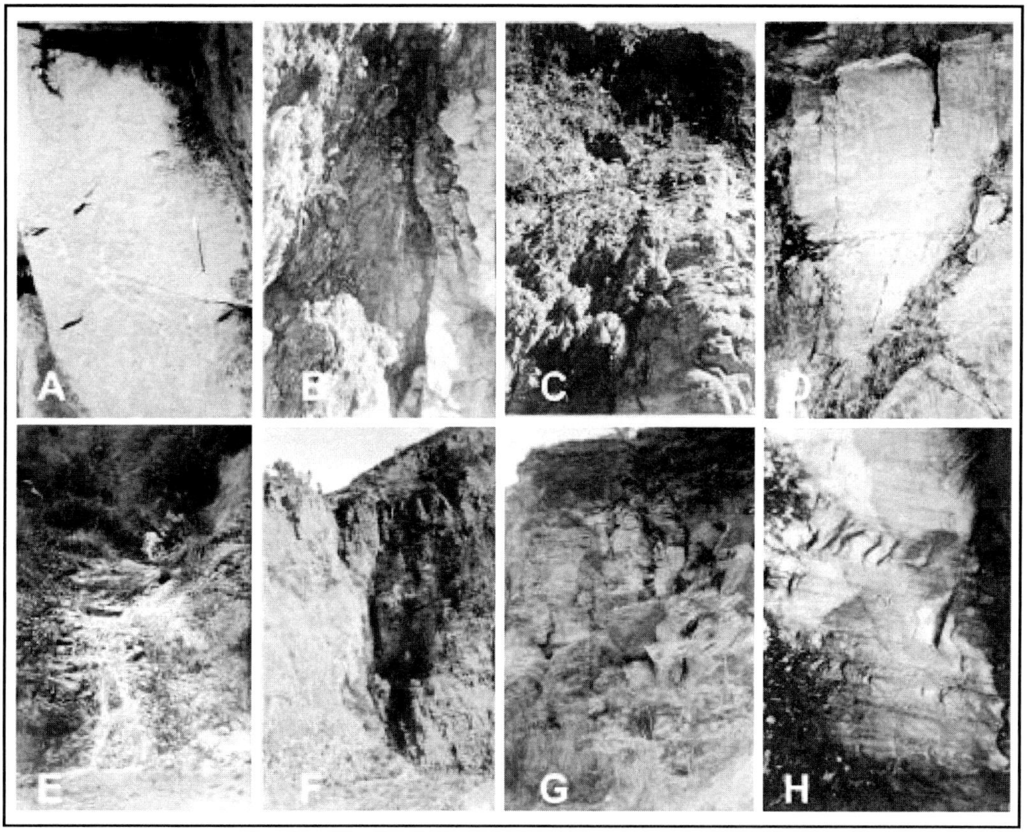

Plate 4.2. **A**- Steeply dipping MCT in Paligarh valley, **B**- Highly crushed and compact Paligarh quartzarenite just near MCT, **C**- high angle Dihargarh reverse fault near Rana, **D**- Development of cataclasite in mylonitised and brecciated granite in Yamuna **E**- Development of waterfall at MCT at Wazri on right bank of Yamuna river, **F**- Development of deep gorge with atleast seven waterfalls on the left bank opposite to photograph E, gorge is running parallel to strike of MCT, **G**- development of crushed breccia in mylonitised granite in Rana village, Yamuna valley, **H**- Development of sub-horizontal mullions in granite near Sayanachatti.

DIMENSIONS OF HIMALAYAN GEOLOGY
A. K. Biyani (2006) : pp.89-159
Satish Serial Publishing House, New Delhi
E-mail: ssp@ndf.vsnl.net.in, hkjain1975@yahoo.com; website: www.satishserial.com

Bhumirapo'nalo vayuh kham mano buddhireva ka
Ahankara itiyam me bhinna prakrtirastadha (Geeta, 7.4)
Earth, water, fire, ether, air, mind, intellect, ego- these constitute
My nature (prakrrti) eight fold divided.

CHAPTER -5

STRATIGRAPHY

The task of establishing age wise hierarchy of beds is a matter of main concern of stratigraphy. This is achieved by using principles of stratigraphy and by utilizing the knowledge of other branches of geology. Usually this exercise is undertaken in a small area. Observations collected from several such small areas are ultimately synthesized so as to produce a geological column applicable to regional level or covering the entire area in which similar rocks are present. The regional picture is drawn with the help of correlation technique. This is a comparatively simple affair if rocks are fossiliferous and undisturbed by tectonic forces. Fossils provide the reliable geological ages of the rocks in many cases. However, if area is unfossiliferous or metamorphosed or tectonically overhauled or thoroughly disturbed, then the construction of stratigraphical column is a highly difficult proposition and often different interpretations are advanced for the same unit of rocks based on the subjective judgment of worker(s) working on local level. The Himalaya provides such kind of examples. Additionally, Himalayan is a physiographically a very tough terrain having steep and forested slopes in which deep furrows are occupied by turbulent streams making field investigations a challenging task and territorially Himalaya is spreading in several countries namely Pakistan, China, India, Nepal, and Bhutan causing extra-difficulty in fieldwork particularly in the border regions.

Progressively increasing geological investigations in last some decades have also drastically revised the earlier held view that all the divisions are developed in belt like fashion of more or less uniform width. Different tectonic belts swell and pinch in wide limits thus in one region full succession is available and in another region a major part of the succession is masked under invading thrust sheet or partially developed due to basinal configuration.

Though this mountain started taking shape in post-Miocene time but its rocks are having long geological history starting from Late Archaean to Holocene with several interruptions of varying time magnitude. These rocks are distributed in different tectonic divisions and these divisions are having their independent stratigraphical identities, which are absolutely different from the surrounding divisions (Biyani, 2005 b).

Among major tectonic divisions of the Himalaya, Outer and the Tethyan divisions being fossiliferous pose no problems in determining the hierarchal arrangement of formations.

More or less it is true for suture zone (s), though tectonically highly disturbed. However, stratigraphy of two important divisions namely the Lesser- and the Higher Himalaya is highly challenging as they are mainly composed of Proterozoic rocks. Lesser Himalaya is by and large unfossiliferous, barring few locations, and tectonically thoroughly overhauled and in some cases some parts are transported to considerable distances from their original positions. Sedimentary rocks dominate in the Lesser Himalaya but also having notable presence of the allochthonous and paraautochthonous ortho- and para metamorphic rocks. The rocks of the Higher Himalaya are metamorphosed to varying degree. These two divisions show not only discontinuity in their extension particularly in western region but also shrinks and swells in a wide limit. Still another problem related with southern part of the Lesser Himalaya where differential translation encountered in the south propagating thrust sheets as well as back thrusting (instead moving to the south, formations translated in the north) in some areas; this differential movement is responsible for different degree of transport and tectonic location of different formations/groups e.g. Kakara Formation occurs in Outer Himalaya in Himachal Pradesh and in the Lesser Himalaya in Uttaranchal. Different views and perceptions of leading workers about stratigraphic setup of particularly in the Lesser Himalaya have yielded complex litho-tectonic stratigraphy of this division by using large number of terms for same rocks unit in different areas; therefore, final stratigraphic column is still elusive.

Plate 5.1. Geological Map of Himalaya

Regional stratigraphy of the Himalaya is usually described in two ways: in first stratigraphy of transverse/latitudinal divisions or regions or state or country can be described. If this methodology is adopted, same stratigraphies of different transverse divisions are described one after another depending upon development with in the format boundary. The drawback of this method is that same formation if developed in more than one transverse division then it is repeated again and again. The other way is to describe the stratigraphy on longitudinal divisional basis, here repetitions do not occur but the same formation may be assigned different names in different regions that quite frequently create confusion and this method being satisfactory is adopted here in brief. Detailed stratigraphies of the entire Himalaya or region/state wise or even of smaller regions are available in several books written by different authors and in journals, memoirs and records published by the Geological Survey of India. Stratigraphic column for Uttaranchal (Biyani 2003 and 2004 a, 2006) is shown in the Table-5.1.

Marine Transgression in Himalaya

Several instances of marine transgression are observed particularly in the southern part of the Himalaya. The Lesser Himalayan terrain is mostly composed of the Late Proterozoic to Cambrian autochthonous to paraautochthonous sedimentary rocks. After this depositional cycle, sea again made incursion in the Late Carboniferous- Permian duration in the Salt Range in Pakistan, Jammu and Kashmir, Garhwal (Binj Formation near Kotdwar), Darjeeling foothills and Arunachal Pradesh. This brief phase of incursion is followed by another one in the beginning of Tertiary i.e. in the Palaeocene (deposition of Kakara Group) and after a minor hiatus another marine transgression occurs in the Eocene (Subathu and Dagshai formations). The regressive Tethys Sea of the north of the Tethyan zone causes the Tertiary transgression in the Lesser Himalaya. No evidences of this invasion occur in the Higher Himalaya located north of the Lesser Himalaya or to the south of Tethyan Himalaya, this could be due to late action of the Main Central Thrust that brought crystallines to the surface or complete erosion of thin veneers of transgressive formations

5.1 Indo-Gangetic Alluvial Plain

The Indo Gangetic Alluvial Plain is made up of alluvium mostly brought about by the Himalayan Rivers. This alluvium was mostly deposited in the foreland basin in Quaternary period. Himalayan Frontal Fault, exposed discontinuously, or alluvial sediments making an unconformable contact with the Siwalik rocks of the Outer Himalaya mark its northern limit and its southern contact is of unconformable nature with peninsular rocks of varying ages. Alluvium is divided into Older Alluvium and Younger Alluvium and in river valleys and in their flood plain areas recent sediment layers of varying thickness also occur. The alluvium has variable thickness, maximum thickness of approximately ~6 km is recorded in the north that decreases gradually towards the south. The alluvium comprises of gravels, sand, silt and clays and below the alluvium, Siwalik, Subathu and progressively older rocks of Himalaya as well as peninsular India are recorded in deep wells drilled for oil by the Oil and Natural Gas Corporation Limited or interpreted by the seismic methods.

Alluvial stratigraphy of the near surface (~50m) Ganga basin sediments at Kalpi near Kanpur is divided into five lithostratigraphic units: the basal floodplain unit comprised of silty clay with dark mottles and scattered carbonate nodules and rhizoconcretions. Unit 2 is composed of fine to medium sands with gravel lag deposits at the base followed by calcareous nodules. Unit 3 is constituted by silty clay deposited in swampy environment. 23m thick Unit 4 is predominantly consists of mud deposited in floodplain to lacustrine conditions and the topmost Unit 5 consists of dipping silty sand, brown mud and lenses of carbonate gravel (Sinha et al., 2005).

5.2 Outer Himalaya

Outer Himalaya, first studied by Medlicott (1864), comprises of partly consolidated to unconsolidated sediments of Late Palaeocene to Quaternary age. It is developed in 10-50 km wide more or less continuous strip from Arunachal to Kashmir that deposited in the Himalayan foreland basin. This Himalayan foothill territory is confined between the Main Boundary Thrust in north and the Himalayan Frontal Fault in the south. In few small locations of Nepal (Fig No.5.1) and Bhutan, the Outer Himalayan rocks are totally concealed under invading thrust sheets of the inner region and the invading rocks are in direct contact with younger alluvium.

This Outer Himalayan foreland basin is largest such basin in the world that contains initially shallow marine to later dominant fluvial deposits. Early workers considered only Siwalik Supergroup in the Outer Himalaya, however, later workers also included pre-Siwalik rocks in the Outer Himalaya e.g. Thakur (1992) included Palaeogene rocks of the Subathu and Muree (Dharmsala) Groups in this division while Valdiya, (1980) considered these rocks as a part of the Lesser Himalaya. The Kakara, Subathu, Dagshai and Kasauli formations belong to early phase of foreland basin and are showing limited development. Outer Himalaya is mainly made up of Middle Miocene to Early Pleistocene Siwalik Supergroup and Pleistocene - Holocene sediments called 'gravels' deposited in 'duns' (longitudinal valleys). However, The broad stratigraphy of the Outer or Sub Himalaya is as follows:

Dun Gravels	Recent
Siwalik Group	
Muree Group = Dharmsala, Dagshai, Kasauli	
Subathu Group	
Kakara Group	
———— MBT ————	
Lesser/Tethyan Himalaya rocks	
———— unconformity/thrust ————	
Outer Himalayan Carbonates	Proterozoic

Proterozoic Outer Himalayan Carbonates

Dolomitic carbonates of the Vaishnodevi (1040-720 Ma) of Jammu and Kashmir and Dharamkot (940-865 Ma), Guma (880-865 Ma), Bilaspur (1065-835 Ma) and Tundapathar (870 Ma) of Himachal Pradesh exposed in inliers are considered as the Outer Himalayan carbonates as these are found to be located to the south of the MBT by detailed mapping for search of hydrocarbon in this area. These carbonates are exhumed by the insequence thrusting. These Proterozoic carbonates are further considered as basement on which Tertiary foredeep sediments were deposited unconformably. Subathu and Dharamsala rocks are unconformably exposed above these carbonates (Dutta, 2003).

Kakara Group

The richly fossiliferous Thanentian to Early Ypresian (Palaeocene) rocks of Kakara Group occur as detached and isolated outcrops over the Mandi-Darla Volcanics, Shali, Deoban,

Shimla, Blaini, Krol and Tal groups/ formations. This group comprises a basal pisolitic laterite/ bauxite, quartzarenite, and shales of various shades and massive to thinly bedded limestone. Neelkanth and Singtali formations of Rishikesh area are considered equivalent to this group. *Ranikothalia, Rotalia, Daviesina, Galeocerda, Pycondus, Operculina* is important fossils form present in this group (Srikantia and Bhargava, 1998).

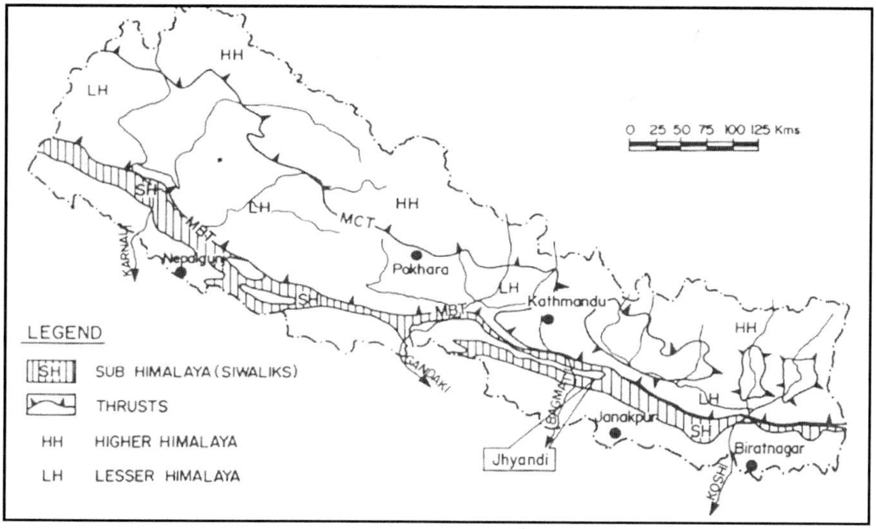

Figure 5.1. Tectonic Zonation map of Nepal showing sub-Himalaya in shaded portion (after Gautam, et.al., 1995).

Subathu Group

The shallow, epicontinental marine transgressive and richly fossiliferous rocks of Subathu Group are exposed as isolated outcrops, in windows or as a part of allochthonous belt from Arunachal in the east to Potwar in the west. Medlicott (1879) first proposed a term Sirmur Group then Sirmur Series for the Subathu, he considered Subathu, Dagshai and Kasauli formations in ascending order. Its thickness varies between 150 m to 1800 m in different places. Subathu rocks are making unconformable contact with the variety of underlying rocks of different formations and ages. Various sedimentary rocks like shale, sandstone and limestone are present in this Group. On the basis of palaeontological evidences in the type area, in Jammu, Upper Palaeocene to Middle to Upper Eocene age is assigned to Subathu Group. The important fossils reported from Subathu Groups are *Isatherium, Neocyprideis, Stephanodus, Nummulites, Lockhartia, Operculina, Assilina, Feddenia, Homotryblium, Pediastrum,* and *Todisporites*. Middle Eocene molar tooth of prosimian primate, first in pre-Siwalik rocks, is reported from Kala Kot area of district of Rajauri of the Jammu and Kashmir.

Muree Group

The rocks of Muree Group are developed between west of Peshawar in Pakistan to Himachal Pradesh in India. Muree Thrust draws its southward contact with Siwaliks rocks in Jammu region while in northern area Muree Group comes in contact with rocks of Shali

Structural Belt along the MBT. The Muree's eastward extension is in Himachal Pradesh is known as Dharamsala Group, Dagshai and Kasauli formations. The Dharmsala, Dagshai and Kasauli are facies variation of Muree Group. Muree is divided into Lower Muree (1000-2300 m) comprising of deep purple to chocolate colored shale, micaceous siltstone and sandstone and Upper Muree (1600m) is made up of purple and green clays, siltstone and sandstone.

The 3150m thick Dharmsala Group comprises of clay, shales, siltstone and sandstones has been divided into Lower and Upper Dharmsala. The rocks of Muree, Dharmsala, Dagshai and Kasauli are poorly fossiliferous e.g. *Unio, Sahal major, Kayea, Poacites, Fious, Persea* and among vertebrate fauna *Nandus, Badis* (pisces), *Palaeochoerus, Microbundon* (aritodactyla), *Teleoceras* (order Perissodactyla) and *Dinotherium* are reported. These rocks were deposited in shallow marine (tidal flat) to continental conditions and are deposited between Upper Eocene to Oligocene duration.

Siwalik Supergroup

This 6-7000 m thick sedimentary succession named after Siwalik Hills of Hardwar in Uttaranchal. In 1925, Pilgrim described Siwaliks as Siwalik System and Siwalik Supergroup is also known as Siwalik Group and it is popularly shortened to Siwaliks. This Supergroup extends from Potwar (in Pakistan) in the west to Arunachal Pradesh in the east. Falconer (1835), Cautley (1836) and Medlicott (1861, 65) were the pioneers who described the Siwaliks initially between Ganga and Ravi rivers. In the East of Kashmir, the northern and southern contacts of Siwaliks are drawn by MBT and HFF respectively. Siwalik sediments are also conformably overlying over the Muree and Dharmsala. Pilgrim divided it on the basis of vertebrate fossils in the type region (NW India and Pakistan) into series and stages i.e. present day's group and formations. The time duration of deposition of sediments in different stratigraphic formations of Siwalik Supergroup is in constant modifications in the light of new data and interpretations and the time base of the same formation in different regions is different.

The stratigraphic succession, on the basis of fossils and magnetostratigraphic records, (with age in Ma in brackets) of the Siwaliks is as follows:

Late Quaternary/ Holocene Alluvium

	Boulder Conglomerate Formation (1.7-0.5)	Pleistocene
Upper Sub Group	Pinjor Formation (2.5-<2.5)	
	Tatrot Formation (5.1-2.5)	
Middle Sub Group	*Dhok Pathan Formation (7.9-5.1)*	*Pliocene*
	Nagri Formation (10.1-7.9)	
Lower Sub Group	Chinji Formation (13.1-10.1	Upper Miocene
	Kamlial Formation (18.3) Middle Miocene	

Muree/Dharmsala

Potwar Plateau in Pakistan is the type area of many Siwalik formations; it is situated between Indus and Jhelum rivers in the Punjab province of Pakistan. Pilgrim's type areas of Kamlial, Chinji, Nagri, Dhok Pathan and Tatrot Formations are present here. The Kamlial Formation is resting over the Muree Formation. Various biostratigraphic zonations of Siwalik are proposed on the basis of abundant presence of vertebrate fossils in the type region. However, in other regions either Siwalik is unfossiliferous or have poor presence of fossils. Lithology of formations of type areas shows large lateral facies variations making identification and correlation of formations of other areas with the type area hazardous. In many areas boundaries between different subgroups are demarcated on the basis of appearance of heavy minerals as well as lithological basis. Recently, these boundaries also are attempted through energy sequence stratigraphy and magnetostratigraphy.

Siwalik is also conveniently categorized into Lower, Middle and Upper Siwalik in Uttaranchal or Lower and Upper Siwalik in Nepal. In Bhutan, Siwaliks are divided into Formation I, Formation II and Formation III. In Arunachal Pradesh four formations namely Kimi, Dafla, Subansiri and Kimin are proposed. In Bhutan a major part of Siwalik is concealed under recent river deposits, its exposures occur at Baxa, Dear Searching and Elephant. In Himachal Pradesh the Lower, Middle and Upper Siwalik are further divided into A and B units.

Chronostratigraphic analysis suggests that Upper Siwalik formations are time transgressive in nature i.e. in different localities the lower time limit of the same formation differs markedly, for example the base of Pinjor Fm ranges from 2.58 to 0.63 Ma and of Boulder Conglomerate ranges from 1.77 to 0.6 Ma (Nanda and Sehgal, 2005)

Lower Siwalik

The 1300 to 2000 m thick Lower Siwalik formations (18.3-13.1 Ma) is comprises alternations of sandstone and clay. Sandstones are gray, to green massive, hard, and coarse and fine to medium grained in nature. Clay, clay stones are red, chocolate and maroon in color. Hard, massive grayish while to green sandstone, medium to fine grained calcareous sandstone and red shales are present in the Kimi Formation of Arunachal Pradesh. Ramnagar area of Jammu and Kashmir in India is considered standard reference area of the Chinji Fm The Kimi Formation is equivalent to Chunabati Formation of Darjeeling Himalaya.

Fossils

Amphicyon, Conohyus, Hipprion, Gazella, Listriodon. Sivapithecus, Giraffa, Unio and *Zizyphs* particularly from Chunabati Formation *Globorotalia, Globogerina* are reported.

Middle Siwalik

The Middle Siwalik (13.1-5.1 Ma) is, in general, 1500-2000 m thick while its equivalent Dafla Formation in Arunachal Pradesh is 3800 m thick. It is made up of sandstone, clay, conglomerate and calcareous nodules. The medium to course grained, massive sandstones are gray to greenish and red in color and are of arkosic to subgraywacke type. These sandstones are interbedded with gray and red clay, clay stones and at many places these are multistoried (Kumar and Nanda, 1989). In some locations, thick horizons of conglomerate also occur. Appearance of kyanite, a heavy mineral, draws the boundary between Lower Siwalik and Middle Siwalik in unfossiliferous areas.

Fossils

Indraloris, Brahmapithecus, Stegodon, Tetraconodon, Hippopotamus Ramapithecus, Sugrivapithecus, Sivapithecus, Sigigium, Trochommina, Bos and *Thalassiphora.*

Upper Siwalik

The Upper Siwalik (5.1- 0.5 Ma), 2300 to > 6000 m in thickness, shows considerable change in lithology from underlying Middle Siwalik. In lower part it is made up of sandstone, mudstone and clay with thin conglomeratic alternations while its upper part is predominantly massive conglomeratic with red and orange clayey matrix and minor sandstone and clay stone layers (Pande and Verma, 1983). In Himachal Pradesh it contains clasts of Mandi-Darla Volcanics. The Upper Siwalik shows characteristic presence of sillimanite and andalusite. In Arunachal Pradesh Kimin Formation constitute the Upper Siwalik Subgroup. *Terminalia* of the family Combretaceae is reported from Upper Siwalik of Arunachal Pradesh. In the Upper Siwalik of Hoshiyarpur, illite, smectite, kaolinite, chlorite and mixed layer phyllosilicate are present which are formed the under moderate chemical weathering condition in cool to temperate climate.

Fossils

Stegodon, Crocodylus, Gavialis, Leptobos, Bubalus, Papio, Sivacobus, Archidiskodon, Equus, Cambaella, Mellivorodon, Hydaspicobus, Granulatisporite and *Neothyrites.*

Outer Nepal Himalaya

Lower Tertiary formations occur in Surkhet, Jajarkot, Dang, Piuthan, Sallajan, and Rolpa and Tansen areas. Among these localities Surkhet area is investigated in detail. The Surkhet Group (Palaeocene to Lower Miocene) is comprises of 160 m thick Melpani Formation consisting of gray to dark gray ferruginous orthoquartzite and sandstone and shale with occasional bands of argillaceous limestone and conglomerate; the Swat Formation (105 m thick) of Eocene age comprises of Nummulitic gray to cream colored limestone and shale and Suntar Formation (550 m thick) of Oligocene to Lower Miocene age, is made up of mainly grayish gray, reddish brown sandstone and shale with bands of marl, pebbly shale, mudstone and conglomeratic sandstone

Siwalik rocks (up to 6000 m thick) are exposed in continuous 20-30 km wide belt in southern parts of the country. Siwalik occurs as fossiliferous and unfossiliferous and cumulative records of Siwaliks are comparable to type sections localities. Koilabas section is richly fossiliferous locality in the Nepal Siwaliks. 5000 m thick Surai khola section of western Nepal contains uninterrupted sequence of Late Lower Siwalik to Upper siwalik of 13 Ma (Corvinus, 2006).

Siwaliks are conventionally divided into Lower, Middle and Upper Siwaliks. These names are proposed to be replaced by Arung (Lower), Binai (Middle) and Chitwan (Upper) Siwaliks. Surahikhola is type locality of Siwalik in Nepal where it attains 5500 m thickness. Here five lithological units namely Bankas Formation, Chorkhola Formation, Surahikhola Formation, Dobatta Formation and Dhankhola Formation are identified. However, in another classification for entire Nepal nine formations namely Bhorlegaon Formation, Bankas Formation, Jangali Formation, Chorkhola Formation, Suraikhola Formation. Dabatta Formation, Dudhaura Formation, Dhankhola Formation and Deorali Formation are proposed (Sah, 1995).

In Tinaukhola area of west central Nepal the earliest Siwalik sedimentation occurred in flood plain and back swamp condition, it was followed by fluvial environment with assemblages of flood and crevasse splay sedimentation by fine-grained meandering river system next change was to braided system and finally gravelly braided system (Ulak and Nakayama, 2001). In Bhutan, Siwaliks occur in detached outcrops of just few kilometers width. In north, Siwaliks are in contact with very disturbed thin intermittent wedges of Damuda Fm of Gondwana and Baxa Fm of the Lesser Himalaya. Siwaliks are divided in to conventional lower; middle and upper groups or Formation I and Formation II.

The Quaternary Diklai Boulder Bed sequence is 2000 m thick and is comprises of dimictite, conglomeratic sandstone, argillaceous sand stone and silty clay stone. The dimictite is constituted of rounded pebbles, cobbles and boulders having ill preserved dicot leaf impressions. In lower Raidak River valley, peninsular Assam Gneiss comprises of hornblende schist occurs below the Siwaliks.

Energy Sequence Stratigraphy

The concept of energy sequence (enseqs) stratigraphy is based on textural characters, particularly grains size of clastic sedimentary rocks. Grains size provides fairly complete history of a sedimentary section or of whole sequence starting from its provenance to final resting of sediments in the depositional basin. It also tells mode of tectonic upliftment of provenance and the distance between provenance and depositional basin. Two ends of clastic rocks namely conglomerate and shale in a stratigraphic section indicate that former was derived from highly uplifted provenance or transported in high gravity gradient condition but sediments for latter rock comes from provenance of low altitude or transported in low gravity gradient. Further presence of conglomerate and shale suggest that proximally (near) and distally (distant) situated places of their depositions in a basin. Thus conglomerate represents deposition by a vibrant or swiftly flowing stream i.e. having high contents of energy and shale represents deposition of sediment by a sluggishly flowing stream having low energy content. In the case of chemically precipitated rocks such as limestone no such transport consideration involves. The energy sequence concept is closely related with the idea of tectonic cycle. A geological column comprising of conglomerate, sandstone, siltstone and shale represents an energy sequence having maximum to minimum every contents. In other words the above sequence represents a tectonic phase or pulsation in which considerably uplifted terrain reduced to peneplain level. In a regional section several energy sequences or phases of tectonic pulsations can be easily recognized. A tectonic cycle in an orogenic belt starts with initiation of depositional basin to its final closing by filling of sediments and its subsequent upliftment. Such cycle begins with deposition of fine clastic or carbonate rocks and ends with deposition of conglomerate or rudaceous sediments. The orogenic movements being episodic in nature so many minor or sub cycles can be identified in a tectonic cycle. The tectonic cycle and sedimentation cycle is hardly synchronous or symmetrical. An energy sequence is an exercise in time series analysis and represents to a stratigraphic unit, which is time parallel. In other word energy sequence represents depth versus mean grain size plot of a stratigraphic section. This time parallel unit is also comparable to biostratigraphic or chronostratigraphic unit. An energy sequence unit is also comparable to litho-stratigraphic formation. Raiverman (1985) applied this concept in the western Outer Himalaya and he recognized eight energy sequence unit in the Outer Himalaya. Each sequence has characteristic ratio of shale, sandstone and conglomerates, their color, stratigraphic characters, and composition of sediments. The eight sequences are Dharampur, Kumarhatti, Makereri, Jawalamukhi, Kalidhar, Sandhot, Batwan and Wahdevi. The Dharampur and Kumarhatti are equivalent to Subathu and Lower Dharmsala; Makereri is equivalent to Lower Dharmsala and rest belongs to Siwalik Group.

Magnetostratigraphy

The principal goal of magnetostratigraphy is to document and calibrate the global geomagnetic polarity sequence in stratified rocks and to apply this geomagnetic polarity time scale for high-resolution correlation of marine magnetic anomalies and polarity zones in other sequence (Ogg and Smith, 2004). The magnetostratigraphy is a tool for correlating the normal and reversed polarity zones of a rock succession to the magnetic anomalies of the standard geomagnetic polarity time scales (GTPS) by the tie- points from independently dated benchmarks (time constrained) stratigraphic horizons, matching of reversal pattern to GTPS and sediment accumulation rates. Magnetic polarity is independent of petrologic or lithologic constraints and uses widely abundant iron oxide minerals that faithfully preserve the prevailing magnetic attributes of the earth at the time of deposition of the sediments. It is a correlation tool rather than absolute dating method and is often used for refinement of biostratigraphy. Magnetostratigraphy basically relied on principle of identification normal and reverse magnetic polarity events and correlating these events with litho or biostratigraphic units. The present day polarity is described as normal polarity, this implies that the magnetic lines of force at the surface of the earth are directed in towards the north magnetic pole and north-seeking pole of a compass needle point in north while in case of reverse polarity the lines of force directed in opposite direction and compass needle would point to the south. The change over time between two polarity events is approximately 5000 years and in one million year duration polarity on an average changes two to three times. The length of polarity interval ranges between 30000 years to several tens of millions of years and the present condition of normal polarity came into existence about 780000 years ago.

In the Outer Himalaya, Keller and coworkers in Potwar Plateau started magnetic polarity stratigraphy study and the study was subsequently undertaken in Himachal Pradesh, Garhwal and in Nepal (Gautam et al., 1995, Kotlia et al., 2005). In the western Himalaya on the basis of the magnetostratigraphy five bench mark levels namely the

A - the base of long normal (C 5n) at 10.95 Ma
B - at the base of C 4n at 8.07 Ma
C - the base of C 3n at 5.23 Ma
D - the Gauss- Matuyama boundary (top of C 2 An) at 2.58 Ma
E - the Brunhes- Matuyama boundary (base of Cin) at 0.78 Ma

are, as most characteristic events so far, reported from the Late Cenozoic Siwalik sequence (Sangode and Kumar, 2003).

In Gabhir Khas section of Potwar Plateau 13 normal and 12 reverse zones were recognized while in Chital Panwali section of the same region 7 normal and 8 reverse zones were identified (Qureshi et al., 1995). In Haritalynagar area of Himachal Pradesh, 8 normal and 7 reverse sections are identified (Ranga Rao, 1993). In Nepal, magnetostratigraphic studies were carried out in Khutiakhola, Dawarkhola, Suraikhola, and Arungkhola and in other regions.

Rate of Sedimentation

In Haripur area of Himachal Pradesh, with the help of magnetostratigraphy rate of Siwalik sedimentation was calculated. Using the rate of sedimentation data, stable tectonic condition interpreted before 2.6 Ma (45 cm/1000 yr) and afterward- unstable tectonic condition (54±8 cm/1000 yr) was inferred. In Kangra sub basin the rate of sedimentation was 37 cm/ 1000 yr between 11.3 to 5.5 Ma age bracket. The rate of sedimentation in Ghaggar was 45-62 cm /1000 yr; in Maginand section it is 26 cm, 60- 94 cm/1000 yr in Patialirao, in Khetpurali 37cm/1000 yr (Kumaravel et al., 2005). In Pinjor Dun, Siwalik rate of sedimentation was 100-150 cm / 1000 yr. The drilling well data in Janauri well records 620 m thickness of Upper Siwalik, 620-2030 m thickness of Middle Siwalik, 2030-3440 m of Lower Siwalik, 3440-4440 of Upper Dharamsala and 4440-4790 m of Lower Dharamsala formations. The rate of deposition of these formations were 463.9 m/Ma, 273.2 m /Ma, 181.8 m /Ma, 155 m/Ma and 60.6 m/Ma respectively.

The generalized sedimentation rate calculated in the Lower, Middle and Upper Siwaliks, over a period ranging between 0.1 to 1 Ma scale, are 0.12, 0.3 and 0.67 mm/yr.

Palaeosols Study in Siwaliks

Study of palaeosoils or palaeosols is a relatively new discipline in geology. Palaeosols entrapped between coarse-grained sedimentary sequences are used to infer palaeoclimate, palaeobotanical interpretation, in palaeoseismology, palaeomagnetism, information on fluvial processes, basin aggradation and in tectonic studies. Carbon and oxygen isotopes of calcretes present in soil, microscopic study and granulometric interpretation of palaeosols are employed for palaeoclimatic interpretations. On the basis of soil morphology parameters such as parent material, color , soil horizon thickness, soil structure or pedality and carbonate content and micromorphological features viz. cutans, b-fabric, concretions etc soil facies or pedofacies are deciphered. Palaeosols studies of Siwalik soils have been done in India, Pakistan and Nepal. In Haripur-Kolar section six pedofacies were identified in the Middle-Upper Siwaliks (Thomas et al., 2005)

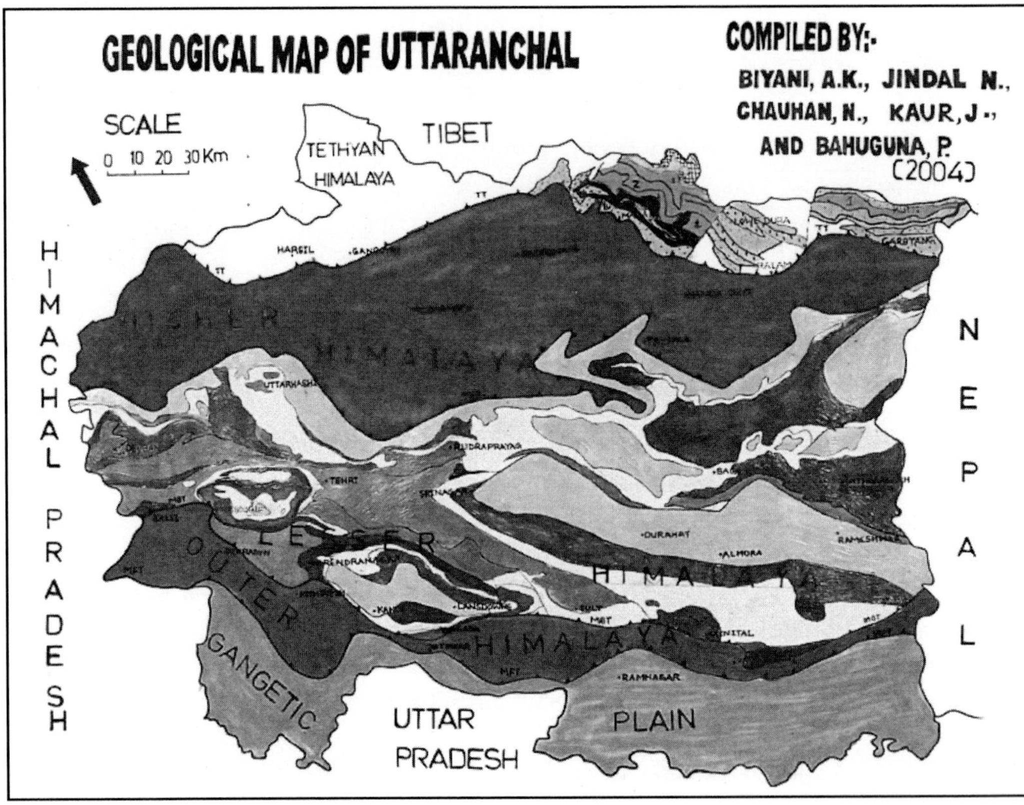

Plate 5.2. Geological map of Uttaranchal. Stratigraphical sucession is given Table No. 5.1

Doon Gravel

Doon Gravel occurs in 80 km long and maximum 29 km wide Doon Valley of the Uttaranchal. This post –Upper Siwalik valley is a tectonic syncline and has been uplifted by 315-420 m. The longitudinally faulted synclinal floor was filled up by the consolidated to unconsolidated sediments derived from the northern Lesser Himalayan formations and southern rising Siwalik Group and are described as Doon/ Dun Gravels. Geomorphologically, three large sized fans namely the Donga-; Doon- and Bhogpur fans are present on the southern valley-facing slope of the Lesser Himalaya. Doon Gravels are divided into Lower Doon Gravel (hard and compact conglomerate) and Upper Doon Gravel (unconsolidated gravelly material). In Doon valley river terraces up to six levels are reported and in different parts of the valley the terraces are uplifted by 0.5 mm/yr to 3.6 mm/yr rate (Singh et al. 2004).

Shortening of 11 km is estimated in the Doon Valley in last 0.7- 0.8 Ma by balanced cross section construction method. This gives shortening rate of 15±1 mm/ yr and in another study based on uplifted fluvial terraces near HFF 11.9±3.1 mm/yr shortening rate was observed. Mohand Anticline, south of Doon Valley, has accommodated 5 km shortening (Mishra and Mukhopadhyay, 2002; Mukhopadhyay and Misra, 2004).

Table 5.1 Stratigraphic column of Uttaranchal Himalaya (after Biyani and Joshi, 2004). Abbreviations: Grp indicates Group and Fm indicates Formation.

ERA	PERIOD		INDO-GANGETIC PLAIN	OUTER HIMALAYA	LESSER HIMALAYA	HIGHER HIMALAYA	TETHYS HIMALAYA	
							GARHWAL	KUMAON
Cenozoic	Holocene	0.001	River Sediment	River Sediment Doon Fan	River Sediment and glacial sediment	River Sediment and glacial sediment	River Sediment and glacial sediment	River Sedimen and glacial sediment
	Pliestocene	1.81		Doon Gravel Siwalik Group				
	Pliocene	5.33		Siwalik Group		Intrusive Granite		
	Miocene	23.03		Siwalik Group			Intrusive Granite	Intrusive Granite
	Oligicene	33.9						
	Eocene	55.8			Subathu Group Nilkanth/Singtali/ (Kakara) Group		(Exotic block) SangchaMalla Group	
	Palaeocene	65.5						
Mesozoic							Guimal sandstone Spiti Shale (Shaligram)	SangchaMalla Group
	Cretaceous	145.5					Sulcacutus Beds Lapthal	
	Jurassic	191.6					Kioto Limestone Kuti Shale	
	Triassic	251					Kalapani Lst Choclate Group	Kuti Shale Kalapani Lst Choclate Group
							Kringkrong Grp Muth Quartzite	Kuling Group Fenestella Shale Kali Group Muth Quartzite Variegated Gr
	Permian	299			Binj Fm.		Variegated Grp Yang Group Siala Group	
	Carboniferous	359.2						
	Devonian							
Palaeozoic		416					Garbyang Fm.	
	Silurian	443.7					Ralam Group Martoli Group	Siala Group Changru Group
	Ordovician	488.3						
	Cambrian	542			Tal Fm. Krol Fm. Infrakrol Blaini Fm. Nagthat Fm. Chandpur Fm. Mandhali Fm. Deoban Fm. Rautgara Fm. Chakrata Fm. Berinag Fm. -Thrust- Ramgarh Group -Thrust- Almora Group			Garbyang Fm.
Proterozoic						Chail Fm. Jutogh Fm. Vaikrita Fm.		Buddhi Schist Vaikrita Group

5.3 Lesser Himalaya

The maximum 100 km wide Lesser or Lower Himalaya is best developed in the western Nepal, Kumaun and Himachal Pradesh. It is also called Peninsular Himalaya as it shows well correlatable similarities with the peninsular rocks of south of Gangetic plain. The Lesser

Himalayan rocks occur in narrow band or detached out crops in Kashmir and Zanskar in India and in Salt Range, Kala Chitta, Naushera and Hazara Region of the Pakistan and in the east as continuous outcrop in the western, central and eastern Nepal, Sikkim, Bhutan and Arunachal Pradesh. The Lesser Himalaya makes tectonic contact (thrust) with Outer Himalaya in south and the Higher Himalaya in north namely by MBT and MCT respectively. However in the western region it is contact with alluvium in south and Tethyan or other sequences in the north. Its northern expanse in India and Nepal is debatable as different workers interpret MCT, which delineates its boundary with the Higher Himalaya, differently. Thus its northern boundary fluctuates in 20 to 30 km wide zone. Rupke (1974) was first to give the comprehensive account of geology of the Kumaun Lesser Himalaya) it was followed by Valdiya (1980) and of Garhwal region by Saklani (1993). Mehdi et al., (1972) and Merh (1977) have initially described the geology of Kumaun region.

Tectonically, it is made up of three types formations namely autochthonous, para-autochthonous and allochthonous nappe and klippe units, some of these are transported from the Higher Himalaya. These rocks are intruded by granites of varying age and have also fair presence intrusive and extrusive basic rocks.

Stratigraphically, it is most complex division of the Himalaya. Firstly the basement of the Lesser Himalaya is nowhere exposed. Secondly its major part is invaded by the crystalline rocks of Higher Himalaya, which are masking the underlying, possibly autochthonous rocks Thirdly sedimentary formations are largely unfossiliferous thus interpretation of their ages are subjective. Fourthly, the sequence is thoroughly deformed and at places autochthonous rocks are mildly metamorphosed. Fifthly, the Lesser Himalaya has long and complex geological history with stratigraphic break(s) of varying magnitude. In the far western Nepal, Kumaun and Himachal Pradesh an overwhelming part of the Lesser Himalaya is comprised of Proterozoic rocks followed by an unconformity of long duration covering the greater part of the Palaeozoic and Mesozoic. Permian and Early Tertiary marine transgression causing deposition of fossiliferous and unfossiliferous Shell Limestone, Subathu and other formations, follow this break. In pocketed areas of Nepal, Kashmir, and Pakistan depositions of Lower Palaeozoic and Mesozoic age occur with in the Lesser Himalaya with several stratigraphical breaks. Lastly, the large number of workers who have established stratigraphical orders in more often than not in small independent areas of studies in fairly accessible regions. Therefore, hundreds of formational terms are introduced in the literature and in many cases stratigraphical columns are erected on the basis of lone or few non-reproducible fossils causing lot confusion in its proper assessment. Classical example of this type of situation, on a regional level, is represented by the Krol Belt of Himachal Pradesh and Uttaranchal; where tens of stratigraphical columns assigned the age of Krol Formation from the Late Proterozoic to Cretaceous The stratigraphical disorders or haze of Krol Belt is cleared by Singh (1980) and Azmi et al. (1981) and provided sound stratigraphical footing, this subsequently helped indirectly to the entire Lesser Himalaya. This situation of regional level particularly in Kumaun Himalaya where immense geological confusion were existed is admirably rectified by Valdiya (1980) through his monumental work in the Kumaun region and his stratigraphical scheme is extended, with age and few other modifications, in the further west by Thakur (1992).

Western Himalaya

The tectonostratigraphic succession of the western Lesser Himalaya (after Thakur, 1992) is as follows:

	Metamorphic Klippe of Jutogh (Munsiari) or Almora and Chail (Ramgarh) Groups
——— MCT ———	
	Subathu Formation Singtali Formation
——— unconformity ———	
Krol Belt/ Super Group	Tal Formation Krol Formation Infra-Krol Formation Blaini Formation Nagthat Formation Chandpur Formation
Tejam Group	Mandhali Formation Deoban Formation
Damta Group	Rautgarha Formation Chakrata Formation Berinag Formation
——— Base Not Exposed ———	

Berinag Formation

Thakur on the basis of available radiometric dates of intrusive Bandal granite (1800 Ma) considered this formation as the oldest Lesser Himalayan formation, however, Valdiya (1980) considered the Berinag Formation equivalent to the Nagthat Formation of Krol belt of Auden (1934,37). It is a very thick sequence of fine-grained quartz arenite and sericitic quartzite. It is mildly metamorphosed and has interbedded chlorite schist and metabasics. Rampur and Banjar formations of Himachal Pradesh and Banku Quartzite of Parchauni Group of western Nepal are equivalent to the Berinag Formation.

Damta Group

Chakrata Formation and Rautgarha Formation constitute this group exposed in Kumaun and Himachal Pradesh.

Chakrata Formation

This turbiditic formation is named after Chakrata town, situated nearly 100 km NE of Dehradun. The flysch or turbidite sequence is made up of greywacke and siltstone of varying shades and show variety of sedimentary structures like flute cast, graded bedding, tool and scour marks and folds formed by penecontemporaneous deformation. Valdiya (1980) considered this is the oldest formation of Kumaun Lesser Himalaya and its base is nowhere exposed. Lower Riphean age is assigned to this unfossiliferous sequence.

Rautgarha Formation

It is named after a village Rautgarha in Pithoragarh district of Kumaun. It is made up of muddy quartzite (subgraywacke to sublitharenite), slates with bands of conglomerate and its extensively intruded or basic sills and dykes and many lava flows have been recognized. Rautgarha is shallow water flysch deposits of Lower Riphean age.

Tejam Group

This group is made up of Deoban and Mandhali formations.

Deoban Formation

It is named after Deoban Mountain situated 5 km north of Chakrata town. This formation consists of stromatolitic blue limestone, cherty dolomite, dolomitic limestone and slates of gray color. *Baicalia baicalica, Kussiella kusgiensis, Minjaria, Uralica, Masdoviella columnaris* suggest Middle to Upper Riphean age of Deoban Formation.

Mandhali Formation

Village Mundali, situated in Deoban Mountain in Chakrata tehsil is type locality of this formation. It is constituted by pyritic phyllite of grayish green and black color and blue color limestone, siltstone, quartzite and para conglomerate.

Krol Super Group

The Krol Supergroup is also known by the other names like Krol Nappe or Krol Belt, is named after Krol peak near Solan in Himachal Pradesh and its geology and allochthonous nature was worked out in detail by Auden (1934) in Himachal Pradesh and Uttaranchal. The more than 6000 m thick Krol Super Group of Thakur consisting of Chandpur, Nagthat, Blaini, Infra-Krol, Krol, Tal and Subathu formations. Several workers have refused to recognized it as allochthonous sheet and proposed it is either paraautochthonous or autochthonous in nature. It is divided into two units, namely the Outer Krol Belt present in southwest and is deposited on the foundation of Shimla Group and Inner Krol Belt situated in northeastern region deposited on foundation of Mandhali Formation. These two units are homotaxial and independently formed (Bhargava, 1972).

Chandpur Formation

Chandpur peak in Himachal Pradesh is the type area of Chandpur Formation. This formation consists of olive green, green and gray colored phyllite interbedded with metasiltstone, quartzite, greywacke and metabasics. Sedimentary structures indicate its flyschoidal nature.

Nagthat Formation

It is named after Nagthat village, situated SE of Chakrata Green, white, purple and fawn color hard quartzite (quartz arenite and sublithic arenite), wacke interbedded with purple slates and conglomerate constitute this formation. It is associated with lava flow of spilitic composition known as Bhimtal Volcanics and younger dykes and sills of basic compositions. The average modal composition of the arenite is quartz 63%, feldspar 5.6%, rock fragments 11.8%, mica 2.5%, heavy minerals 1.2% and matrix 11.3%. Sediments for this formation is inferred from the granitic source possibly from the Banded Gneissic Complex of the Aravalli- Bundelkhand craton (Islam et al., 2002).

Blaini Formation

The name Blaini is given after Baliana River of Himachal Pradesh. This formation consists of conglomerate with siltstone, carbonaceous shale, varved argillite, purple or red limestone and minor quartzite. The conglomeratic horizon was considered equivalent to Upper Carboniferous Talchir Boulder Bed of Gondwana sequence of peninsular area. However, this concept got invalidated in recent time due revision of age of Krol Belt.

Infra-Krol Formation

The Infra-Krol sequences exposed at the base of Krol Formation is comprised of gray-green siltstone, shale and greywacke, pyritiferous shale and slate and thins bands of quartzite. The Infra-Krol Formation is clubbed with Blaini Formation by Valdiya (1980) while Srikantia and Bhargava (1998) considered this formation as basal member of Krol Formation. The fossil assemblage reported from this includes cynobacterial filaments, coccoids, acritarchs, red algae, sponge spicules like structures, vase shaped microfossils suggesting Vendian age (Tiwari and Pant, 2004).

Krol Formation

This formation, named after Krol Hill (2278 m) of district Solan, Himachal Pradesh, is exposed in several synclines like Krol, Pachmunda, Kamlidhar, Nigalidhar, Korgai, Mussoorie and Nainital synclines. This dominantly calcareous formation also consists of red and green shale and sand stone. Its base is constituted by a sandstone horizon designated as Krol sandstone exposed in Krol and Kamlidhar synclines of Himachal Pradesh and Dehradun (Biyani and Joshi, 1997). In type area, Auden (1934) divided Krol formation into five members which he named as Krol A (dolomite, oolitic limestone, calcilutite, shales), Krol B (red shales with calcilutite and thin bands of dolomite), Krol C (massive, dark blue dolomite), Krol D (mainly cherty limestone with minor occurrence of conglomerate, sand stone and shale) and Krol E (calcareous sandstone, quartzarenite and calcilutite interbeds with red and black shales). This distinction of five subdivisions is not clear in Kumaun and Nepal.

This Neoproterozoic formation has shown presence cynobacteria (*Tetraphycus, Syphonophycus, Oscillatoriopsis Myxococcoides, Bavlinella*), vase shaped microfossils, VSM, (*Melanocyrrillium*), algae (*Wengania, Thallophyca*), acritarchs (*Satka,*

Margominuscula, Paracrassosphaera), oncolites, edicaran, coelenterates and stromatolitic fossils.

Tal Formation

This formation is named after the Tal River situated near Rishikesh and has more than 1000 m thickness. It is divided into Lower Tal Formation comprising of black bedded chert, phosphorite and black shale, silty shale and sandstone, ferruginous, siliceous or sandy limestone and quartzite, and Upper Tal Formation consists of quartzite, arkose, grit and thin bands of gray, green and red shales and siltstone. Azmi et al. (1981) assigns the Lower Cambrian age to chert-phosphorite Member of Lower Tal Formation on the basis of conodont fossils.

Age of Krol Belt: Riddle Solved

Geology of Krol Belt is one of the monumental contributions of Auden (1934,37), who investigated this more than 6000 m thick, then unfossiliferous, mainly sedimentary Lesser Himalayan nappe (?) in the Kumaun Himalaya. Auden equated Blaini Formation (mainly conglomeratic formation), a middle formation of the Krol Belt, with the Talchir Boulder Bed, a basal formation of glacial origin in fresh water Gondwana sequence of the peninsular India. Auden assigned the Permian age to Blaini Formation and the Jurassic-Cretaceous age to overlying Krol and Tal Formations and the same was accepted by Wadia (1937), Gansser (1964) and Krishnan (1982). However, in the post sixties period several workers like Sitholey et al. (1954), Sab et al. (1968), Tewari (1969), Valdiya (1969, 1980), Bhargava (1972, 79), Ghosh and Srivastava (1972), Mithal and Chaturvedi (1977), Kalia (1972), Sinha (1975, 1979) Prasad and Bhatia (1975), Kumar (1978), Patwardhan (1978), Tewari and Singh (1979), Tiwari and Joshi (1995) Singh and Shukla (1981) and many others referred in Singh (1980), Azmi et al. (1981), Azmi and Joshi (1983), Azmi (1983), Singh and Rai (1983 a and b) and Shukla et al. (2005) reported single or few fossil assemblages of many families like nannofossils, oncolites, mollusca, trilobites, stromatolites, branchiopod, annelids, ichno fossils, plant fossils etc. from different formations of the Krol Belt. Different ages varying from Proterozoic to Cretaceous were assigned to that formation from which fossils were reported. For example L. Proterozoic, Carboniferous to Cretaceous ages has been assigned to the Krol Formation. Using this information as base, attempts were made to establish the stratigraphic succession of the entire belt. Thus the real age of Krol Formation became a mystery or riddle and no straight answer to this riddle was in sight. The age problem of the Krol Belt is critically reviewed by Singh (1980) and he stated that most of the reported fossils are either poorly preserved, fragmentary in nature, non-reproducible or reported from only one locality, lacking precise locations with vague account of geology and invariably lithologs are missing; therefore, these finding should be discarded. He assigned the Precambrian age to Krol and Tal formations and also supported Auden's contention that entire belt is an extension of peninsular Vindhyan basin.

In 1981, Azmi, Joshi and Juyal made a significant breakthrough in biostratigraphy of the Krol Belt by reporting large number of reproducible conodont fossils – *Amphigeisina, Cordylodus, Drepanodus, Furnishina, Hirsutodonties* etc. from phosphoritic horizon situated at the Krol-Tal contact of Mussoorie Syncline near Dehradun. Azmi (1983) assigned the Late Proterozoic to the Krol Formation and Lower Cambrian age to the Tal Formation. Tiwari and Parcha (2006) reported presence of trace fossils like *Monomorphichnus, Dimorphichnus, Diplichnites, Neonerites* etc. of Early Cambrian age from the lowermost Quartzite member of Upper Tal Formation of the Mussoorie Syncline. With the amount of work done in Krol-Tal boundary, this section should be considered as one of the type localities for Precambrian Cambrian boundary.

Conodontic age of the Krol Formation is further reinforced by reports of brachiopods trilobites, mollusca and ichno fossils from different parts of Mussoorie Syncline (Singh and Rai, 1983 a and b). Shukla et al. (2005) assigned Terminal Neoproterozoic age to basal Krol sandstone of Chambaghat Formation of Kamlidhar Syncline of Himachal Pradesh.

The age Gangolihat dolomite (= Deoban Fm) is recently revised from Mesoproterozoic (~1600-1000 Ma) to Vendian- Early Cambrian on the basis of protoconodonts (organophosphatic grasping spines with deep inner cavity of oldest chaetognaths, arrow worms) species like *Protohertzina* and *Proonetodus* (Azmi and Paul, 2004).

Crystalline Klippen and Nappes

Several nappes and detached inliers klippen units of Higher Himalaya overlie the Lesser Himalayan sedimentaries in Himachal Pradesh, Uttaranchal, Nepal, Sikkim, Bhutan and Arunachal Pradesh. These eroded crystalline rocks are present in cores of major synclines or synformal structures. In Kashmir and further west the Lesser Himalaya as well as the Higher Himalaya are hidden under Tethyan rocks. The important crystalline nappes and klippen units comprising metasedimentary and granitic rocks of Uttaranchal and Himachal Pradesh are:

1. *Banali : Satengal and Lansdowne Klippe of the Garhwal Nappe* : Garhwal Nappe is comprised of slate, phyllite, quartzite, siltstone, and conglomerate. Two thrust sheet less traveled Bijni nappe and more traveled Amri nappe are recognized by Auden while in Lansdowne Klippe Bijni Nappe is further divided into Lower and Upper Bijni units. The Lansdowne Klippe is intruded by muscovite biotite granite showing gneissic and mylonitic textures.
2. *Ramgarh Nappe* : Ramgarh Nappe in Nainital district is comprised of medium to coarse-grained granite, granodiorite and adamellite and locally described as Amritpur Granite. The other units of Ramgarh Nappe are Rampur porphyry or Debguru porphyroids (gneissose quartz porphyry and mylonitic granite gneiss and Nathuakhan Formation (Crystalline limestone, sheeny phyllite and sericite phyllite and quartzite). It is equivalent to the Chail Nappe of Higher Himalaya.
3. *Almora Nappe* **:**The extensively developed Almora Nappe occupies Dudatoli Ranikhet-Almora and Champawat range of Kumaun. Almora Nappe comprises of schist, micaceous quartzite and quartzite. Metamorphosed in amphibolite facies and also have batholithic granitoids like Champawat Granodiorite, Almora and Dudatoli Granites.
4. *Baijnath Klippe***:**This klippe is developed between Bageshwar and Nandprayag; chlorite schist, sericite schist, graphite schist, mica schist, quartzite, augen gneiss, granite

gneiss and amphibolites constitute the Baijnath Crystallines. Gwaldam Granite (1307±79 Ma) occurs as intrusive mass within the crystallines.

5. *Askot Nappe* **:** The Askot Crystallines are resting over Berinag Formation in the south east of Baijnath Crystallines in Kumaun region and is comprises of granite gneiss, garnet mica schist and bands of amphibolite. The granite gneiss has yielded the 1983±80 Ma age.

6. *Shimla Klippe* **:** It is also known as Jutogh Klippe and occurs around Shimla area in Himachal Pradesh. This composite klippe is comprised of Chail Formation (low grade schist and phyllite, slates and limestone) and Jutogh Formation (medium grade metamorphic rocks namely quartzite, phyllite, garnet schist, staurolite, schist, kyanite schist, amphibolite bands and gneissic granite. The Shimla Klippe is resting over rocks of Jaunsar Group.

Himachal Pradesh

The Lesser Himalayan rocks of Garhwal-Kumaun of Uttaranchal continue in same disposition in eastern Himachal Pradesh. However, in the west beyond the river Beas and the river Ravi correlation becomes difficult particularly in the lower part of stratigraphical column as the base of lowest formation is not exposed and has different structural and tectonic setting. The Bandal-Jeori-Wangtu Granitoid Gneiss is the oldest (2000 Ma) rock unit of western Himachal Pradesh. It is followed by quartzarenite, phyllite slate and metabasics of Rampur Group. This is equivalent to Berinag Formation. The Mesoproterozoic Larji Group, containing of quartzarenite, phyllite, slate, shale, dolomite and limestone, succeeds it. The Shali structural Belt was defined as the belt of carbonate rocks with the underlying orthoquartzite-volcanic suit (Mandi-Darla volcanics), structurally delimited by the belt of paraautochthonous Palaeogene formations along the western boundary and by the Jutogh regional nappe in the eastern side (Srikantia and Sharma, 1976). The belt comprises of Lower Sundernagar Group of Palaeoproterozoic age comprises with slates, phyllite, quartzite, shale and having mildly metamorphosed basics designated as Mandi-Darla Volcanics. The Sundernagar Group is equated with Rautgarha Formation. Chatterjee et al. (2005) have identified two magmatic episodes as Mandi Complex (pre-Shali) and Tattapani Complex (post-Shali/pre-Shimla) on the basis of field and geochemical evidences. Earlier this was identified as single Sundernagar volcanics/Mandi-Darla volcanics. The succeeding Shali Group (>3100 m thick) is made of red shales, siltstone, quartzarenite, dolomite and limestone. Its basal part, in Mandi, shows presence salt bed, salt grit and marly salt layer called Lokhan. Shali Group is equated with Deoban Formation. The Deoban Structural Belt of Himachal Pradesh is constituted by Lower Dharagad Group (quartzarenite, slates, shale siltstone and limestone with basic lava flows) and Upper Deoban Group having lithology same as Deoban Formation. The remaining formations namely Shimla Group, Jaunsar Group and Krol Belt match with Kumaun Himalaya in their petrological composition and structural and tectonic setup (Sharma, 1977).

Folded Autochthonous Belt of Kashmir

The Shali Structural Belt of Himachal Pradesh joins with Folded Autochthonous Belt of Wadia (1928,31) in Kashmir. In Kashmir, the Lesser Himalaya is represented by Gamir

Formation (quartzite, purple and green shale slate and limestone); Baila Formation (limestone with shale and slate); Ramban Formation (shale, slate quartzite and gypsum); Bhimdasa Formation (shale, slate, siltstone, quartzite) and Sincha Formation (Dolomite, limestone, black chert and gypsum) of Late Proterozoic age. The Gamir and Baila Formations are correlated with the Shali Formation; Ramban and Bhimdasa Formations with Shimla Slates and the Sincha Formation is equivalent to Tal Formation. In the Poonch area, these formations are overlain by the Tethyan sequence comprising of Agglomeratic Slates of Upper Carboniferous to Lower Permian, Panjal Volcanics of Lower Permian, Zewan Formation comprising of slates, shales, quartzite and limestone of Upper Permian, Poonch and Mandi formations constituted by shale, slate, limestone and fine grained quartzite of Triassic Jurassic age) and Rajpur Formation having Lower Nummulitic limestone member and upper Variegated shale member of shale and slate, of Palaeocene and Lower Eocene age.

Shali Structural Belt and the Autochthonous Folded Belt terms have been criticized on the ground that former does have true character of a tectonic belt and latter is neither folded nor autochthonous in nature (Thakur, 1992, pp. 86), instead a more appropriate term Panjal Imbricate Zone is proposed for the Lesser Himalayan rocks occurring between the Panjal (=Chail) Thrust and MBT. The westward extension Panjal Imbricate Zone occurs in the NW syntax in Hazara district of NW Pakistan. Stratigraphically, Late Proterozoic to Cambrian sequence comprising of Hazara Formation (slate, phyllite, shale and thickly bedded greywacke and limestone beds and gypsum of deep water turbidite origin), Tanwal Formation (metaquartzite and conglomerate with intrusive Masher Granite), Tanakki Boulder Bed Formation (Conglomerate at the base and red and maroon shales, sandstone and quartzite). Abbottabad Formation (dolomite, phyllite, quartzite conglomerate and phosphatic marble of Lower Cambrian age) and Hazara Formation (quartzite, shaley siltstone and hematite of Cambrian age) have been identified. Further in the south of Hazara, in the NE-SW trending elongated Garhi-Habibullah Syncline west of Muzaffarabad town, Jurassic age Datta Formation (sandstone and shale of Early Jurassic age), Samaner Sukh Limestone (Late Jurassic limestone) and the Chichali Formation (black Shale and sandstone) are overlying the Hazara Formation.

Lesser Himalaya West of NW Syntaxial Bend

Across the northwestern syntaxial bend, the Lesser Himalayan rocks occur in several independent localities such as Salt Range, Attock-Cherat Range, Kala Chitta Range and Naushera area. Part of Attock-Cherat Range, Kala Chitta Range, Salt Range and Naushera area are situated west of the Indus River (Thakur, 1992).

Salt Range

The Late Proterozoic to Middle Cambrian, Permian and Triassic to Eocene rocks of the Salt Range represent the southern most part of the western Himalayan orogenic belt. The main range is developed between Jhelum and Indus rivers and a part of this range is also developed west of the Indus River. Its southern boundary with alluvium is demarcated by the Salt Range Thrust equivalent to HFF and in the north it is in contact with Muree or Siwaliks.

The stratigraphic sequence of the Salt Range is as follows:

Surghar Group	Late Jurassic - Early Cretaceous
——— unconformity ———	
Baroch Group	E-Middle Jurassic
——— unconformity ———	
Musakhel Group	Triassic
——— unconformity ———	
Zuluch Group	L. Permian
Neelawahan Group	E. Permian
——— unconformity ———	
Jhelum Group	E-M Cambrian
Salt Range Formation	Eocambrian

Salt Range Formation was earlier described as Saline series or Panjab Saline Series (Wynne, 1878). Salt, marl, gypsum, dolomite and clay constitute this formation. The Jhelum Group is comprised of Khurpa Sandstone, Kussuk Formation (= Neobolous Bed; sandstone, shale, siltstone, dolomite, gypsum and asphalt), Jutana Formation (=Magnesian Sandstone; dolomite) and Bhaganwala Formation (= Salt Pseudomorph Bed; red shale, clay, sandstone and salt pseudomorph casts); Kishor Formation or Gypsiferous Series is equivalent to Bhaganwala Formation.

The Neelawahan Group of Early Permian age is constituted by Tobra Formation (Salt Range Boulder Bed, (illite, shale, clay, siltstone and sandstone), Dandate Formation (Olive Series; Eurydesma Beds or Conularia Beds or speckled sandstone or Dandate Group; shales of various shades), Warcha sandstone (sandstone) and Sardhani Formation (= Lavender Clays, bluish and greenish gray clay and minor sandstone and siltstone).

Amb Formation (sandstone, limestone and shale), Wargal limestone (limestone and dolomite) and Chhidru Formation shale, sandy limestone and calcareous sandstone) constitute the Zuluch Group. The Triassic Musakhel Group is divided into Mianwali Formation (marl, sandstone, sandstone and dolomite), Tredian Formation (micaceous sandstone and shale) and Kingriali Formation (dolomite, limestone and shale).

The succeeding Baroch Group is comprised by Datta Formation (= Variegated Stage; variegated sandstone, shale, siltstone), Shinawari Formation (limestone, marl, shale and sandstone) and Samna Sub-Formation (limestone, marl and shale). Glauconitic shale, phosphorite nodules, sandstone of Chitchat Formation and thick current bedded sandstone and sandy and glauconitic shale of Lumshiwal Formation make the upper most Surghar Group.

Attock Cherat Range

The Attock Cherat Range is situated between Peshawar and Nizampur basins. Three faults namely Khairabad, Cherat and Hissartans delineate this range into northern, central and southern blocks. Argillite, quartzite, limestone, shale and diabase of Proterozoic, Palaeozoic, and Neogene ages occur in this Range.

Kala Chitta Range

In between the Main Boundary Thrust and Attock Cherat Range, the Kala-Chitta Range is situated on the both sides of Indus River. Marl, sandstone, mainly limestone, siltstone, dolomite and shale of Mesozoic-Palaeocene-Eocene age are divided into several formations namely the lower most Mianwali, followed by Chale Jabbi, Samna Suk, Kiawah Marl, Lockhart and the uppermost Magalia Hill Limestone formations (Qureshi at al., 1995).

Naushera Region

In hilly terrain of Peshawar basin, Palaeozoic rocks of the Lesser Himalaya are exposed. Stratigraphically, on the basement of Proterozoic Manki Formation, Palaeozoic Amber dolomite probably of Cambrian age, Misri Banda quartzite (Ordovician), Panjpir Formation (phyllite and limestone of Ordovician to Silurian age) and Naushera Formation (quartzite, limestone, marble and dolomite of Early Devonian age) occur with an unconformity between Amber dolomite and Misri Banda quartzite.

Lesser Himalaya of Nepal

This 80-100 km wide division is situated between two thrusts namely the southern MBT and the northern MCT Zone. Stratigraphically, the Lesser Himalayan formations of India continue, with different nomenclature, in the farwestern Nepal (Upreti, 1990) and the Lesser Himalayan of eastern Nepal passes into the Sikkim of India. Two physiographic regions the southern Mahabharat and the northern Midland Range having combined width of nearly 100 km in the west and less than 20 in the east are delineated in this division. Thus the Lesser Himalaya covers wide area in the farwestern and western regions while occurs as continuous narrow strip of few km widths in the central and eastern regions of Nepal. The Lesser Himalayan also occurs in two windows and is surrounded with Higher Himalayan Crystallines namely between Dudha Koshi and Melamichi rivers and across Arun river where it attains maximum length 110 km and 95 km and width 55 km and 25 km respectively.

On the MBT, the Siwaliks, the older Tertiaries and the Quaternary alluviums of south come in contact with northerly situated diverse group of rocks belonging to the Gondwana, Proterozoic to Mesozoic sedimentaries and metamorphics of the Lesser Himalaya, crystalline and sedimentary nappes and klippen of the Higher and Tethyan Himalaya respectively and igneous rocks of various ages at different places.

Stratigraphy of Lesser Himalaya is worked out in small independent localities at different places thus same horizon is named differently in different regions (Figs. No. 5.2, 5.3 and 5.4). However, on regional scale a relatively simple picture appears in this division and is constituted by autochthonous or para-autochthonous Proterozoic to Mesozoic age Nawakot and Tansen Groups; and intrusives granitoids of Proterozoic to Early Cambrian and Ordovician age, augen gneisses nepheline syenite and basic sills and dykes and thrusted high to medium grade crystallines of the Higher Himalaya and fossiliferous sedimentary rocks of Tethyan division. In some areas autochthonous Nawakot and Tansen Group rocks are thrust over itself.

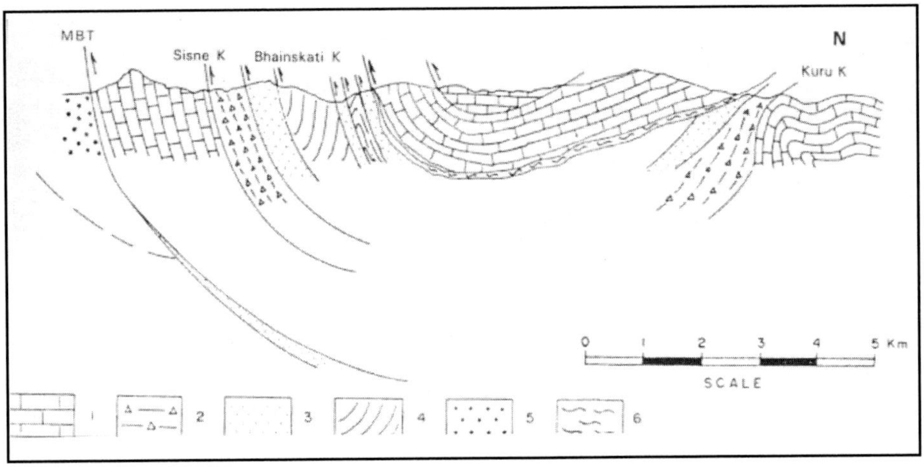

Figure 5.2. Geological cross-section of Tansen synclinorium, abbreviation; 1- Proterozoic carbonates and low-grade metasedimentary rocks, 2- Late Paleozoic Gondwana diamictite of Sisne Formation, 3- Cretaceous-Eocene Taltung, Amile and Bhainskati formations, 4- Early Miocene Dumri Formation, 5- Neogene Siwalik, 6- Rock underlying Palpa Klippe (after Sakai, 1983).

The western Nepal Lesser Himalayan area is best-investigated one and the stratigraphic succession proposed by Upreti (1996) is as follows:

Nappe of Higher Himalaya and Tethys Himalayas

Granitoids, nepheline syenite and intrusive sills and dykes of basic rocks

	Dumri Formation
Tansen unit (Lower Tertiary)	Bhainskoti Formation
——— (Palaeozoic to Early-Tertiary) ———	
Gondwana	Upper Gondwanas
	Lower Gondwanas
——— unconformity ———	
	Robana Formation
	Malekhu Limestone
	Nawakot Unit
——— Upper Nawakot	Benighat Slates
	Dhading Dolomite
	Nourpul Formation
——— Unconformity ———	
	Dandagaon Phyllite
——— Lower Nawakot	Fagfog Quartzite
	Kuncha Formation

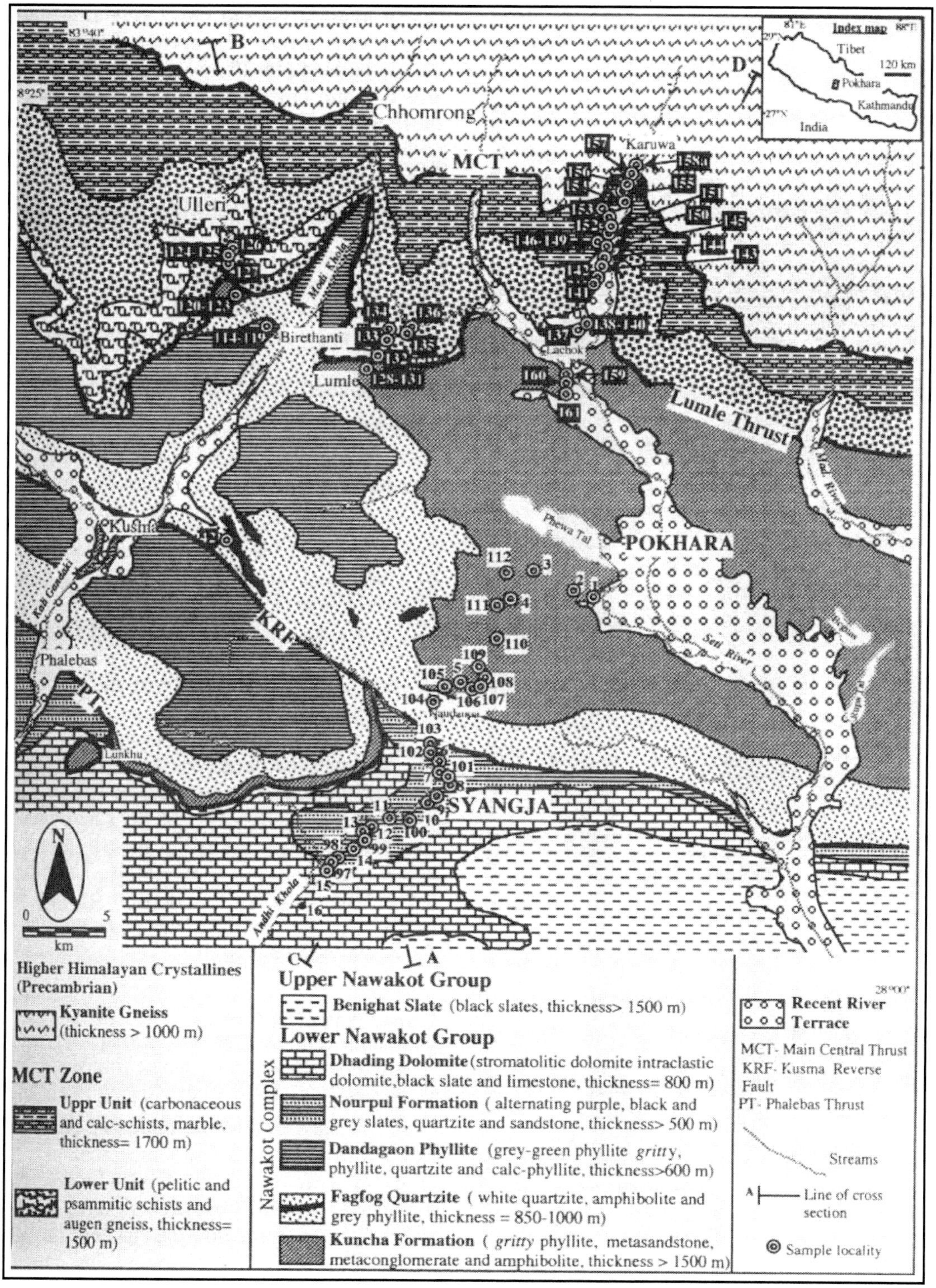

Figure 5.3. Geological map of Pokhra region, western Nepal Lesser Himalaya (after Paudel and Arita, 1998)

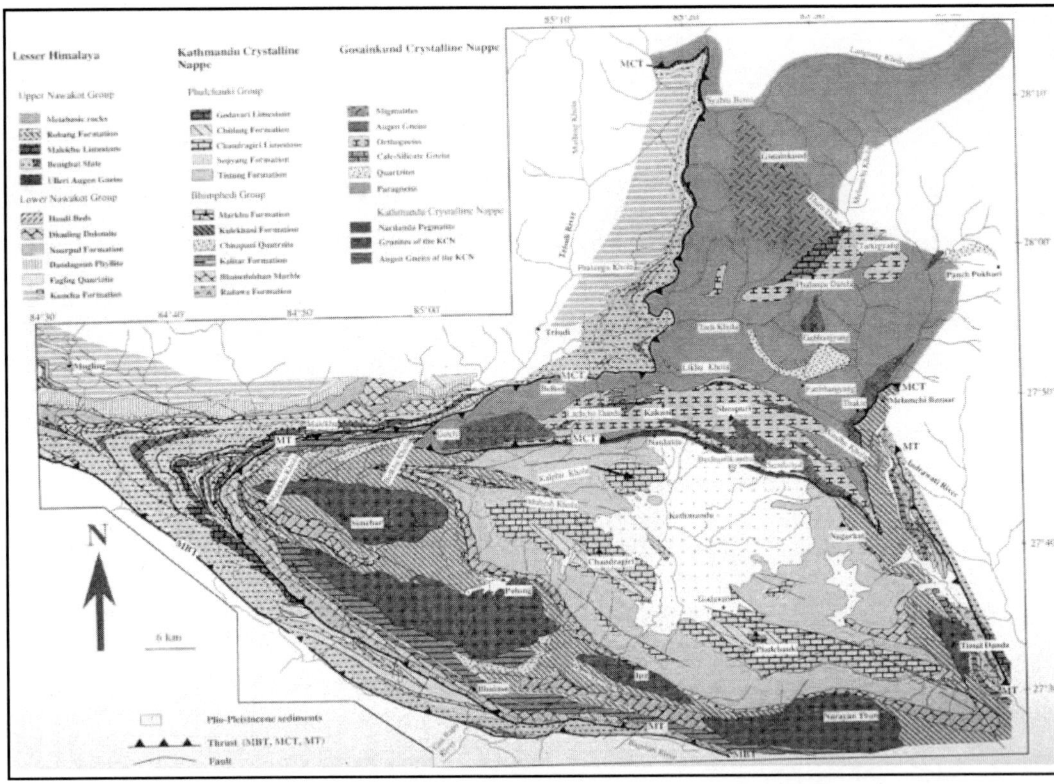

Figure 5.4. Geological map of central Nepal Himalaya (after Rai, 2001).

This sequence of western Nepal has been recognized with modification and under different names in other regions. However, the nature of overlying nappes and klippen are vastly different.

Nawakot Unit

Hagen (1959) was first to propose the term Nawakot Series, after a small township in Trisuliganga Valley of the central Nepal. Later different workers designated it as Nawakot Complex, Nappe, Series or Group. It is equivalent to Suparitar Formation or Suparitars or Midland Sediment Group. The Nawakot rocks are widely distributed in the entire Nepal Lesser Himalaya. The Lower and Upper Nawakot is divided on the basis of unconformable contact, however, exact location of contact in stratigraphical column is disputed one.

Kuncha Formation

Kuncha Formation, having maximum length of 300 km and width of 40 km, is named after Kuncha Village is Gorkha district, characteristically developed in Midland Range and occurs widely from west to east Nepal. The oldest Proterozoic formation makes the base of Nawakot Unit. The base of this formation is not exposed anywhere in the Nepal. This formation consists of flysch sediments i.e. phyllite, phyllitic quartzite and gritstone and quartzite along with metabasics, augen gneisses of Ulleri and nepheline syenite. It is equivalent to

Budar Group of farwestern Nepal where it is classified into Lower Budar Quartzite and Upper Rupaskanra Phyllite, in northern Dang area it is named as Dangri Formation and in the eastern Nepal it is described under Tribeni Supergroup. In central Nepal it divided into the Lower Banspani Quartzite Member and the Upper Labdi Phyllite Member.

Fagfog Quartzite Formation

It is named after Fagfog village situated in Thaplikhola Valley. The formation is mainly made up of fine to coarse grained, thick to very thick, white quartz arenite or metaquartzite. Profusely rippled and cross-bedded quartzites also have arkosic arenite and intercalations of phyllite, conglomerate and metabasics. It is called Balle Quartzite in northern Dang area and is equivalent to Berinag Quartzite of Kumaun. This quartzite formation is absent in the eastern Nepal.

Dandagaon Phyllite Formation

The formation is named after Dandagaon village situated east of Dhading township. The 1000 m thick Dandagaon Phyllite Formation is mainly consists of dark blue to green argillaceous to finely quartzitic phyllite with less than centimeter scale intercalation of chlorite-sericite quartzite. Its weathered surface shows reddish tint. This unit has also been described by other names such as Kochhap Formation, Harichur Formation and Heklang Formation in other areas.

Nourpul Formation

This 800 m thick formation is named after village Nourpul located south of Dhading and is well exposed in the Tansen and Pithuan areas. It consists of three units; the lowest one lying above Dandagaon Phyllite is designated Purebesi Quartzite constituted by thin to coarse grained, pure white to greenish and occasionally yellowish and pinkish arkosic quartzite. This is followed by dominantly purple phyllite with quartzite unit, which is dark green to blue-gray in color. The upper most part is made up of dolomite and dolomitic quartzite. The Nourpul Formation is also described Purple Phyllite, Gulang Suite and Khamari Formation.

Dhading Dolomite

The 500 to 1000 m thick dolomitic formation is named after town Dhading around which dolomites are exposed in ridges. The blue gray dolomite is massive, fine crystalline to dense and shows splintery fracture on the surface. It also contains chert nodules and a 20 m thick zone black argillaceous and siliceous slate in the Burhi Gandaki River. The dolomite shows presence of collenia and conophyton structure. This formation is also named as Surtibang Dolomite, Kerabari Dolomite and Chappani Formations. It is equivalent to Krol Formation of the Kumaun,

Benighat Slate

The term Benighat Slate is coined after type locality around Benighat village situated at the confluence of the Burhi Gandaki and the Trisuliganga rivers. The Benighat Slate is 500 m to several kilometer thick formation comprises of dark black to bluish grays slates, phyllite and subordinate quartzite, conglomerate, calc-phyllite, phyllitic limestone and dolomite. This formation occurs extensively in the central and western Nepal (Adhikari, 1991). The term Benighat Slate is equivalent to Ramedighat Formation and Kerabari Formation in Tansen area, Hatiya Formation in Kali Gandaki and Sirchaur Formation in Dang area.

Malekhu Limestone

The 800 m thick mainly limestone formation is named after village Malekhu situated at the junction of Malekhukhola and Trisuliganga on Kathmandu-Pokhra road. It is youngest formation in western Nepal. The formation consists of fine-grained to dense, hard, light yellow limestone and dolomite in lower part and thickly bedded dolomite or dolomitic limestone of dark gray color in upper part. This carbonate horizon also contains chert, slates and conglomerate. It shows presence of poorly developed long columnar stromatolite.

Robang Phyllites and Dunga Quartzite Beds

The phyllite is named after village Robang situated in the Hugatikhola. This strongly tectonized formation is developed in the Central Nepal consisting of green gray color sericite phyllite. The phyllites are associated with metadiabasics and gabbroids and greenish white soapy talcose phyllite. Dunga Quartzite is named after Dunga tributary in Manharikhola. The quartzite is light green to white in color.

Tansen Unit

This Late Palaeozoic to Early Tertiary age fossiliferous unit, named after city Tansen, is nearly 2 km thick. The rocks of this unit are developed in the west of the Marsyangdi and Narayani rivers in western and far western Nepal. The outcrops of this unit occur near MBT as well as over Nawakot unit in the form of long, narrow and highly elongated outcrops at different places. The Tansen rocks are described from the Tansen, Piuthan, Surkhet, Tosk, Sallyan, Jajarkot, Rukun, Barikat, south of Darchula, Baitadi, Chainpur and Libang areas. The age of rocks varies between carboniferous to Eocene. The Tansen Unit includes following:

Gondwana

The Lower and Upper Gondwana are deposited in non-marine condition, however, at some places marine nature is also recognized. Gondwanas are situated unconformably over the older Nawakot unit.

Lower Gondwana

It is also called Sisne Formation and is best exposed in Tansen Synclinorium and in the Tansen, Phalabang and Sallyan areas. This formation rests disconformably over the Robang Formation and consists of mainly dimictite, clay stone, mudstone, sandstone and conglomerate. The dimictites are unstratified and contains clasts of sand to boulder up to two meter in size embedded in clayey matrix. *Fenestella, Polypora, Asansthoclasdia* and *Pamirella* fossils of Middle Permian age are reported from this formation. The Sisne Formation in equivalent to Talchir Boulder Bed or Agglomeratic Slate of India and in Nepal its other equivalents is the Sallyan Formation. Phalabang Formation, Daban Group, Sattim Formation, and Tosh Group.

Upper Gondwana

The Upper Gondwana covers a larger area than the Lower Gondwana of Nepal and is divided into lower Taltung Formation and upper Amile Formation.

Taltung Formation

The Taltung Formation is resting disconformably over the Sisne Formation and is comprised of conglomerate and sandstone. Near Aulis, exposure of a basic volcanic body of 13 km length and 30 to 200 m thickness occur. This volcanic unit is called Aulis Volcanics and is equivalent to Rajmahal Trap of India. *Ptillophyllum, Pterophyllum, Cladophrebis, Sphenopteris* of Upper Gondwana (Upper Jurassic Lower Cretaceous) are reported from Taltung Formation.

Amile Formation

In Tansen area, the Amile Formation is lying unconformably over the Taltung Formation. Sand stone mainly constitutes this formation; however, subordinate sequence of sandstone, shale and limestone are also reported. Besides this, it also contains calcified wood and thin coal seams. The limestone contains lamellibranchia, gastropods, echinoids, corals and foraminifera fossils of Cretaceous to Palaeocene age.

Lower Tertiary

The marine transgressive Lower Tertiary rocks occur in very narrow outcrops at many places. These are divided into the Lower Bhainskoti Formation and the Upper Dumri Formation (Kayastha, 1992).

Bhainskoti Formation

The marine Bhainskoti Formation composed of black, green, reddish colored shales and fossiliferous limestone, mudstone and sandstone. It is correlated with the Middle Eocene Subathu Formation of India and contains *Nummulites, Assilina* and *Asteracantus*. Swat Formation of Surkhet is equivalent to this formation,

Dumri Formation

The 775 m thick, Dumri Formation of Oligocene - Early Miocene age is made of medium grained quartzose sandstone and red, purple green colored shales with occasional presence of conglomerate and marl beds. It is correlated with Dagshai or Muree Formations of Western Himalaya. In Surkhet area its equivalent is Suntar Formation.

Nappes Over Lesser Himalaya

Over the Lesser Himalayan rocks of Nepal, presence of several nappes has been recorded. The brief stratigraphy (Rai et al., 1997) of important nappes is as follow:

Kathmandu Nappe

Kathmandu, the capital city of Nepal, is lying in an intermontane valley. Valley of 30 by 25 km size covers an area of 650 sq km. The piggyback basin lies in Kathmandu Nappe consists of metamorphic and sedimentary rocks and 650 m thick Late Pliocene- Quaternary basin fill sediments.

The Kathmandu Nappe is also designated as Kathmandu Series or Kathmandu Complex and is developed around the city. This nappe is divided into the older Bhimphedi Group and the younger Phulchowki Group (Sychanthavong and Sah, 1980; Sharma, 1981).

Bhimphedi Group

It named after Bhimphedi town and consists of relatively high-grade metamorphic rocks of Proterozoic age. The Bhimphedi Group is divided into several stratigraphic formations, namely:

Formation	Rock Types
Tistung	Slates, phyllite, metasandstone and sandstone.
Markhu	Schist, quartzite, phyllite, marble
Kulikhani	Biotite - sericite quartzite and schist
Chisapani	Fine-grained pale green quartzite with intrusive Palung Granite and Narayanthan Granite.
Kalitar	Strongly micaceous quartzite, biotite schist, dolomite and conglomerate.
Bhainsedo bhan	White, coarse, crystalline marble with intercalations of biotite garnet schist in the lower and upper part.
Raduwa	Lying unconformable over Upper Nawakot Unit and composed of biotite - garnet schist and gneisses and subordinate micaceous quartzite.

Phulchowki Group

It is named after Phulchowki Hill situated south of Kathmandu. This 4300 m thick group of fossiliferous Early-Middle Palaeozoic rocks lying conformably over Tistuna Formation of the Bhimphedi Group of the Kathmandu Group. This group shows affinity with Tethyan rocks. It is divided into following formation:

Formation	Rock Types
Chandragiri	Yellow to brown weathered limestone layered with quartzite. Most prominent formation of Phulchowki Group.
Sopyang	Phyllitic slates, marls argillaceous limestone and limestone.
Chitlang	Dark purple, soft slates with intercalations of quartzite and limestone.
Godawari	Lower part consists of bedded green and purple argillaceous crinoidal limestone and upper part is white to light brown coarsely crystalline dolomite.

In the south central region, the Phulchowki Group is divided into oldest Tistung Fm of Cambrian age containing meta-siltstone, siltstone, quartzose sandstone, arenaceous limestone, pebbly mudstone of 5000 m thickness. The Sopyang Fm is grouped with lowermost Chandragiri Fm, This is followed by Ordovician Chandragiri Fm (2300 m) comprising of gray to light-gray, finely crystallized arenaceous and argillaceous limestone; 1100 m Chitlang Fm containing alternate fine sandstone and siltstone with hematite bed, white quartzose sandstone and purple to dark purple siltstone and topmost 400 m thick Godawari Fm contains siltstone, sandstone and crystalline fossiliferous arenaceous limestone (Funakawa, 2001).

Quaternary of Kathmandu Valley

On the top of Bhimphedi Group in and around Kathmandu intermontane valley 650 m thick lacustrine-fluviatile deposits of Plio-Pleistocene age occur (Fig No. 5.5). These Quaternary sediments are divided into following formations (Paudal and Sah, 1997).

Formation	**Composition**
Kalimati	Black clay and silt sands, silt and clays.
Thimi	Sands, silt and clays
Gokarna	Massive to thinly laminated dark gray to brown clay
Shankhu	Pebbly sand, gravel and clay
Sunkoshi	Dark brown to black clays and micaceous sand
Lukundal	Pebbly conglomerate, black to gray clay and occasionally lignite.
Tarebhir	Conglomerate with lenses of sandstone.

The fluvio-lacustrine sediments form several depositional terraces around Kathmandu. These terraces are named 19-11 ka old Patan (1300-1320 m), 28-24 ka old Thimi (1330-1340 m), and ~28 ka old Gokarna (1350-1390 m) terraces in central part of the basin, and Boregaon (1410-1430m), and Pyangaon (1480-1520 m) in the southern part. These terraces represent delta-plain, delta front and prodelta deposits (Sakai et. al., 2001). The rate of sedimentation in Kathmandu Bain varied between 20-200 mm/ka with an average of 70 mm/ka.

Figure 5.5. Geological map of of the inner Lesser Himalaya between Kusamand Ramdighai (after Dhital et al., 2002)

Jajarkot Nappe

It is 100 km long and 10-20 km wide in dimension and joins with Karnali Nappe in the west. It occupies the core of east-west trending synclinorium (Arita, Sharma and Fuji, 1984). It is divided into three formations:

Chaurjhari Formation -gray to greenish mica schist, garnetiferous schist and quartzite.

Thabang Formation- consists of crystalline impure marble and mica schist.

Jaljala Formation- fine to medium grained calcareous sandstone, siltstone, gray phyllite and limestone.

Karnali Nappe

The high grade Higher Himalayan Crystalline thrust over low grade/sedimentary rocks of Jajarkot Nappe in the western and far western Nepal over a large area. The rocks of Karnali Nappe are kyanite-sillimanite schist and gneisses. Carbonate rocks of Tethyan Himalayan unconformably overlie these rocks.

Dadeldhura Nappe

This nappe is situated in farwestern Nepal and is a direct continuation of Almora Nappe of Kumaun. The highly folded rocks of Dadeldhura Nappe show tectonic contact with Baitadi Group in the north. Stratigraphically, three formations are identified in the nappe:

Gaira Formation: it is constituted by augen gneisses, feldspathic schist, garnet schist, quartzites, minor metacarbonates and famous Dadeldhura Granite; *Saukharak Granite* - composed of gneisses of varying mineralogical composition and granite and *Pokhra Phyllite Formation* -this formation is situated in the core of Dadeldhura Synform and the phyllites are mainly lead gray to black in color and also occur in light gray shades and quartzite.

Lesser Himalaya of Darjeeling and Sikkim

A major part of Darjeeling of west Bengal and Sikkim is covered with the Lesser Himalayan formations. Gondwanas, Daling Groups Darjeeling Group and Lingtse Gneiss constitute the Sikkim Lesser Himalayan (Pawde and Saha, 1981).

Gondwana Group

The Gondwana Group is comprised of older Rangit Pebble Slate and the Younger Damuda Formation. The Rangit Pebble Slate Formation comprises pebbly gritty slates, lithicwacke, quartzite, carbonaceous argillites, rhythemites, volcano clasts and dimictites. It is correlated with the Talchir Boulder Bed. The Damuda Formation is constituted by sandstone, shale, and thick coal seams of Permian age (Raina, 1982).

Daling Group

Daling Group is equivalent to Nawakot unit of Nepal. Initially this term was proposed by Mallet for typically greenish fissile slates and green quartzite. Later it was realized that a complete inverted metamorphic sequence from low to high- grade rocks exists in Sikkim (Srikantia, 1973). Daling Group term is restricted to Late Proterozoic low grade metamorphic and sedimentary rocks and divided into three formations: (1) Gorubathan Formation- quartz-chlorite-slate or schist, phyllite and quartzite, (2) Reyong Formation- characterized by quartzite, calcareous phyllite and thin strips of carbonates and (3) Baxa Formation- dolomite, limestone and phyllite (Acharyya, 1989).

Darjeeling Group

The Daling Group gradually passes into the garnetiferous gneisses, designated as Darjeeling Group. The high-grade two mica gneisses contain quartz and feldspars and also calc silicates, marble, graphite schist and quartzite called Chungathang gneiss.

Lingtse Gneiss

The streaky, sheared 1050 Ma old granite gneiss developed with in the Darjeeling Group with tectonic contact.

Lesser Himalaya of Bhutan

The development of the Lesser Himalaya in Bhutan has taken place in a narrow bend in southern part of the country. Much of the Lesser Himalayan is concealed under the crystalline rocks of Thimpu Group. The unfossiliferous rocks are situated between MBT in south and Thimpu Thrust (=MCT) in the north (Jangpangi, 1978, 1989).

The stratigraphic sequence in Bhutan after Bhargava (1995) is as follows:

Tethyan Himalaya	sedimentary succession
	Unconformity
Higher Himalaya	Thimpu Group
	Thimpu Thrust
	Jaishidanda formation
	Jaishidanda Thrust
	Shumar Formation
Lesser Himalaya	Shumar Thrust
	Diuri Formation
	unconformity/ locally tectonized
	Baxa Group
	Baxa Thrust
	Damuda Subgroup
	MBT
Outer Himalaya	Siwalik Group /Formation I, II & III.

Lesser Himalaya of Arunachal Pradesh

Stratigraphic formations of Western Bhutan continue in the Arunachal Pradesh and in the east the Lesser Himalayan rocks abruptly terminate against the Tidding suture. The Lesser Himalayan geology (Kumar, 1997) is as follow:

Bomdila Group

The Middle Proterozoic Bomdila Group's rocks are extensively developed in the Lesser Himalayan. Medium grade garnet-mica schist, calc-silicate, marble, quartzite, para amphibolite, phyllite, metavolcanics and dolomite constitute the Bomdila Group. These rocks are divided into the lower Khetabari Formation; middle Tenga and the upper Chillepam formations. The Bomdila Group contains intrusives of various types, for example, lit-per-lit quartzofeldspathic gneisses, biotite granite, mafic intrusives and tourmaline pegmatite etc.

The Baxa Group of the Lesser Himalaya comprises of dolomites, limestone, cherty stromatolitic oolitic- intraclastic dolomite, black carbonaceous shale, calcareous quartzite and phyllite and is equivalent to Chillepam Formation or Menga Limestone (Tewari, 2003).

Dirang Formation

Rocks of Dirang Formation are situated unconformably over the Bomdila Group and comprise of low-grade metamorphics such as garnet biotite muscovite schist, phyllite, quartzite, golden yellow dolomite, tremolite marble and intrusive biotite granite.

Gondwanas

Gondwana rocks of Lower Permian age are making unconformable contact with underlying Bomdila Group particularly in the southern part of the Lesser Himalaya. These rocks are stratigraphically divided into Miri, Bichom, Bhareli and Abor-Volcanic formations. The Miri Formation is constituted by a sequence of conglomerate, sandstone, shale and dimictite. The Bichom Formation is made of dimictite conglomerate, mudstone, shale and phosphatic and cherty nodules while the Bhareli Formation is characterized by a very thick (4050m) sequence of feldspathic sandstone, gray to black carbonaceous shales and coal beds. The Abor Volcanics comprises amygdaloidal basalt and andesite, acidic tuffs and volcanic agglomerates.

Yinkiong Group

The Lower Tertiary rocks are described as Yinkiong Group. This group is divided into lower Geku and the upper Dalbuing formations. The Geku Formation is sequence of ignimbrite, siltstone, shale, mafic volcanics, quartzite and sandstone while Dalbuing Formation is basically composed of foraminiferal limestone and shale.

5.4 Higher Himalaya

The Himalayan crown i.e. Higher Himalaya is a spectacularly uplifted geanticlinal terrain of crystalline nature. Its height is ranging between 3000-8000 m. This part of the Himalayan Orogenic Belt is also known by several other superlative terms like Inner, Great, Central or Himadri Himalaya. This NW-SE trending discontinuous or dismembered zone particular in western Himalaya is frequently shrinks and swells shows maximum width of more than 100 km in western sector (Fig. No.5.6). This continuous entity from Arunachal to eastern Himachal Pradesh; disappears in Lahaul and reappear in the Zanskar Range of western Himachal and Eastern Kashmir, once again hide itself under the Tethyan cover in Central and Western

Kashmir Valley and reemerge in a NE-SW trending zone between Astar to Hazara area in the Pakistan. Rocks of Higher Himalayan characters also occur in the north of Tethyan Zone in Tso Morari area of Himachal Pradesh and Jammu and Kashmir.

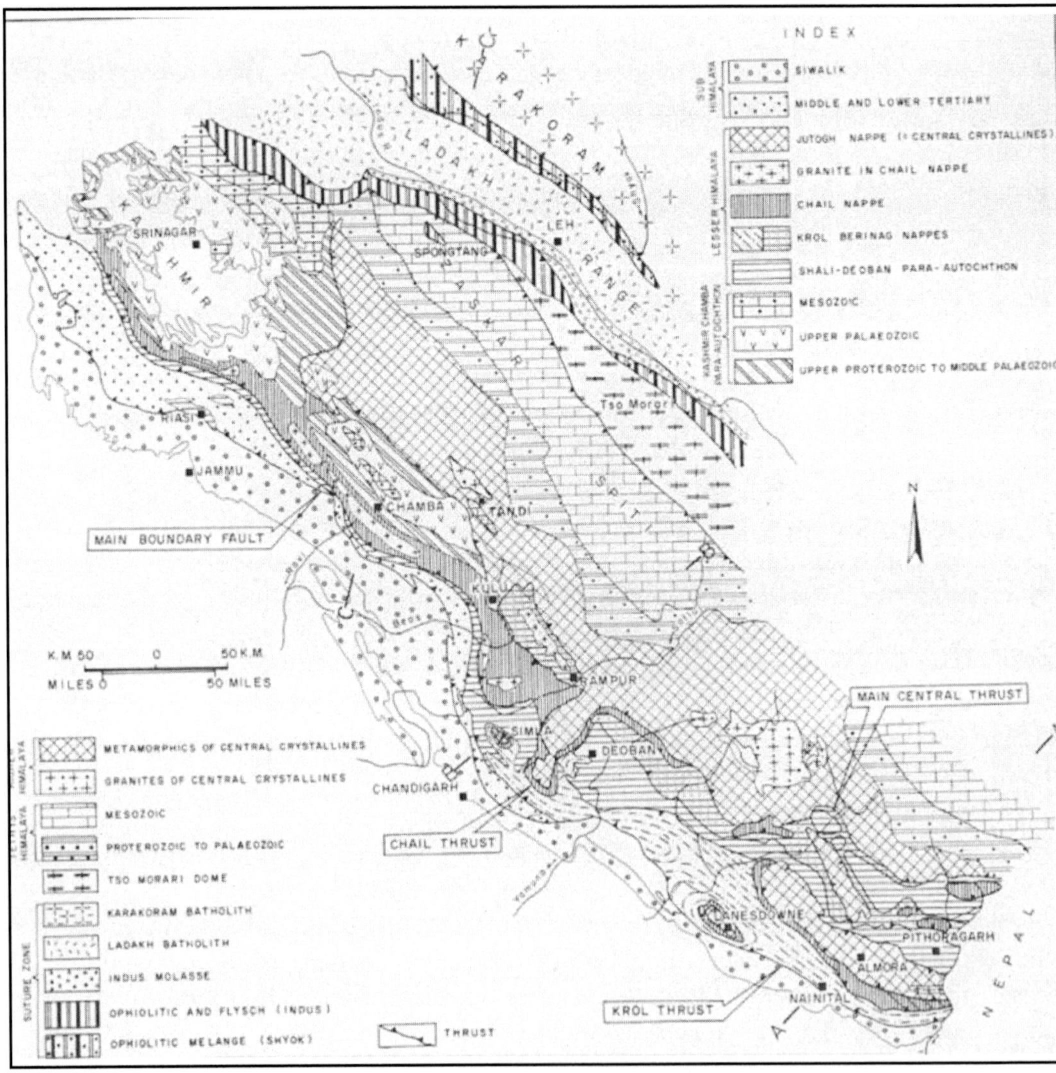

Figure 5.6. Geological map of western Himalaya (after Thakur, 1992)

The 15-20 km thick metamorphic sequence is intruded by granites of Proterozoic to Neogene age. It is separated from its southern neighbour the Lesser Himalaya along a tectonic contact described as MCT, a sharp plane or 5-15 km wide region also named MCT Zone. From Arunachal to eastern Himachal its northern contact with Tethyan rocks is of tectonic nature and is called Tethyan Thrust, which is locally named as Malari Thrust, Trans-Himadri Thrust, and Jhala Normal Fault etc. However in the western region it is with the Tethyan rocks. The Higher Himalaya is a remobilized cratonic terrain which is made of highly deformed, polyphase, inverted metamorphic rocks of Proterozoic or possibly Late

Archaean age. The Higher Himalayan rocks also occur as nappes and klippen over the Lesser Himalaya.

The Higher Himalaya is made of varieties of metamorphic rocks ranging from low grade green schist facies rocks like slate, phyllite, mica schist, garnet schist to upper amphibolite facies rocks like kyanite schist, sillimanite gneiss, migmatites, layered gneisses besides these amphibolite, granite gneiss, mylonitic gneisses, quartzites, marbles, serpentinites metadolomite etc are also present. These rocks are formed in shallow to mid-crustal level or even in the upper part of lower crustal level. Granulites of lower crustal origin are reported from eastern and western extremities of the Himalaya. These rocks have been brought on the surface by a pack of nappes during the Himalayan orogeny and show horizontal displacement estimated between 80-150 km. During peak metamorphic duration anatectic melting had widely taken place along the axis of geanticlines and is resulted into the formation of white colored S-type granite over an extended period of time i.e. between 30 to 3 Ma duration in different region.

Stratigraphy

The Higher Himalaya tectonically represents a giant pack of thrust or nappe sheets showing distinctive petrological characters. In Himachal and Kumaun three nappes are recognized and named as Chail Nappe, Jutogh Nappe and Vaikrita Nappe. A consensus is slowly emerging to adopt these terms in other parts of the Himalaya also or equate the locally designated rocks with these units. The individual thrust or nappe sheet is locally divided into two or more formations. Like Lesser or Tethyan Himalayan divisions large number of local formational terms are proposed in the geological literature for Higher Himalayan rocks. The local stratigraphic framework of the Higher Himalaya is also revised time to time and even forceful proposals are available to abandoned the established term e.g. in Himachal it is proposed to discard the Chail Nappe and added new units or in the Kumaun some authors have proposed only Vaikrita should be considered in the Higher Himalaya and the Chail and Jutogh be treated as parts of the Lesser Himalaya. In the foregoing description, following Thakur (1992), first the salient characters of the Chail, Jutogh and the Vaikrita are described and this is followed by the local or regional set up of the Higher Himalaya in different regions.

Chail Group

The rocks of Chail Group or Nappe are making contact with sedimentary or low-grade metamorphics of Lesser Himalaya in the south and medium grade metamorphics of Jutogh in the north along the Chail (= MCT) and Jutogh Thrusts respectively. The Chails are low-grade green schist facies metamorphic rocks and comprises green or gray color phyllite, black slates, quartzites, chlorite to garnet grade schists, marbles and amphibolites. They are intruded with two mica granites. In many areas the frontal part is mylonitised. The mylonite zone in some regions is up to 15 km wide. With in mylonitic zone granite (Gupta, 1987; Gupta et al., 1988), quartzite, amphibolite and low-grade schists are present. In some areas Chails are absent and the Higher Himalayan sequence begins with Jutogh metapelites.

Jutogh Group

This group is sandwiched between the Chail and Vaikrita rocks. In Kumaun, the Jutogh is designated as Munsiari Formation. The Jutogh is constituted by lower amphibolite facies rocks namely quartzite, marble, garnet to kyanite and staurolite grade pelitic schists and gneisses, graphitic schist, amphibolites and granite gneisses of mid-crustal origin. Granites in Jutogh have yielded 1800 and above million years ages. In many regions Jutogh shows break in metamorphic grade. Some workers believe that along the northern boundary (Vaikrita Thrust) of Jutogh the Lesser Himalaya is subducted in the Higher Himalaya.

Vaikrita Group

The Vaikrita represents the upper amphibolite facies rocks like migmatite, leucogranite, sillimanite schist and gneisses, amphibolites, quartzites, calcsilicates, amphibolites gneisses and granulites and aplite veins. The leucogranites are of Neogene age (28-10 Ma).

Uttaranchal

The Higher Himalaya of Kumaun is best-investigated one and has been studied in all the major river valleys. In eastern Kali, Goriganga and Ramganga valleys three tectonic units namely the Chipalkote Crystallines –augen gneiss, mica schist, chlorite schist, mylonite and marble; Sirdana unit -quartzite, phyllite, limestone and amphibolite and the Central Crystallines -Munsiari Formation and Vaikrita Group (Misra and Bhattacharya, 1976). In Nanda Devi area Vaikrita is divided into the Lata and Ramani formations (Meruo, 1979). In the Alaknanda Valley the Higher Himalaya is divided into Tapovan, Joshimath and Pandukeshwar formations. In Mandakini Valley, it is divided into the Chail, Jutogh and Vaikrita groups. Gupta and Dave (1979) described crystalline geology of Bhagirathi Valley.

In the Yamuna Valley the lower region of the Higher Himalaya is described under the Yamuna Formation (Figs. No. 5.7 and 5.8) while in the further east in the Tons, Rupin and Supin valleys the tectonostratigraphy is grouped into Naitwar Group, Imbricate Zone and Jutogh Group.

Roy and Valdiya (1988) proposed that Proterozoic Vaikrita Group situated between Vaikrita and Malari Thrusts constitutes entire Central Crystalline Zone. They divided Vaikrita Group into Joshimath Formation (streaky kyanite and banded gneisses, garnet kyanite biotite schist and calc-silicate gneiss), Pandukeshwar Formation (mica schist, kyanite-garnet-mica schist and psammatic gneiss), Pindari Formation (lime-silicate gneiss, biotite gneiss, migmatites, augen gneiss included with aplite veins, pegmatite and granite) and Budhi Schist (biotite porphyroblastic calc-schist, mica schist and phyllite).

Himachal Pradesh

A concise account of geology of Higher Himalaya under Proterozoic Crystalline Group is given by the Srikantia and Bhargava (1998). They identified four lithotectonic units namely the Vaikrita, Salkhala, Jutogh, and Kulu groups. These units are mainly distributed in centrally located NW-SE trending region of the Himachal Pradesh and as independent regional inliers in the NW, SSE, SW and NE areas of Himachal Pradesh.

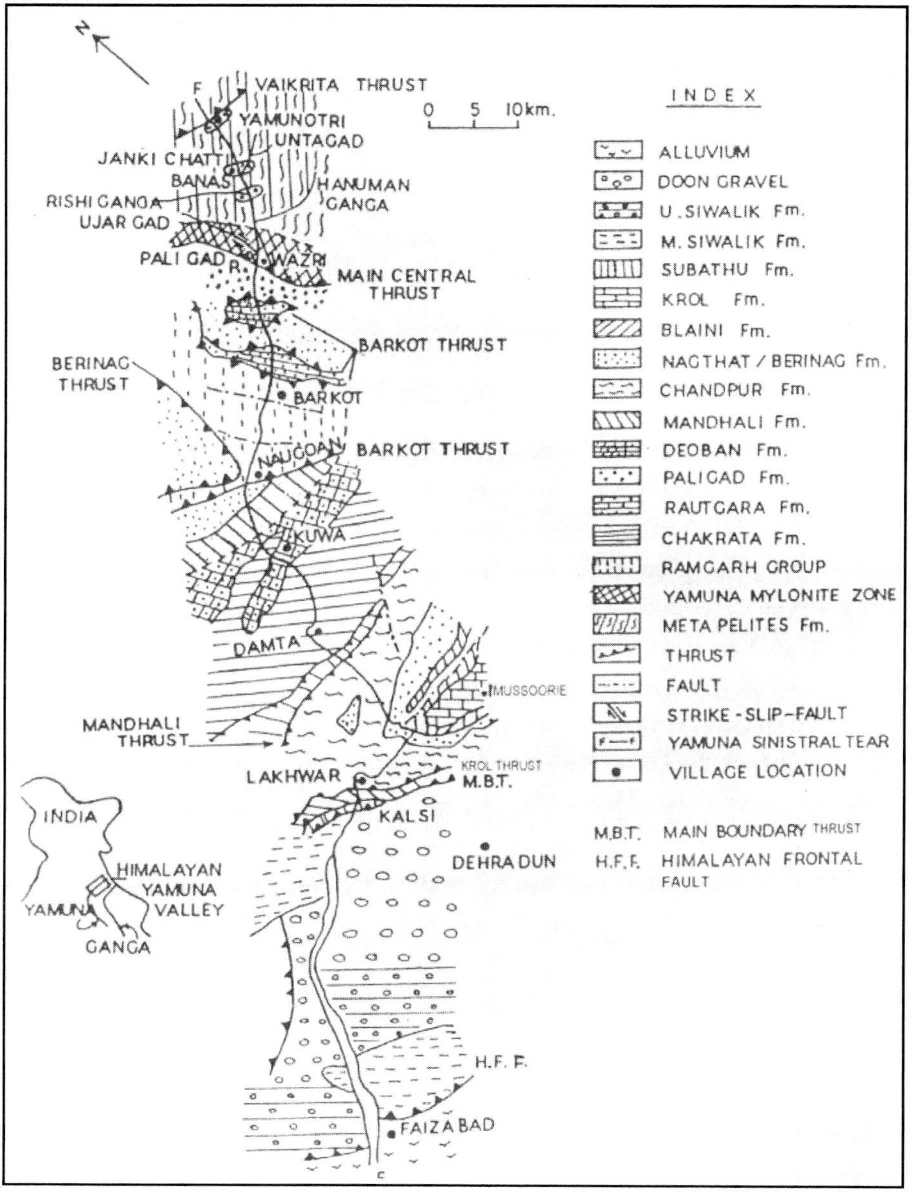

Figure 5.7. Geological map of Himalayan Yamuna river (after Biyani, 2002)

The Vaikrita Group comprises of high-grade layered gneisses, augen gneisses, calc gneisses, kyanite-sillimanite schist and gneiss, quartzite, marble, amphibolites, granite gneiss and granites. The Salkhala Group is divided into Lower Bhalai Formation comprising of grayish green talcose phyllite, garnet biotite schist, micaceous quartzite and psammatic gneiss with sporadic crystalline limestone bands, and the Upper Chamba Formation of metagraywacke, slate, carbonaceous slate and quartzite sequence. Salkhala is intruded by granitoids namely Dalhousie-Dhauladhar, Mandi and Karsog granitoids. Kulu Group is the newly proposed unit constituted by Khamrada Formation (carbonaceous to graphic schist and locally garnetiferous phyllite, and locally occurring limestone and calc-schist), Garh-Manjrot Formation (streaky kyanite mylonitic gneiss and banded and augen gneisses), and Khokan Formation (low grade pelitic schists, phyllite, slates, garnetiferous schist and quartzite). The Jutogh Group is very thick succession of low to medium grade metamorphic rocks namely carbonaceous limestone, carbonaceous phyllite and schist, slate, mica schist, garnetiferous schist, psammatic gneiss, quartzite, marble, amphibolite and granitoids of Chaur, Kainchawa and Pabar. The Jutogh Group is divided into eleven formations amazingly occurring in the normal order of superposition and without even single tectonic break. These formations in ascending order are: Panjerli, Manal, Bhotli, Khirki, Taradevi, Kanda, Naura, Badrol, Rohru, Chirgaon, and Jaknoti. These formations have combined thickness of nearly 10 km.

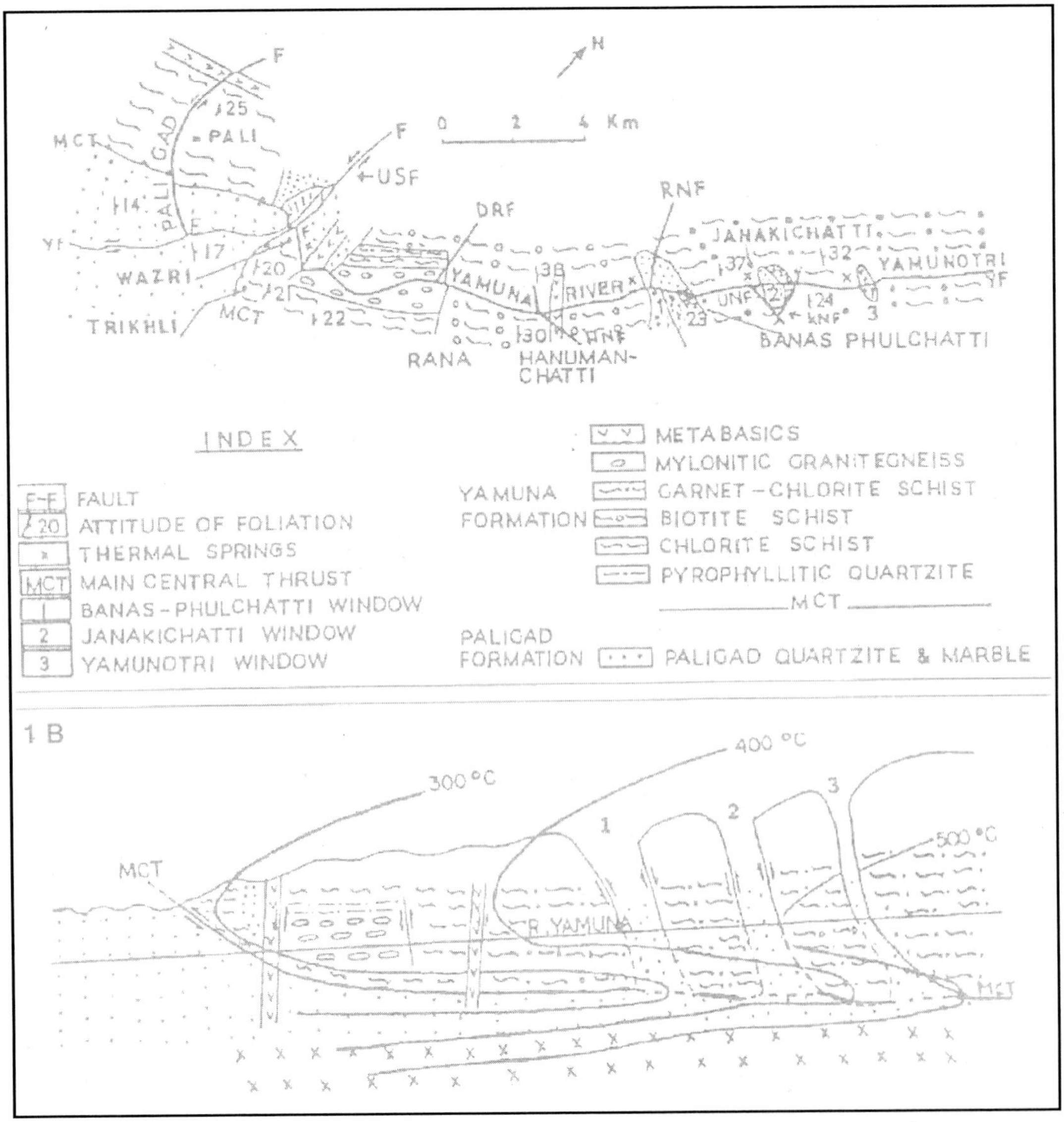

Figure 5.8. Geological map of Higher Himalya Yamuna valley (after Biyani, 1998)

Jammu and Kashmir

The crystalline rocks of Himachal Pradesh continue in the Jammu and Kashmir state along the narrow Pir Panjal-Dhauladhar Range and as western continuation of Lahaul Valley and of Zanskar area.

The narrow Pir Panjal-Dhauladhar continuation is situated north of the Panjal Thrust (MCT) and is constituted by Salkhala Formation (=Chail Group). The Salkhala comprises of mica schist, phyllite, quartzite mylonitic gneisses and early Palaeozoic granites of the Dhauladhar, Kund, Kaplas and Ramban areas. On the foundation of Salkhala, Tethyan deposition took place (Srikantia and Bhargava, 1976). The Central Crystallines of Zanskar region is also known as Higher Himalayan Crystallines (HHC) or Zanskar Crystallines (ZC). The ZC has considerable development in the highly rugged terrain of eastern Jammu and Kashmir (Choudhuri, 1987; Kunding, 1989). These rocks have been investigated in river valleys like Suru, Kishtwar-Chenab and Padam. In the Suru Valley ZC are overlain by Tethyan rocks. The green schist to upper amphibolite facies rocks are divided into the Sankoo Formation (phyllite, chlorite biotite schist, quartzite, limestone, amphibolite and foliated granite), Sangra Formation (garnet-staurolite schist, amphibolite and two-mica Karpokhar granite) and the Sangra Formation (kyanite-sillimanite schist and gneisses and leucogranite, this is equivalent to Vaikrita). In the Kishtwar-Chenab valleys besides occurrence of famous Kishtwar window, amphibolite facies rocks are reported from several areas. These rocks contain staurolite, kyanite, and sillimanite minerals. In Doda-Ramgarh-Rajgarh areas these high-grade rocks are thrust over by the low grade Salkhala. In Padam-Darcha and Shingo La-Gumburanjan areas, the Zanskar Crystallines are composed of gneisses containing biotite, garnet, staurolite, kyanite, sillimanite and clinopyroxene minerals, migmatite, granite gneiss, metabasics and leucogranites (Thakur et al., 1990).

Pakistan Occupied Kashmir and Pakistan

The Higher Himalayan Crystallines of Zanskar and eastern Jammu and Kashmir area are completely overlain by Tethyan rocks in Kashmir Valley. However, concealed crystallines resurface again in the Pakistan occupied Kashmir and in Pakistan in NE-SW trending region from north of Nanga Parbat to Abbottabad in the south and Swat area in the east. The geological descriptions of crystallines are available from the Nanga Parbat, Hazara-Manshera, Swat-Malakand and Kaghan Valley areas. In Nanga Parbat region low-grade green schist facies to upper amphibolite facies are present and they are called lower, middle and upper crystalline units; these units are correlatable with the Chail, Jutogh and Vaikrita groups of rocks of Himachal Pradesh. In the Hazara-Manshera area, Tanwal and Salkhala formations show low to medium grade metamorphism and are correlated with the Chail and Jutogh groups. The Cambrian age Manshera Granite is intruded in Tanwal and Manshera formations. In Swat-Malakand region, following tectonostratigraphy, in the ascending order is reported:

Manglaur Schist (mainly quartz feldspar and quartz-mica-garnet/kyanite schists with intrusive Swat Granite gneiss complex, locally called Chogu Granite, Ilum Granite, Kot Granite, Kalangai Granite etc. Swat Granite is equivalent to Manshera Granite); Alpuri calc-mica-garnet schist (calcareous quartz-mica schist, quartzofeldspathic gneisses and marbles); Said calcareous graphitic schist (calcareous, graphitic schist of black color) and Younger Granites namely tourmaline granite of syntectonic origin and biotite granite of post-tectonic nature of Neogene age. The 70 km wide Swat-Kaghan area is situated between the Panjal Thrust and Main Mantle Thrust. In this region a pack six internally imbricated nappes stacked one over the other is recognized by Treloar et al. (1989). These are: Besham Nappe (granite, biotite gneiss, marbles and amphibolite), Swat Nappe (Manglaur schist, calcareous schist, marble, amphibolite and Swat Granite), Hazara Nappe (metapelites and quartzites of Tanwal Formation and Manshera Granite), Lower and Upper Kaghan Nappes (varieties of metapelites, psammites, granites etc.) and the Banna Nappe (chlorite bearing slates and interbedded limestone).

Nepal

A continuous spread of low to high-grade crystalline rocks constituting of the Higher Himalaya occurs in Central to Northern part of Nepal. The crystalline sheet is much wider in the east, nearly 120 km; shrinks to width of 10 km in Central or Western Nepal particular in the south of Churen Himal or Dhaulagiri locations and again widens in the farwestern sector. Its klippen are present in many areas in western and farwestern sectors. Deep erosion of crystalline rocks particularly in the eastern sector has exposed the underlying Lesser Himalayan lithology in several windows like Arun, Tamur or Okhladunga. Its contact with the Lesser Himalaya is described as Main Central Thrust Zone, which is 5-20 km wide belt of highly tectonized rocks. The crystallines of Nepal Higher Himalaya are broadly divisible into three parts namely the lower part comprising of kyanite-sillimanite gneiss with dominant argillo-arenaceous fractions and well bedded quartzite; the middle part is constituted by the calcareous gneisses with interbedded marble and intruded with pegmatite and aplite and the upper part contains augen gneisses, granitic gneiss and migmatite (Le Fort and Lanord, 1995). Despite highly rugged nature of the terrain, the crystallines have been thoroughly investigated along the several river valleys or in considerably large areas.

In the Mt. Everest region the crystallines are divided into three groups a lower complex of migmatised paragneiss with minor metabasites, calc-silicates and marble, the middle one is of migmatised orthogneisses and the upper most is of black color paragneiss. In the Chomolonga-Makalu region near Mt Everest, the Higher Himalayan Crystallines are classified by Lombardo et al. (1993) into Barun Gneiss (garnet-biotite gneiss, amphibolite and marble); Namche Migmatite Orthogneiss (augen gneiss and granite gneiss); Black Gneisses (biotite gneiss, biotite-sillimanite gneiss, biotite-sillimanite-cordierite gneiss and biotite-sillimanite gneiss) and the Rangbuk Formation (biotite gneiss, biotite-muscovite-sillimanite, cordierite gneiss and marble). In Somdang-Lari area of Ganesh Himal region the crystallines are: chlorite schist, quartzite, mylonitised quartzite, biotite gneiss, dolomitic marble, biotite-kyanite schist and amphibolite. In Manang area of Marsyangdi Valley the stratigraphy is described in terms of Formation I (gneisses), Formation II (metacarbonates) and Formation III (augen gneiss). The same pattern is followed in Kali Gandaki Valley where 20 km thick sequence of crystalline rocks are exposed; the MCT zone is followed by Formation I (kyanite bearing paragneiss and mica schist), Formation II (calc-silicate gneisses with minor presence pelitic and psammatic rocks) and Formation III (granitic orthogneisses of Cambro-Ordovician and

Late Eocene-Oligocene age). In Langtang Valley (Reddy et al., 1983), MCT zone is at least 5 km wide and is followed by the Syabru Unit (12 km thick metamorphosed kyanite and sillimanite gneisses are present which in lower part contains migmatite); Kyangjin Unit (1 km thick augen gneiss contains K-feldspar, plagioclase, quartz, biotite, muscovite, garnet and sillimanite); Langshisa Unit (1.5-2 km thick mainly metapsammites and subordinate metapelites and carbonate bands) and the Langtang Lirung Unit is petrologically same as the Langshisa Unit but contains in nearly half volume granitic material). In the western Nepal Himal gneiss are divided into garnet-biotite gneiss, calc-siliceous gneiss, migmatitic gneiss and augen gneiss.

Bhutan

Bhutan has widespread development of crystalline rocks. These rocks show continuity with crystallines of Sikkim in the west and also with the crystallines of Arunachal Pradesh in the east. Gansser (1983) has observed presence of several crystalline units, showing low to high- grade metamorphic nature and leucogranites in Bhutan. These units are:

1. *Chasilakha Crystalline Unit.*
2. *Sure Crystalline Unit* - comprises migmatitic muscovite - biotite granite gneiss, biotite-kyanite schist, sericite phyllite, psammatic gneisses, and biotite - garnet schist, quartzite and amphibolites.
3. *Tashgang Crystalline Unit* - its components are migmatites, sillimanite gneiss and biotite gneiss and schist.
4. *Paro Metasediments* - comprises augen gneiss, high to low-grade pelitic schist and gneiss, amphibolites, quartzites and varieties of marbles. This unit is reported from several areas of Bhutan.
5. *Takhtsang Crystallines* - It is also considered equivalent to Darjeeling Gneisses and contains banded sillimanite bearing garnet-biotite gneisses.
6. *Leucogranites* - These Tertiary granites are reported from Chomolhare, Kankar and Punzum areas.

Arunachal Pradesh

Two units constitute the Higher Himalaya, north of MCT, namely the Se La Group and the Dirang Formation (Kumar (1997). Polyphase deformed metasedimentary rocks of green schist to amphibolite facies constitute the Early Palaeoproterozoic Se La Group. The Se La Group is divided into the Lower Taliha (named after Taliha village in the Subansiri valley) and the Upper Galensiniak formations. The Taliha Formation comprises graphitic schist, sillimanite gneiss, calc-silicate marble and quartzites. The Galensiniak Formation (named after Galensiniak village in the Subansiri valley) contains migmatites, kyanite-sillimanite schists and gneisses, quartzite, lit-par-lit gneisses, granite gneiss and tourmaline granite. Fig. No. 5.9 shows geological map of the Siang Dome area.

The Mesoproterozoic Dirang Formation comprises garnetiferous schist, muscovite-biotite schist, calcschist, phyllite, quartzite and biotite granite. Its exposures in the Tethyan zone is called Lum La Formation containing pelitic schist having biotite, garnet staurolite, calc silicates, marble, quartzite, graphite schist and amphibolite.

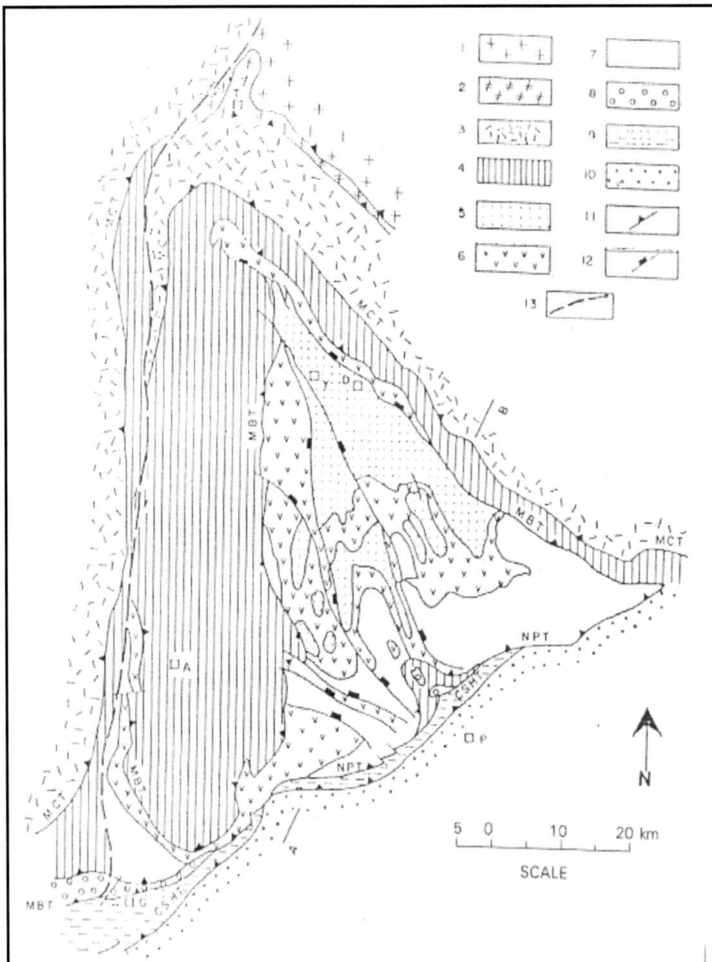

Figure 5.9. Geological Map of Siang Dome abbreviation: 1-Trans Himalayan granitoid migmatites, 2-basic and ultrabasic rocks, 3-high grade metamorphic rocks, 4-low grade metasedimentary rock, 5-middle Eocene Yinkiong formation, 6- upper volcanics, 7- Palaeocene Eocene quartzite, 8- late Proterozoic sedimentary rock, 9- Deformed neogene sediment, 10 – Southern Neogene sediment, 11-major thrusts of Himalaya, 12- Thrust within Siang window, 13- Tuting – Garu strike slip fault; D-Dalbuing, G- Garu, P- Pasighat, T- Tuting, Y- Yinkiong (after Acharyya, 1998).

5.5 Tethyan Himalaya

E. Sues first used the term 'Tethys' for a Palaeozoic - Mesozoic sea or geosyncline existing between India and Eurasia. Tethyan Himalaya is also known as Tibetan Himalaya or Tethys Himalaya or in short as Tethys.geographically this division is developed to the east of Main Mantle Thrust i.e. western Kashmir to Arunachal Pradesh in the east. Autochthonous part of Tethyan Himalaya is situated between the Higher Himalaya and Indus Suture Zone.

Its many nappe and klippen, covering small to large areas, occur over the Higher and Lesser Himalaya. Its southern contact with the Higher Himalaya is described as tectonic (thrust or fault), unconformable or gradational. The tectonic contact is known by various names like Malari Thrust, Tethyan Thrust or Jhala Normal Fault. The northern contact with Indus Suture Zone is called Counter Thrust, or Zanskar Thrust or Indus Suture Thrust.

The more than 10-12 km thick mainly sedimentary sequence of the Tethyan rocks were deposited in a single continuous basin or epicontinental sea or geosyncline during Late Proterozoic to end of the Cretaceous and in some regions till the Eocene. This continuous record of at least of 700-800 Ma shows small time breaks in type region in Lahaul and Spiti (Himachal Pradesh) and Garhwal-Kumaun (Uttaranchal). However in many areas the stratigraphical sequence show much late beginning of sedimentation, large time gap between the stratigraphical column and relatively early end of deposition. The depositional continuity of sediments were later broken down during Tertiary upheaval and now there rocks occur in detached manner from the Hazara in Pakistan, Kashmir, Himachal Pradesh, Uttaranchal, Nepal, Sikkim, Bhutan, and in Arunachal Pradesh and also continue in north i.e. in Tibet. Despite their lost continuity, they can be correlated rigorously in convincing manner due to fairly richly fossiliferous nature of strata. Tectonic movements also translated the Tethyan rocks to long distance in south in the form of nappe and klippen. The base of the Tethyan rock is metamorphosed in low-grade condition. The impact metamorphism in some sections occurs in Late Proterozoic Tethyan rock only while in some other sections metamorphism affects even Cambrian and Ordovician rocks.

Starchy's (1851) collection of fossils was reported in 1865 by Salter and Blanford and at same time Staliczka (1865) report of Spiti of Himachal Pradesh and subsequent Griesbach's (1891) and Hayden (1904) work in Spiti shot this richly fossiliferous Tethys into unfading limelight.Correlation of various Tethyan area has been given in Table-5.2.

Stratigraphy of the Tethyan rocks are described from many sections or valleys and as usual different workers have given local names to stratigraphic formations of the same age.

Himachal Pradesh and Uttaranchal

The lithostratigraphic succession in the type area i.e. Spiti-Lahaul area of Himachal Pradesh and Uttaranchal regions are as follows:

Group	Formation	Age
	Chikkim Shale	
	Chikkim Limestone	Cretaceous
Lagudarsi	Guimal Sandstone	
	Spiti Shale	
	Kioto	
	Nunuluka	Jurassic
	Alaror	

	Hangrang	
Lilang	Sanglung	Triassic
	Chomule	
	Kaga	
	Mikin	
Kuling	Gungri	Permian
	Gechang	
	Ganmachidam	
Kanawar	Po	Carboniferous
	Lipak	Muth
Muth	Devonian	
Sanugba	Takche	Silurian
	Thango	L. Ordovician
Haimanta	Kunzum La	L. Cambrian
	Batal	
	Terminal Proterozoic	
Salkhala/Vaikrita Group	Proterozoic	

In other areas for example Zanskar the age of uppermost rocks belong to even Eocene age. The lithostratigraphies of type locality along with other sections of Tethyan zone are described below.

Meso to Terminal Proterozoic

Base of Tethyan Zone

The Tethyan rocks are deposited on the lap of Higher Himalayan crystallines i.e. either on the Vaikrita group (in Himachal Pradesh, Uttaranchal or high grade rocks of Nepal) i.e. high grade kyanite-sillimanite schist and gneiss, migmatites, quartzites and garnetiferous schists, and comparative low grade Salkhala i.e. low to medium grade schist and quartzites in Kashmir and Zanskar basins. Along a tectonic contact, Haimanta Group forms the base of Tethyan sequence in Himachal Pradesh and Uttaranchal. Metamorphosed sediments constitute the lower part of this group. Thus a gradual decrease in metamorphism is observed between Vaikrita to Upper Haimanta Group. In Kumaun, Budhi Schist constitutes the base of Tethyan sequence.

Haimanta Group

Greisbach (1891) used the term Haimanta System (Sanskrit = snow covered) for the rocks overlying the lithology of Vaikrita Group. The Haimanta Group is divided into two formations; the lower is designated as Batal Formation of Terminal Proterozoic age and the upper one Kunzum La of Lower Middle Cambrian age.

Batal Formation

This basal part of Tethyan succession shows conformable contact with Vaikrita Group in Spiti-Lahaul region. It is mainly composed of gray green color phyllite, carbonaceous phyllite and quartzite; however, occasionally migmatites and schist and granitoids are also present. Martoli Formation (low-grade metamorphic) and Ralam Formation (conglomerate, quartzite and shale) are equivalent of Batal Fm in Kumaun and Garhwal.

Cambrian

Kunzum La Formation: It is named after Kunzum La in Lahaul. It is constituted by greenish gray siltstone, slate, phyllite, quartzite, sandstone and dolomite in lower part and pebble bed in upper part. The Early Cambrian trace fossils present in the basal part of this formation in Spiti are *Phycodes, Pedum, Skoliths, Planolites, Bergaueria, Cruziana, Gordia Nereites, Chondrites, Diplichnites, Dimorphichnus, Rusophycus, Taphrhelminthopsis* (Parcha et al., 2005) while slightly in the upper part brachiopod, trilobita and conodont namely *Ptychoparia, Olenus, Redlichia, Lingulella, Onetodus* and many other forms are reported. Garbyang Formation (flysch, limestone, dolomite, siltstone, sandstone and shale) in Kumaun-Garhwal, Hilap Formation (Slates and quartzite with acidic tuff and marble) in eastern Kinnaur, Phe Formation (gray, green slates, quartzite, sandstone and greywacke and limestone) belong to Late Proterozoic to Early Cambrian age and Karsha Formation (arkose, siltstone, dolomite and shale) and Kurgiakh Formation (pelites, dolomite and basalt flow of Cambrian age are reported from southern Zanskar region are equivalent to Kunzum La Formation.

Ordovician – Silurian: The Sanugha Group is divided into the lower Thango Formation of Ordovician age and the Upper Takeha Formation of Silurian age in the Spiti area.

Thango Formation: The Thango sequence rests unconformably over the Kunzum La Formation. It consists of conglomerate, purple quartzite, sandstone, shale and dolomite. Reported fossils are *Pentameras, Propara, Raphistones, Bifungites, Spirophys* and other genus.

The Changru Formation (limestone and shale) and Shiala Formation (shale, sandstone and limestone) in Garhwal and Kumaun, Suti Formation (calcareous phyllite and ferruginous quartzite), Tiwri Formation (quartzite and purple-green shale), and Nakdum Formation (slate, quartzite and conglomerate) in eastern Kinnaur are equivalent to the Thanga formation.

Takche Formation: The richly fossiliferous Takche Formation comprises green and purple shales, quartzite, dolomite and limestone. The important fossils are *Fauosites, Halysites, Calymene, Hallopara, Chandrites* etc. *Calostylis* (Yong limestone of Silurian and *Leiofusa* and multiplicispaeridium = acritarcha), *Rafinesquina, Monotrypa, Orthis* (coral and brachiopods) reported from the Shiala Fm, of Ordovician age and *Teradinium, Domasia and Baltisphaeridium* = acritarcha are recently reported.

Yang Formation (green limestone) and Variegated Formation (purple limestone shale and sandstone) in Garhwal-Kumaun, Yamrong La Formation and Maroon Mandrop Formation (sandstone, shale and coral limestone) in eastern Kinnaur, Thaple Formation (Conglomerate, quartzite and shale) in the southern Zanskar are equivalent to Takche Formation.

Muth Formation: The poorly fossiliferous Devonian Muth Formation a characteristic marker horizon developed from Kashmir to Nepal and was earlier designated as Muth Series or Muth System. It is making sharp contact with underlying Takche Formation. The formation is known after occurrence of fine to medium grained granular white quartzite, which is mottled at places. Besides quartzite, dolomite, doloarenite are also present in middle part while in the upper part it contains olive green shales and shale and slaty quartzite.

Carboniferous

Kanawar Group: The Kanawar Group lying between the Muth Formation and the Kuling Group consists of Lipak, Po and Ganmachidam formations in ascending order.

Lipak Formation: The constituents of Lipak Formation are dark gray to black limestone, pink limestone, argillaceous dolomite, quartzite, shale and gypsum. The important fossils present in the Lipak Formation are *Syringothyris, Neospirifer, Productus* and *Phillipsia*. Kali Formation (Sandy dolomite and pink to gray shale) in Kumaun is of L. Carboniferous age while in southern Zanskar the same describes it name i.e. Lipak Formation.

Po Formation: A thick sequence of white, medium grained quartzite, sandstone, siltstone, shales of various color and chert comprise this formation. *Rhacopteris, Fenestella, Rhizocorallium* and *Lepidodendropsis* are notable fossils report from this formation. Fenestella Shale Formation (black carbonaceous shale with subordinate limestone) in Kumaun is of M. Carboniferous age is considered equivalent to the Po formation. Hercynian break is observed in the eastern Kinnaur and in the Sarchu Chharap and Pang region across the Bara Lacha Pass across the Great Himalayan range in Lahaul.

Ganmachidam Formation

This unfossiliferous formation comprises of pebbly mudstone, sandstone and siltstone, conglomeratic sandstone and shales and basic sills and dykes. The upper age of this formation is believed to be earliest Permian.

Permian

Kuling Group: This group unconformably overlies the various older formations and is divided into two formations.

Gechang Formation: Gray, pale gray, coarse-grained sandstone and localized occurrence of conglomerate and shale at the base constitute the Gechang Formation. *Eurydesma, Deltopection, Zoophycos and Waagenophyllum* fossils are present in it.

Gungri Formation: This formation comprises of black shale, nodules of chert, phosphates and calcareous materials and limestone. Fossils present in this formation are

Marginifera, Productus, and Atheyris Cyclolobus. Kringkrong Formation (carbonaceous and micaceous shale and limestone) and Kuling Formation in Kumaun, Garhwal, eastern Kinnaur and Sarchu Chharap and Pang region of Himachal Pradesh, Panjal Trap of Permian and Kuling Formation of Late Permian from southern Zanskar region are equivalent to this formation.

Triassic

Lilang Group: The earlier named Lilang Limestone shows conformable contact with Kuling Group, is now divided into several formations on the basis of rich fossils presence.

Mikin Formation: This key formation of Lilang Group shows alternate presence of carbonate and shale units. The cherty dolomite is dark gray to gray in color.

Kaga Formation: Calcareous shale of dark gray and green color and lenticular beds of limestone and dolomite make up this formation.

Chomule Formation: This formation shows the presence of light to dark gray dolomite and subordinate calcareous shale and marls.

Hangrang Formation: Middle Norian age Hangrang Formation is composed of light to dark gray massive dolomite.

Alaror Formation : This formation comprises of shale with subordinate limestone and sandstone.

Nunulaku Formation : Limestone, shell and sandstone sequences constitute this formation.

Kioto Formation : It was earlier called Meglodon Limestone. It is made of gray, pale, cherty thick-bedded dolomite and limestone.

Fossils

Large numbers of ammonoid fossils are reported from the Lilang Group. The important species are: *Xenaspis, Otoceras, Ophiceras Meekoceras, Hedenstroemia, Rhynchonella, Daonella, Ptychites, Dielasma, Juvavites, Spiriferina, Megalodon, Belemnopsis.*

Its equivalent formations are Kalapani Limestone and Kuti Shale and Kioto Limestone Formation in Garhwal and Kumaun. In eastern Kinnaur, limestone and shale sequence is divided into A to E units and described as Spiti Formation. In the Sarchu Chharap and Pang area the limestone and shale sequence is divided into several beds and rocks have been locally metamorphosed to upper green schist facies condition. Zanskar Carbonates in the northern Zanskar belongs to Late Triassic to Middle Jurassic age.

Jurassic

Lagudarsi Group : The Jurassic-Cretaceous age Lagudarsi Group is divided into three formations namely the Spiti Formation of Jurassic and Guimal and Chikkim formations of Cretaceous age.

Spiti Formation : The earlier designated as Spiti Shale, begins with thick gray ventricular sandstone and shales followed by fossiliferous black shale, sandstone, siltstone and limestone of Chikkim Formation Fossils: *Lima, Mayaites, and Inoceramus* etc. Sulcacutus Member (ferruginous) and Saligram Member (black shales) in Garhwal are equivalent of Spiti Formation.

Cretaceous

Guimal Formation : This formation is made of interbedded sandstone and black shales as well as siltstone limestone and glauconitic sandstone in upper part. The fossils include *Cardium, Ostrea, Gryphaea, Tellina, and Perisphinctes*. This formation is also reported from eastern Kinnaur and Chharap area of Himachal Pradesh and Garhwal region of Uttarañchal.

Chikkim Formation : the Chikkim Formation in Spiti area represents the upper most Mesozoic. It is constituted by gray to bluish gray limestone and dolomite in basal part and marl and shale member in the upper part. Fossils: *Globotruncana, Textularia, Nodosaria, Dentallina, Belemnites, Rudistes* etc.

Its equivalent in other areas are Fatu La Formation (gray, red and green color limestone) of Albian age and Kagila Formation (flyschoidal sandy and silty slates; siltstone and marl) belongs to Campanian to Early Maastrichtian age in the southern Zanskar area and Shilla Kong Formation (multicolored limestone, dolomite, slate and phyllites and calc schist) of Late Campanian age and Lamayuru Formation (black slates and silty shales, and subordinate micaceous sandstone and quartzite) of Maastrichtian age are present in the northern Zanskar.

Paleocene Eocene

In most of the places Tethyan sedimentation stopped in the Cretaceous but in a few places like in Garhwal and Zanskar regions development of Palaeocene – Eocene Tethyan is reported.

Sangcha Malla Formation

This deep marine formation consists dark greenish gray, purple shales, greywacke, foraminiferal ooze and radiolarian chert in Malla Johar area of Garhwal. On the basis of radiolaria, fucoids markings, *Globotruncana, Eouvigerina, Archaeoglobigerina, Rositafornicata, Globotruncana. Heterohelix, Psuedotextularia* and other fossils Late Cretaceous to Early Eocene (Turonian Maastrichtian) age is assigned to Sangcha Malla Formation (Juyal et al.,2002)

In southern Zanskar, Tethyans are divided into Spanboth Formation (limestone, quartzite and calcareous sandstone) of Palaeocene to earliest Eocene age and Chulung La Formation (purple green slates, siltstone, fine grained quartzite with minor ophiolitic detritus) of Early Eocene age. In the northern Zanskar Lingshet Formation of Paleocene age contains quartzite sandstone and limestone and Kong Slate of Early Eocene age contains slates, schistose marl and limestone.

Tethyans of Kashmir and Chamba Valleys

The Phanerozoic Tethyan rocks are present in the NW-SE trending Kashmir Valley with further eastern extension in the Chamba Valley of western Himachal Pradesh. These valleys are physiographically located between the northern Zanskar Range of High Himalaya and the southern Pir Panjal - Dhauladhar Range. According to Wadia, the Tethyan succession represents to allochthon Kashmir Nappe whose base is constituted by the Proterozoic high to low grade metamorphic rocks of Salkhala Formation, which is also described by many other names such as Zanskar Crystallines, Chamba Formation, Bhalai Formation, Sewa Paragneisses or Bhadarwah Formation. The root of Kashmir Nappe lies in the northern Zanskar Range. Several workers in different isolated region of the Kashmir-Chamba Belt have studied the Tethyan succession; therefore, different stratigraphic names have been used for formations of same age. It is also observed that the Tethyan sedimentation did not start uniformly in all regions but began in different period of times in different regions. For example, the base of Tethyan sediments in the western Kashmir is constituted by Terminal Proterozoic Dogra Slates while in the Lidar Valley by the Ordovician Gauran Beds. Several break in sedimentation of small to large time gaps are recorded in these regions and in some areas fairly extensive igneous activities have also been reported. The stratigraphic position of formations is also being revised time to time in the light of new data and observations.

The stratigraphic set up in Kashmir valley is as follows (Singh and Bhat, 2002):

Panjal Volcanics: Basaltic and andesitic vesicular to massive lava flows Early Permian.

Agglomeratic Slates: Slates, phyllite, quartzite with angular to rounded fragments in slates Late Carboniferous – Early Permian.

Fenestella Shales: Shales, phyllites and sandstone quartzite containing *Fenestella*, brachiopods, trilobita - Middle Carboniferous

Syringothyris Limestone: Fossiliferous limestone, sandstone and marl- Early Carboniferous

Muth Quartzite: white to gray quartzite- Devonian

Chamalwas Formation: Greenish-gray slates, sandstone, siltstone and shales Cambrian-Silurian.

Basement of Tethyan Rocks

The basement rocks in Kashmir Valley is designated as Salkhala Formation that comprises of kyanite-staurolite schist and gneiss, garnetiferous biotite schist, phyllites, slates, quartzite and black limestone of Mesoproterozoic age. Sewa Paragneiss in Bhadarwah area is 1400 Ma old mylonitic gneiss and Bhalai Formation comprises biotite and garnet schist, phyllite, quartzite, limestone, basic volcanics and Dalhousie Granite constitute the basement of Tethyan rocks in the Chamba Valley.

Terminal Proterozoic to Cambrian

Salkhalas are unconformable overlain by Dogra Slates, Attock Slate and Hazara slates in western Kashmir Valley and Haimanta Formation and Ramsu Formations in eastern part of the valley and Bhadarwah Formation in Bhadarwah-Bhalesh area. These Terminal-to Neo Proterozoic formations comprise slate, phyllite, and quartzite and are intruded by granitic and basic rocks. The Cambrian in NW Kashmir Valleys are represented by fossiliferous clay shales, limestone, sandstone and quartzite, in the eastern part of Valley by Ramsu Formation (phyllite and limestone) and Sunbain Formation (coarse grained quartzite, quartzitic slate and conglomerate). In Chamba area the revised Tethyan stratigraphy indicates Manjir Formation (conglomerate) of Neoproterozoic age and Batal Formation (= Katari Gali Formation) is of Terminal Proterozoic age.

Ordovician and Silurian

The Ordovician and Silurian rocks are poorly developed in the Kashmir-Chamba Belt and are only reported from northwestern Kashmir and described as Ordovician beds of Marhaum and Silurians of Sham Abri area and in the Lidar Valley. Fossiliferous shales and limestone are of Late Ordovician age and Silurians are divided into Early Silurian unfossiliferous shales and Late Silurian and Early Devonian into Harpatnar beds (mudstone and shale) and Naubug beds (shales and siliceous limestone) respectively.

Devonian

Muth Quartzite Formation of Middle to Late Devonian age is present in the western Kashmir Valley, Liar Valley and Kishtwar-Marble Pass area. These quartzites contain *Atheyris, Schizophoria, Cyrtina Cathophyllum* and other fossils in Lidar Valley.

Carboniferous

Carboniferous rocks are widely developed in this region. In northwestern Kashmir and Hazara they are described as Tanwal Formation (thick sequence of moderately metamorphosed arenaceous rocks) followed by Agglomeratic Slate (slate, sandstone and conglomerate), Plant beds of Nagmarg and Panjal Traps. In the Lidar Valley, Kishtwar, Marble Pass and Bhadarwah-Bhalesh area Syringothyris Limestone, Fenestella Limestone (limestone, shale and conglomerate), Agglomeratic Slates (slate, shale and conglomerate), Gangamopteris Bed (novaculites and siliceous shale) and Panjal Traps (andesite and basalt extending into Upper Triassic) are present. Earlier workers have put Chamba Formation, Pukhri Formation and Manjir Formation of Chamba and Manjir Formation and Salooni Formation of Bhadarwah area in the Ordovician but now they have been placed differently.

Permian

Permians also show extensive development in this area. Interbedded Limestone in developed in the Panjal Traps and in the Hazara area. Panjal eruptions continue in the Kashmir area. Late Permian fossiliferous limestone and shale bearing Zewan Formation is reported from Lidar Valley and Kishtwar area. Salooni Formation of Chamba-Bhadarwah area comprises dark gray to black carbonaceous pyrite slates with thin bands of limestone and lava flows of basalt to andesite composition.

Triassic

Triassic rocks are reported from the western Kashmir, Chamba and Bhadarwah area. In Kashmir fossiliferous very thick evenly bedded limestone with interbedded black slates and quartzites are present. The Kalhel Formation of the Chamba comprises gray, yellow and bluish gray limestone and dolomite with interbedded quartzite and shale. Dalmon Formation of Bhadarwah contains shale and limestone.

Quaternary

A major part of Kashmir Valley, of more than 5000 sq km, is covered with 1.2 km thick Plio- Pleistocene glaciolacustrine Karewa Formation. This sediment was deposited in the then Karewa Lake that formed around 4 Ma ago due to upliftment of the southern Pir Panjal range. The Karewa deposits are divided into Lower Karewa (1000 m thick conglomerate, sands, mud and lignite) and Upper Karewa of 200 m thickness comprises sands and clays. In Khaigam section five horizons of volcanic ash are reported (Kotlia and Sanwal, 2005)

Tethyans of Nepal

The Tethyan rocks in Nepal occur in detached locations. Two prominent regional occurrences are one between Trisuli and the Karnali rivers in the western Nepal and other between Saipal to Nepal-India border in the farwestern Nepal. In case of farmer the maximum length and breadth dimensions are 350 km and 120 km while in case of latter it is 100 and 35 km. Besides these, Tethyans also occur in Mount Everest, Mount Choyu and the east of Mount Lhotse areas. The Klippen of fossiliferous Tethyans occur in the Phulchowki (south of Kathmandu), north of Libang, south of the Jumla or east of the river Karnali and between Dipayal-Dadeldhura areas.

Very limited published literature is available on the geology of Tethyan sequence of the Nepal; this may be due to extreme ruggedness and remoteness of the Tethyan terrain. Tethyan rocks are described from Mt. Everest and Thhakhola region.

In Mt. Everest area, lower calcareous layer along with dark green gneisses and phyllites are followed by the Everest Pelites that comprises dark greenish gray argillaceous sandstone, sandy limestone and siltstone and the Permocarboniferous Everest limestone which is light in color, fine grained, thin bedded and fossiliferous in nature. Tertiary Mabales Granite has intruded at the base of the Everest Mountain.

The Palaeozoic-Mesozoic of Nepalese Tethyans are best studied in Thhakhola and adjoining western Dolpo and Nyi-Shang areas. Proterozoic migmatite, kyanite schist, quartz-mica schist calcareous gneiss and marble form the base of Tethyan sequence. The Dhaulagiri and Neelgiri Limestone of Cambrian-Ordovician age is nearly 5 km thick and comprise impure limestone, micaceous marble, dolomite and calcareous siltstone. The limestone is metamorphosed in low-grade marble in the Nyi-Shang region. In Thhakhola and Nyi-Shang areas the, 560 m thick bed of quartzite of arkosic nature, siltstone and calcschist are designated as North Face Quartzite Formation of Late Ordovician are reported. The North Face Quartzite Formation is followed by Silurian age Dark Band Formation which is made of dark gray limestone, dolomite blackish calc siltstone and shale and blackish crystalline limestone and silty shale. The Devonian age Tilochopass Formation is equivalent to Muth Quartzite and comprises of micaceous and calcareous sandstone, arenaceous limestone and dolomite. Tilicho Lake Formation of Carboniferous age makes a sharp contact with underlying formation and is consists of fossiliferous limestone. Nearly 700 m thick sequence of slate, sandstone, dolomite, limestone and pelitic rocks constitute the Thinichu Formation of Permian age.

The Mesozoics begin with an unconformity at its base. The transitional Permotriassic is called Scythian bed and 1000 m thick Triassic is divided into Lower Mukut Limestone (thin bedded limestone), Middle Tarap Shales and Upper Quartzite Formation. The Early Jurassic (Rhaetic) is called Kioto Limestone is the western Dolpo region and Jomson Limestone (dolomitic limestone) and Lumachell Formation of Lower Dogger contains calcareous sandstone, limestone and calcilutite. The Late Jurassic is called Spiti shales in the western Dolpo area while in the Thhakhola area Jurassic begins with Ferruginous Beds containing oolitic limestone followed by the famous Saligram Formation consists of black slates intercalated with sandstone. Chert or slate coated spherical to spheroidal ammonoid fossils believed to represent Sudarshan Chakra (ultimate weapon) of Lord Vishnu, the sustainer of the world, are worshipped by the Hindus, come from this formation. The Thhakhola Formation comprising of 150 m thick light green sandstone and 750 m thick succession of siltstone, shale and glauconitic sandstone represents Cretaceous.

Tethyans of Eastern Himalaya

Tethyan rocks in the eastern Himalaya occur in north Sikkim and Bhutan. However, main development of the Tethyan sequence has taken place in the Tibet about which systematic geological description is poorly reported. In Sikkim, the Permian Lachi Formation and the Triassic Tso Lhamo Formation follow Early to Middle Palaeozoic Everest Pelite and the Everest Limestone Formations. Tethyan sequence in Bhutan was first described by Hayden (1907). These rocks cover a major part of the country and occur in four-isolated synclinoria, namely Lingshi, Gurpola, Black Mountain and Saktong and north of the Lunana Lake region. These rocks occur over Thimpu Group. Tethyan rocks are also reported from the Bhutan-Tibet border region but due inaccessible nature of terrain, details are lacking. The lithostratigraphy of Tethyan rocks worked out (Tangri and Pande, in Bhargava,1995) is as follows:

The oldest Chekha Formation of Late Proterozoic is showing tectonic, gradational and unconformable contacts with underlying Thimpu Gneisses in differenr regions. It is mainly made of argillo-arenaceous sequence and subordinate limestone. The Chekha Formation is unconformably followed by Pele La Group, which is divided into Lower Singhi Formation comprising of basic flow low-grade metamorphosed amygdaloidal basalt and pyritiferous phyllite; Deshichiling Formation constituted by arenite, phyllite fine grained limestone and oligomictic conglomerate; Maneting Formation contains monotonous sequence of green to greenish gray laminated phyllite, fine grained quartz wacke and arenite; and the upper most Quartzite Formation having 750 m thick bed of fossiliferous quartzite. Terminal Proterozoic to Mid Cambrian age is assigned to Pekla Group. The succeeding Tang Chu Group of Lower Devonian to Lower Permian age is divided into Wachila Formation of more than 250 m thick fine grained limestone, dolomite, calcareous phyllite and flaggy quartzite; Ripakha Formation, contains richly fossiliferous thin bedded limestone, dolomite, shale and slate sequence; Bari Shong Formation is having white, granular quartz arenite and calcareous quartzite in lower part and pure and impure limestone in upper part and the upper most Shodug Formation shows presence of thinly bedded limestone and slates and pebbly conglomerate. The unconformably overlying Lingshi Group has been divided into richly fossiliferous sequence of greenish gray to light gray quartzite and interbedded slates; Chebesa Formation contains 1500-2000 m thick succession of fossiliferous dark carbonaceous, pyrite slate and minor arenite and the youngest one is Nile Formation comprising of carbonaceous shale and quartzitic sandstone. The Lingshi Group yields fossils of Middle Jurassic to Cretaceous age.

In Arunachal Pradesh, Tethyan rocks occur in the inaccessible northern part. The geology of the northern region is not well known. The pale brown ferruginous Upper Permian Yamune Formation has presence of *Aviculopecten, Etheripecten, Mysidioptera, Schizodus, Productus, Spirifer, Loxonema, Neilsonia, Meekospira* etc.

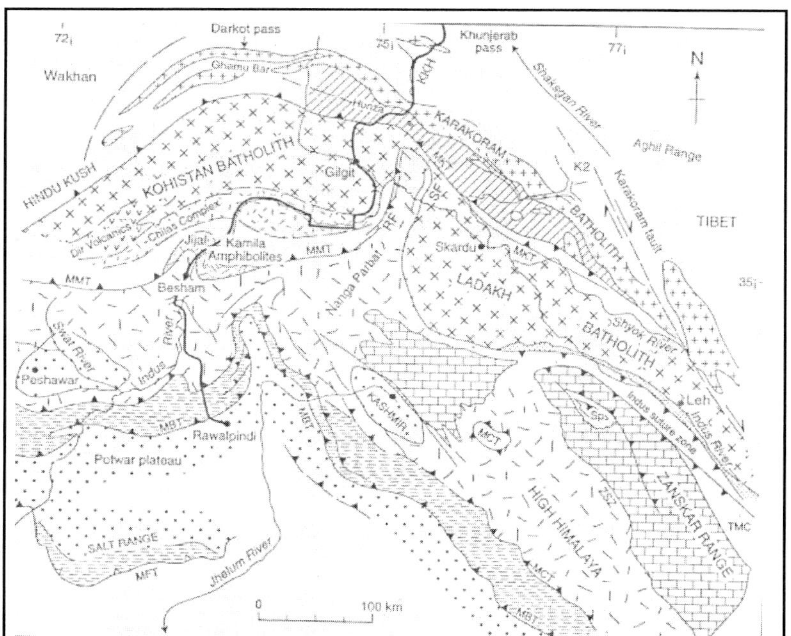

Figure 5.10. Geological map of Kohistan and Karakoram (after Searle et al., 1999) abbreviations:- SF- stak fault, RF- Raikot fault, SP- Spontang ophiolite in Ladakh, TMC- Tso Morari complex, ZSZ- Zanskar shear zone, MCT- Main Central Thrust, MBT- Main Boundary Thrust, MFT- Main Frontal Thrust, Thick line shows the route of Karakoram Highway(KKH).

Table.5.2. Correlation of Geological Formations of Tethyan Himalaya (compiled by Divya Dudeja & A. K. Biyani)

Age		Kashmir			Himachal Pradesh				Zanskar			Uttaranchal		Nepal		Bhutan	Arunanchal Pradesh
		Lidar Valley	Kishtwar	Northwest	Spiti	Chamba	Sarchu Charap	Kinnaur	Northern	Eastern	Southern	Malla Johar,Garhwal	Kali valley, Kumaon	Thakola Area	Western		
Eocene	L										Chunglung la Fm						
	M								Kong Slate								
	E								Lingshat Fm		Spanboth Fm	Sangcha Malla Fm					
Paleocene	L									Lmst & Marls							
	M																
	E																
Cretaceous	L				Chikkim Fm Giumal Fm		Giumal fm Spiti shale Lapthal beds	Giumal Fm	Shillakong Fm Lamayuru Fm	Khumak Fm	Fatu La Fm Kangila Fm	Chikkim Fm Giumal Fm	Sangcha Malla Fm	Thakola Fm	Kagbeni Gp	Chebasa Fm	
	M																
	E								Giumal Fm		Giumal Fm						
Jurassic	L	Coral Lmst			Spiti Shale		Kioto Lmst Lungun Fm Coral Lmst	Spiti shales		Lama y-uru Fm	Spiti shales	Spiti Shale Ferruginous Oolite Sulcacutus Fm		Saligram Fm Ferrugino-us Beds	Saligram Fm Ferrugino-us Beds Lapthal Fm	Lingshi Fm Mochu Fm	
	M								Kioto Fm	Kioto Fm	Kioto Fm	Lapthal Fm					
	E	Kido Fm			Kioto Fm									Kioto Fm	Jomson Lmst		
Triassic	L			Trias Lmst	Lilang Gp	Kalhel Fm	Juvavites beds	J Lilang Group T b				Kioto Lmst Kuti Shale	Kuti shale Kalapani Lmst	Upper Quartzite Fm			
	M			Trias Lmst with Traps					Lilang Gp		Lilang Gp	Kalapani Lmst		Middle Trap shale	Thinigaon Fm		
	E			Infra trias lmst			Tropites beds Halobia beds							Lower Mukut Lmst			

Period																
Permian	L	Zewan Fm	Zewan Fm		Kuling Fm	Salooni Fm	Kuling fm	Kuling Fm	Kuling Fm	Silt-Quartzite-limestone Fm	Kuling Fm	Kuling Fm	Kringkrong Fm	Thini Chu Fm	Shodung Fm	Yamune Fm
	M										Panjal Trap					
	E	Panjal Traps	Panjal Traps		Gamnachidam Fm				Gamnachidam Fm		Gamnachidam Fm				Tong Chu Fm	
Carboniferous	L	Gangamopteris beds Aggolomeratic slates	Agglomeratic slates	Khunmu Panjal trap Agglomeratic slate					Lipak Fm		Po Fm Lipak Fm					
	M	Fenestella Shale	Fenestella Shale	Tanawal series	Po Fm								Fenstella Shales			
	E	Syringothyris Lmst	Syringothyris Lmst		Lipak Fm								Kali Fm	TilichoLake Fm		
Devonian	L M E	Muth Fm	Muth Fm	Muth Fm	Muth Fm			Muth Fm	Muth Fm		Muth Fm	Muth Fm	Muth Fm	Tilicho Pass Fm	Nakechu Fm	
Silurian	L	Naubug Beds		Shams Abri	Takche Fm			Manchop Fm Yamrung La Fm	Thaple Fm		Thaple Fm	Variegated fm Yong Fm	Variegated fm	Dark Band Fm		
	M															
	E															
Ordovician	L			Marhaum Fm	Thango Fm			Nakdum Fm Tiwri Fm Suti Fm				Shaila Fm Changru Fm	Shaila Fm Changru Fm	North Face quartzite		
	M	Harpatnar beds Gauran beds												Nilgiri Carbonates		
	E														Maneting Fm	
Cambrian	L															
	M		Ramsu Fm		Kunzun La Fm		Haimanta Fm		Kurgiakh Fm Karsha Fm		Kurgiakh Fm Karsha Fm	Garbyang Fm	Garbyang Fm	Dimphu Gniesses		
	E															
Proterozoic	L		Haimanta Fm	Dogra slates	Batal Fm	Batal Fm		Hilap Fm		Basal Flysch		Ralam Fm Martoli Fm			Chekha Fm	
	M															
	E		Salkhala	Salkhala	Vaikrita Fm	Manjir Fm Pukhri Fm Chamba Fm	P		Phe Fm	Nimaling Crystalli-nes	Phe Fm	VaikritaFm	Buddhi schist			
						Dalhousie Granite							Vaikrita Fm		Thimphu Gniesses	

Note:Kunzum La and Batal formations were designated as Haimanta Formation by Hayden (1904)

5.6 Trans Himalaya

The area situated north of Tethyan zone is generally described as Trans Himalayan region. The Trans Himalaya includes Indus Tsangpo Suture Zone that joins the Indian and Eurasian plates together and small part of Eurasian plate in Kohistan and Karakoram regions, because in this region continuity of Himalayan structure is present Figs. 5.10. and 5.11.

The Trans Himalayan region is intimately associated with the Himalayan orogeny. The Indus Suture Thrust demarcates the southern boundary of Trans Himalayan region from the Tethyan zone of Himalaya. This region, evolved in the Mesozoic and Tertiary, shows highly complex geological set up having varieties of tectonic regimes and the geology of this region is put into proper perspectives by commendable work done by several research, technical and academic institutions.

Workers have identified many tectonic units in Trans Himalayan region. The Trans Himalayan, Indus-Tsangpo Suture running parallel to Tsangpo river in the east and central regions but bifurcates into two sutures namely the Indus and Shyok Sutures in north west besides sutures the other identified tectonic units are Tso Morari Crystallines, Karakoram Zone and Kohistan Zone. The Trans Himalayan territory falling into the Ladakh of India, and in the Pakistan and is geologically well known while the southern Tibet is geologically little known.

Indus Suture Zone

The Indus Suture zone runs parallel to the Indus and the Tsangpo rivers and its 600 km long and 40-100 km wide stretch exposed in the Ladakh is well studied. It is bordered by the Indus Suture Thrust or Dras or Southern Thrust and the Nubra-Shyok or Northern Thrust in south and north respectively. This region has short geological history ranging between Triassic to Oligocene. With in this suture zone basic volcanics, deep to shallow water flyschoidal sediments, ophiolitic mélange, turbidite, calc- alkaline volcanics and granitoids are present (Thakur and Misra, 1983). The Ladakh Granitoid exposed in 600 km by 20-80 km, area is most conspicuous unit in the suture zone. Following tectonostratigraphic sequence is reported from the east of Leh region of western Ladakh (Gergan and Pant, 1983; Rai, 1983, Upadhyay, 2002).

——— Nubra-Shyok Thrust ———	
Khardung Formation	Oligocene
——— Intrusive Contact ———	
Ladakh Plutonic Complex	Cretaceous-Miocene
——— Intrusive Contact ———	
Dras Volcanics	Jurassic-Cretaceous
——— Thrust ———	
Kargil Formation	Late Oligocene-Lower Pliocene
——— Thrust ———	
Indus Formation	Early Cretaceous-Oligocene
——— Thrust ———	

Shergol Ophiolitic Melange		Jurassic-Cretaceous
	Thrust	
Nindam Formation		Cretaceous
	Thrust	
Lamayuru Formation		Triassic-Jurassic
	Indus Suture Thrust/Dras Thrust	

In the west of Leh, the Lamayuru Nindam, Shergol Ophiolitic Melange and the Dras Formations are present. Besides these, Tso Morari Crystalline complex and Nidar Ophiolite are also exposed in the eastern region.

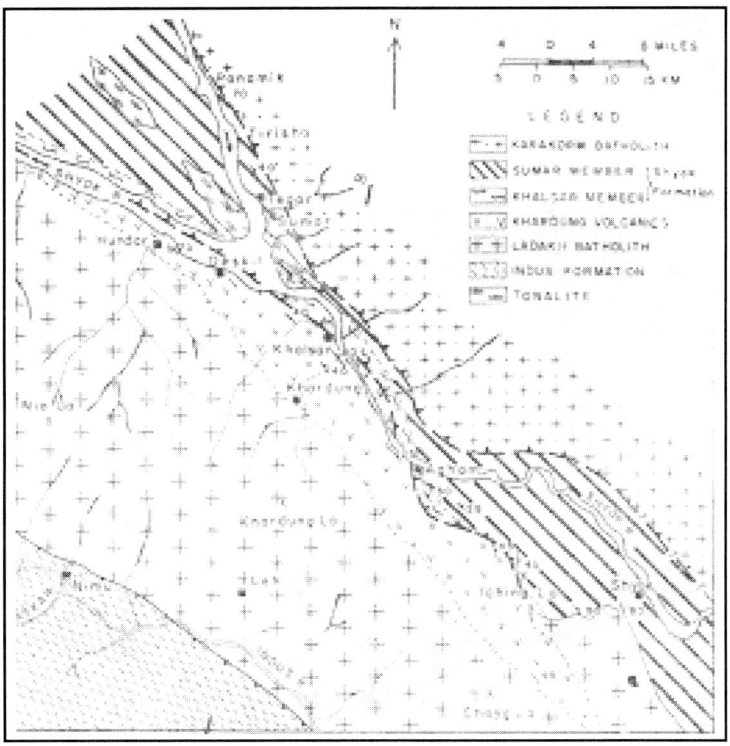

Figure 5.11. Geological map of Nubra shyok area (after Thakur, 1981)

Lamayuru Formation

This 3000 m thick NW-SE trending formation of continental slope to deep marine flysch sediments occurs along Zanskar crystallines in a narrow strip. The formation comprises shale, siltstone and graded sandstone of distal turbidite facies deposited on the passive continental margin. The exotic olistostromal limestone blocks of transported origin of couple of km length yielding carboniferous and Permian fossils are embedded in the Lamayuru Formation. On the basis of *Daonella, Psiloceras* and *Meekoceras* Late Triassic to Middle Jurassic age and foraminiferal fossils of Late Cretaceous- Early Eocene is assigned to the Lamayuru Formation (Bose et al., 1983).

Nindam Formation

The maximum 5 km wide Nindam Formation occurs as a wedge between the southern and northern Shergol Ophiolitic Melange. Earlier it was considered as lateral facies variation of Dras Volcanics but now it is recognized as a separate tectonostratigraphic unit. The 3000 m thick formation is constituted by thinly as well as thickly bedded sandstones, siltstones, red and green shales and bedded tuffs. The flysch sediments were originated from basic to intermediate volcanic source of Dras islands arc.

Nidar Ophiolite

The NW-SE trending Nidar Ophiolitic Belt is exposed between the northwest-southeast trending Zildat Ophiolitic Melange in south and the Indus Formation, Kargil Formation and Ladakh Granitoid of the north. Three rock units namely ultramafics, gabbro and volcanics associated with chert and clastic constitute this maximum 10 km wide ophiolitic belt. Pyroxinites, peridotite and ultramafic lenses are the main constituents of ultramafic unit. Ultramafics are associated with chromite mineralizations. The dark colored gabbro is coarsely crystalline in texture and the volcanic rocks show pillow structure. Thickly bedded chert, jasperoids, and conglomerate, grit and sandstone occur above the pillow lava. Recently few cm to a few m thick veins of tonalite and trondjhemite variety of oceanic plagiogranite are also reported and these veins are believed to be formed by fractional crystallization and filter pressing from the gabbroic magma within the ophiolite sequence (Rao et al., 2004).

Shergol Ophiolitic Melange

The Shergol Ophiolitic Melange unit occurs between Khalsi and Shergol. This unit is developed in less then 5 km wide area across the Nindam flysch in north and south regions. Besides the main occurrence, its thin lenticular exposures are reported from other places. The Spontang Klippe and Shilakong Nappe are considered equivalent to Shergol Ophiolitic Melange Formation.

Haphazard assemblage of chert and jasper bearing shales, sandstone, conglomerate magnesite, limestone, blue schist metamorphics, phyllite, serpentinite and peridotite constitutes this formation. The blue schist is formed around 400°C at 9-11 kb pressure range. This mélange zone formed in the trench.

Indus Formation

The Early Cretaceous to Early Oligocene age sediments of the Indus Formation was deposited in the arc-trench gap (Thakur and Bagati, 1983). This 4- 5000 m thick formation occurs along a strike length 500 km and has maximum width of 15 km. Indus Formation consists of massive to thickly interbedded sequence of conglomerate, sandstone, siltstone, shale and limestone. The Indus Formation is divided into the Khalsi Flysch representing deep sea fan deposits and an assemblage of flysch, conglomerate and marl units designated as Miru Unit (Mathur, 1983). In south, this formation is in contact with the Dras Volcanics, Ophiolitic Melange Zone and Tso Morari Crystalline Complex along the steep dipping thrust and in the north, Ladakh Batholith and Kargil Fm are in thrust contact with Indus Fm.

Nummulites, Orbitolina, Assilina, Globotruncana Hippurites, Ostrea along with algae, miliolids and rotalids of Early- Late Cretaceous age are reported from the Indus Formation.

Kargil Formation

The 4000-5000 m thick of Kargil Formation is resting over the eroded surface of Ladakh Granitoids. It comprises boulder bed, conglomerate, arkose, sandstone, shales and coals seams. Various ages starting from Palaeocene have been assigned to this formation but now it believed to be of Late Miocene to Late Pliocene age. Liyam Formation and Kailas Conglomerate are considered equivalent to the Kargil Formation.

Dras Formation

This formation is named after the village Dras. The 5000 m thick formation occurs along a strike length of 400 km and has maximum width of 15 km. Volcanics, pyroclastic material, radiolarian chert, limestone and serpentinite lenses constitute the formation. The volcanics are basaltic, andesitic to dacitic in composition and belong to island arc tholeiite and calc-alkaline series typical of island arc environment. The magma is generated by incipient or dehydration melting of Tethyan oceanic crust which was subducted into the mantle.

Ladakh Plutonic Complex

It is also known by several other names such as Ladakh Granitoids, Ladakh Granite, Ladakh Intrusives or Ladakh Batholith. It is nearly 600 km long and 20-80 km wide and its exposure has 5 km thickness. It is mainly composed of varieties of granite like I-type calc alkaline granites, quartz-diorite, monzodiorite, tonalite granodiorite, gabbro, norite, anorthosite, pyroxenite and other igneous rocks. This I-type granite is believe to be formed by incipient or dehydration melting of subducted oceanic crust into the mantle. This plutonic complex is formed in several phases and the oldest and the youngest reported dates are 103±3 and 60.8+4 Ma respectively. This age is obtained from the Kargil, however, 235±13 Ma is also reported from Gaik-Kiari area and the youngest is around 20 Ma.

Khardung Formation

The Khardung Formation is named after village Khardung. The Khardung Formation mainly comprises acidic volcanics overlies northern margin of Ladakh Granitoid and in further north overlain by rocks Shyok Suture along the Shyok Thrust. The formation comprises of 2000 m thick succession of acid and intermediate volcanic rocks like massive flows of rhyolite, dacite, andesite and pyroclastic and tufaceous material. The pyroclasts comprise of volcanic ash flow, tuff beds, volcanic breccia, ignimbrite and welded tuff. The volcanic activity had taken place between 38-26 Ma duration.

Spontang Klippe

The klippe is named after Spontang village and is composed of ultrabasic and gabbroic plutonic igneous rocks. Its petrologic composition is similar as Shilakong Ophiolitic Nappe. Evidences of mylonitisation and ductile shearing are present in the peridotite. The hornblende age of the ophiolitic rock is 130-140 Ma. This klippe was subdivided into lower Photang subnappe comprises of pillow lava, purple chert, sepentinites, volcanogenic sediments and carbonate rocks. The upper Spong subnappe is composed of varieties of altered ultrabasic rocks. This nappe is also interpreted as an obducted slab of upper mantle and oceanic crustal rock of Neotethys onto subducting passive Indian margin during Upper Cretaceous.

Shyok Suture

The NW-SE trending Shyok Suture or Northern Suture zone is situated between Ladakh and Karakoram granitoids. Its northern limit follows the Nubra Valley, a tributary of Shyok River, and the southern limit is located a few km south of the Shyok River. The tectonic limits are drawn by the Shyok Thrust and Main Karakoram Thrust present in the south and north respectively. The highly rugged and glacier covered terrain of Shyok Suture is physiographically known as Saltoro Hills where world's highest battle ground, Siachen Glacier, is located. The suture further extends up to Nanga Parbat in the NW. Granitoids volcanics, metasediments, sediments and Shyok Ophiolite constitute this suture zone. The Shyok Ophiolites representing the Shyok Suture is considered by one group of worker as product of pre-Cretaceous subduction while the other group considers that of Late Tertiary subduction. Shyok Suture is also considered as extension of Indus Suture of south as well as back arc basin. The geology of Shyok Suture is reported from Nubra-Shyok and Chushul Koyul areas. The tectonostratigraphy of Shyok Suture is as follows:

———————— Main Karakoram Thrust ————————

Saltoro Molasse

Saltoro Andesite

Nubra/Shyok Ophiolitic Melange

———————— Thrust ————————

Hundri Formation

———————— Thrust ————————

Shyok Volcanics

———————— Thrust ————————

Khalsar Formation

———————— Shyok Thrust ————————

These formations occur in highly disturbed condition and do not show regular order of superposition.

Tirit Granites

This granite is aligned in the WNW-ESE direction, south of the Shyok Suture is mesocratic medium to coarse-grained granodiorite- tonalite to gabbro diorite in nature. The granites are intruded by vertical dolerite and aplite dykes.

Khalsar Formation

The Cretaceous age Khalsar Formation is composed of high to low grade metamorphics and sedimentary rocks. The metamorphics include migmatitic gneiss, augen gneiss, mica schist, phyllite, quartzite, and amphibolite. Among sedimentary rocks, limestone, feldspathic quartzite and shales are present. However, the fossiliferous limestone member has yielded L. Carboniferous to Early Permian age.

Shyok Volcanics

Early to Middle Cretaceous age high Mg basaltic and andesitic rocks are exposed in the Shyok Valley and Saltoro Hills. In the south, these volcanic rocks are separated from the Ladakh Granitoids by lenticular bend of slate and serpentinite and in the north they are thrust over Soltaro Molasse and limestone, phyllite and quartzite of Hundri Formation. Fine to medium grained basalt is metamorphosed in very low–grade condition. In comparison to basalt, fine-grained, porphyritic andesite has wide distribution. These rocks give evidence of submarine origin. *Orbitolina* bearing limestone overlies the volcanics.

Hundri Formation

Exposed near Khalsar and Hundri was earlier designated Saltoro Flysch or Saltoro Molasse. The Late Cretaceous to Early Eocene Hundri Formation contains phyllite, slate, calcareous shale, limestone and quartzite. Turbidites of this unit show profuse presence of current ripples, graded bedding, flute casts and convolute laminations. Gastropods, echinoids, corals, foraminifera and rudists are reported from the 200 m thick limestone formed in shallow marine condition. Luzarmu Formation (andesite and dacite interbedded with chert and limestone) and Diong Formation (limestone, quartzite, sandstone and volcanics) of eastern Ladakh are equivalent to Hundri Formation.

Nubra Ophiolitic Melange

The Nubra Ophiolitic Melange was earlier described as Shyok Ophiolite or Shyok Ophiolitic Melange. The few km wide mélange zone occurs in three considerably separated beds namely: between Dongpolas to Margi and Chameshan to Tirit, Skam to Sonamarg and between Dongpolas to Bukidu. Melange is constituted by red, greenish, gray black shales, grading into slate and phyllite, limestone, calcareous shales, green sandstone, boulder bed and serpentinite rich in chromite, schistose volcanic greenstone, calc schist, quartzite, polymictic metaconglomerate.

Saltoro Molasse

The 50 km long and 3000 m thick molasse zone of post-Oligocene age resting unconformably or thrust upon Shyok Volcanics and the Hundri Fm contains conglomerate, siltstone, breccia, sandstone and shale was deposited in alluvial to deeper sub aquatic environment. Within the molasse, an andesite exposure of 500 m thick and 40 km length occur. This andesite is called Saltoro Andesite. The Kole Molasse of eastern Ladakh is equivalent Saltoro Molasse.

5.7 Tso Morari Crystallines

The Tso Morari Crystallines occurs as independent tectonic unit between the southern part of the Indus suture Zone and Tethyan Himalaya in the eastern Ladakh and western

Himachal Pradesh. Tectonically Tso Morari Crystallines exhibits regional doubly plunging anticlinal or domal structure. A similar structure with same lithological setup named as Gurla Mandhata Crystallines, is present in the southern Tibet. The Tso Morari Crystallines is composed of three tectonostratigraphic units namely the Puga Formation of the Late Proterozoic to Lower Palaeozoic age, the Taglang La Formation of Carboniferous -Permian age and Eocene-Oligocene granites of Rupshu and Polokong. The Late Proterozoic Puga Formation is composed of kyanite - sillimanite bearing schists and gneiss, augen gneiss and amphibolites. This formation is believed to be equivalent to Higher Himalayan Crystallines. The metamorphosed sediments of the Taglang La Formation are correlated with Neoproterozoic to Cambrian age Haimanta Formation and Palaeozoic - Mesozoic Tethyan sequence. Calcareous, marly and argillaceous sediments are metamorphosed in green schist facies condition and sills of amphibolites. The Rupshu and Polokong granites, occur as stocks, are concordant intrusions in the Taglang La Formation. These 500 Ma Old granites are coarse grained in nature and contain quartz, feldspar, hornblende, muscovite, biotite and tourmaline. The Polokong Granite is foliated in nature while Rupshu Granite is massive and non-foliated type. Contact metamorphism is also observed around granitic masses. Within the crystallines coesite-bearing eclogite is also reported which is similar to eclogite of Kaghan Valley of Pakistan (Thakur, 1983 a and Upadhyay, 2002).

5.8 Kohistan Region

The western continuation of Ladakh in Pakistan is known as Kohistan. This territory of 36000 sq km area is situated to west of the Nanga Parbat and makes the NW end of Himalaya. Though it is in continuation of Ladakh here the petrological and tectonic dissimilarities are more pronounced than the similarities. Tectonically, Kohistan region is situated between the southern Main Mantle Thrust (= Indus Suture Thrust of Ladakh) and the northern Main Karakoram Thrust (= Northern or Shyok - Nubra Thrust of Ladakh). Island are tectonic setting is inferred in the Kohistan region.

The Kohistan rocks are thrust over along a low angle fault on the Higher and the Lesser Himalayan rocks in the south and on the north of MKT, a part of the Karakoram zone of Eurasian plate is also present in the Kohistan region. This region is composed of highly diverse group of rocks such as garnet granulite of lower crustal origin, blue schist, amphibolites, diorite, granites, volcanics and metasedimentaries. Rocks of Kohistan region is divided into two tectonic groups namely the Kohistan Sequence and the Northern Suture. The tectonostratigraphy of the Kohistan region, after Tahirkheli Khan (1979), is as follows:

Main Karakoram Thrust (= Shyok Nubra Thrust)	
Northern Suture	Yasin Group Melange
	Chalt Volcanics
Kohistan Sequence	Kohistan Batholith
	Chilas Complex
	Jigal Complex
	Sangla Blue Schist
Main Mantle Thrust (= Indus Suture Thrust)	

Kohistan Sequence

Sangla Blue Schist

This is actually an ophiolitic melange zone consisting of dismembered fragments of metavolcanics, and phyllite and also to small extent of serpentinite, metagabbro, metadolerite, metamorphosed greywacke, chert and limestone. Metalimestone has yielded Jurassic to Middle Cretaceous age and is inferred to be formed at 380° C temperature and 7 kb pressure (Khan et al., 1997).

Jigal Complex

The Jigal Complex occurs as wedge of 200 sq km, comprises garnet granulite and ultramafics. The garnet granulite possibly representing the lower crustal origin is formed at 670°-770°C and 12-14 kb temperature-pressure condition. The garnet granulite shows heterogeneous petrology and includes metamorphosed norite, olivine gabbro, hypersthene diorite, anorthosite, pyroxenite, and other magmatic rocks having higher alumna content. Ultramafics like dunite, peridotite, diopsidite and webesterite also occur as independent bodies. The garnet granulite and ultramafic rocks have been originated at different time. These rocks underwent two-phase metamorphic history the earlier one is of high temperature and low pressure (800°C/7-8 kb) and later one is of low temperature and high pressure (670°-770°C/12-14 kb).

Kamila Amphibolite

The Kamila Amphibolite consists of amphibolites, leucohornblende gneiss and flaser gabbro. 300 km long and 50 km wide amphibolite band comprising of hornblende, plagioclase, epidote, quartz and opaque extends from north Pakistan to Pakistan- Afghanistan border. In Khwazakhola area, this amphibolite is formed at 500°-600° C and 10-12 kb pressure.

Chilas Complex

The arcuate shape 300 km long and 40 km wide Chilas Complex is located between the Kamila Amphibolite of south and the Kohistan Batholith in the north. Feldspathic gabbro and norite are main components of this zone besides these also have minor presence of ultramafics, anorthosite, gabbro, diorite, pegmatite and amphibolite dykes in the Complex. This complex is interpreted as root zone of magma chamber of Kohistan arc or early-stage diaper produced during intra-arc rifting.

The Kamila Amphibolite is overlain by low to medium grade graywacke, chert, phyllites, chlorite schist, calcareous sediments, amphibolites and low grade pillow basalt belonging to Kalam Formation. The Kalam Formation is overlain by calc-alkaline volcanic rocks like andesite, dacite, rhyolite and ignimbrite composition and is combinedly called Dir Volcanics.

Kohistan Batholith

The Kohistan Batholith composed of diorite, granodiorite, tonalite and aplite is evolved in five phases. The earliest phase is 102±12 Ma Old and the youngest one 24±8 Ma old.

5.9 Northern Suture

The Northern Suture is extension of the Shyok Suture. Two units namely the Chalt Volcanics and the Yasin Group constitute this suture.

Chalt Volcanics

A four thousand meter thick Cretaceous sequence of volcanics and metasedimentaries belonging to Chalt Volcanics is situated north of Kohistan Batholith. The volcanic suit is constituted by basalt, andesite dacite and volcanic tuff along with lenses of pyroxinites, hornblendite, diorite and gabbro. The metasedimentaries contain quartzite, phyllite and schist. The entire sequence is metamorphosed in green schist to amphibolite facies condition.

Yasin Group

The Early Cretaceous Yasin Group is also described as Northern Melange and is composed of several km long randomly distributed fragments of mafic to ultramafic rocks, limestone, sandstone and conglomerate in a matrix composed of chloritoid slates. Besides these fragments of biotite-garnet-staurolite schist and gneiss are also reported in the melange.

5.10 Karakoram Block

The Karakoram region, 600 km long and 150 km wide, is situated north of Ladakh and Kohistan areas of India and Pakistan respectively. This mountainous terrain is one among the several mountain ranges radiating out from the famous Central Asian Pamir Knot; the other being the Pamir and the Tienshan trending in the north; the Aghil and the Kunlun ranges trending to the east; westward trending Hindukush and the southeast trending Himalaya. The Karakoram region is geologically part of the Eurasian plate. This Central Asian region is evolved from two juxtaposed orogenic movement namely the Hercynian and the Himalayan. The Karakoram mountain, the Aghil, the Hindukush and the south and Central part of Pamir region are considered as part of the Alpine-Himalayan Orogen as this zone is also evolved during the Himalayan orogenesis.

The Karakoram zone is thrust over the Northern Suture and Shyok Suture along the Main Karakoram Thrust. Several workers have described the geology of Karakoram region from the eastern, western and the northern regions. An unified geological account reveals that Karakoram zone is constituted by three tectonic units: the Proterozoic low to medium grade crystalline rocks, the main part is occupied by a granitoid batholith and a fossiliferous sedimentary zone of interrupted Carboniferous to Permian age. These units have been assigned many local names, a broad account of these units are as follow:

Proterozoic Crystalline Unit

This unit is also known as Pangong Tso Group in the eastern Karakoram and the western Karakoram Metamorphic Complex. The deformed crystallines include migmatites, kyanite-sillimanite gneiss, garnet-staurolite schists, phyllite, marble, quartzite, amphibolite, mylonitised granite gneiss, augen gneiss and granite intrusions.

Karakoram Batholith

The Karakoram Batholith is situated to the north of the Shyok Ophiolitic Mélange zone, along the Karakoram Fault. A 50 m wide mylonite zone and a chain of the hot springs are present along the contact. This mainly granodioritic batholith is also called Karakoram Axial Batholith and occurs in a length 600 km and has maximum width nearly 100 km. Besides granodiorite and granite it also contains tonalite, aplite, pegmatite and migmatite. At many places batholith has acquired gneissic characters due to metamorphism. The batholith is intruded in the crystallines of Proterozoic age as well as in the younger sedimentary rocks. It is also recognized by several local names like Masherbrum Migmatitic Complex, Hunger Plutonic Complex, and K 2 Gneiss etc. The batholith is evolved through a series of intrusive pulses the earliest one is probably of Mesozoic (204 Ma, Jurassic) age and the youngest is of Pliocene (4-6 Ma) age. Geochemical interpretation of the data suggests that granites are volcanic arc and syn collision (I and S types) origin; therefore, granites are compared with continental- Chilean- arc type.

Karakoram Sedimentary Unit

Several detached exposures of fossiliferous sedimentary zones are reported from Karakoram zone. These rocks have yielded Carboniferous and younger fossils. The depositional history in this region begins in Ordovician and continued till Pleistocene. In between this duration several sedimentation breaks have been recognized. In the eastern region it is described as Karakoram Super Group and is divided into following formations.

Saser Brangsa Formation

This 1000 m thick Carboniferous-Early Permian formation comprises of fossiliferous black slates, sandstone quartzite and limestone. The important fossil species are *Spirifer, Productus, Fenestella* etc. a 30 m thick bed of pyrite, chalcopyrite, pyrhhotite along with malachite and magnetite is present in this formation.

Chhongtash Formation

This 1500-2000 m thick Permian formation mainly composed of black shale, siltstone, greywacke, dimictite, conglomerate and pillow lava. The lava is interbedded with argillaceous sandstone and limestone containing *Parafulsina. Schwagerina. Noeggrathiopsis, Gangamopteris* and *Equisetalean* etc are Gondwana plant fossils.

Aqtash Formation

This mainly purple and green shaley, polygenetic conglomerate, gray-yellow limestone, chert lenses and volcanics formation is 500 m thick and is of Middle to Upper Permian age.

Morgo Formation

This 300 m thick formation mainly contains limestone and subordinate calcareous shales intercalated with red gray shale and chert. The faunal assemblage includes *Meglodon* and corals of Triassic age.

Burtsa Formation

Limestone and black and pink shales of 1000 m thickness constitute Burtsa Formation of Jurassic-Cretaceous age. The formation yields *Nodosaria, Dentalina Ogmoconcha, Ledahia* and other fossils.

Qazil Langer Formation

The 500 m thick, Cretaceous age Qazil Langer Formation comprises of red conglomerate, black shales, sandstone and limestone. *Globigerina* and *Textularia* forams have been used to decide its age.

Khumdan Formation

It is a constituted by rhyolite flow spreading over couple of hundred square kilometers area. It may be similar to Khardung Volcanics.

Depsang Group

The Quaternary glacial, fluviatile and lacustrine sediments define this formation and are spread over in 30-50 km wide flat region at 5000 m altitude. Roches moutonne types hillocks are present in this plain

The geology of western Karakoram region is continuation of eastern region. However, in the northern region Permian to Jurassic age stratigraphy is reported from Shaksgam and Upper Hunza Valleys and Ordovician to Cretaceous rocks from the Chitral and Western Gilgit area of Pakistan - Afghanistan. Permian rocks are described as Shaksgam Formation containing limestone, impure sandstone and marl and mudstone and cherty limestone. Formation containing impure sandstone in Shaksgam Valley while the Gircha Formation comprising of shales and laminated sandstone in Hunza Valley. The Aghil Limestone Formation contains 1000 m thick dolomitic limestone in Shaksgam Valley and the Panjshah Formation (arenite, and siltstone), Kundil Formation (calcarenite and white chart), Boram Formation (limestone and black chert) and the Aghil Formation (limestone and dolomite) in Upper Hunza Valley belong to Triassic age. The Tekri Formation (mudstone and wackestone), Morpo Sandstone Formation (sandstone) and the Bongo La Formation (limestone, siltstone, shale and sandstone) in Shaksgam Valley and the Yashkuk Formation (sandstone, limestone and calcarenite) and the Reshit Formation (impure sandstone and mudstone) in Upper Hunza Valley are of Jurassic age.

The Karakoram Batholith and Northern Karakoram sedimentary Unit continue in the western Gilgit and Chitral areas situated partly in the Pakistan and partly in the Afghanistan. Sandstone quartzite, shale, slate, limestone of Ordovician and Permian age, and the slate, chlorite schist, limestone, sandstone and conglomerate of Cretaceous age are reported from Chitral and Western Gilgit region. The Karakoram Batholith is exposed in the form of small sized granodioritic to granitic plutons at Dobargar, Zogar Umalsit and Kesu-Buni Zom. The Zagar - Umalsit pluton yielded 56 and 40 Ma radiometric (K-Ar) dates.

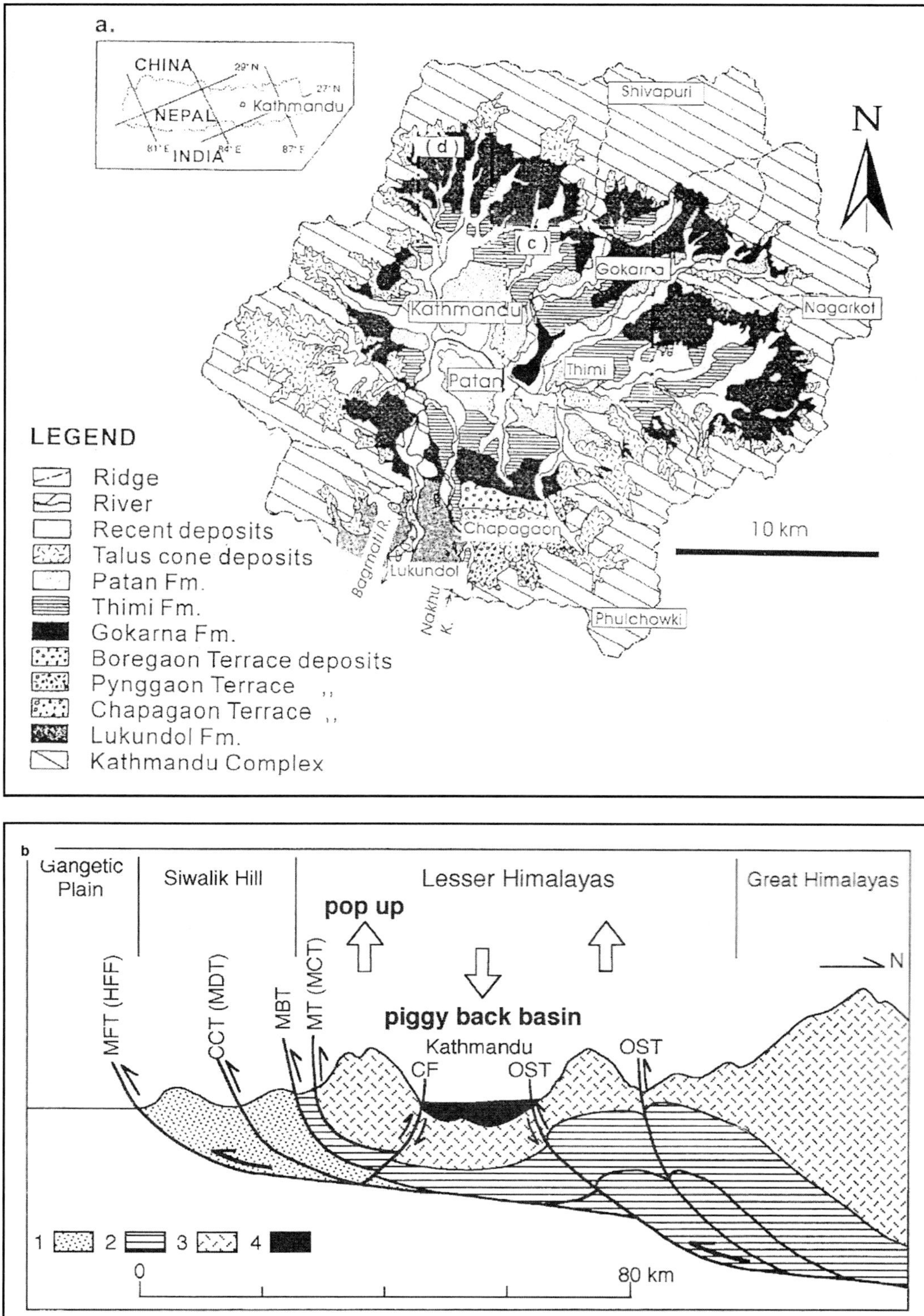
a.
CHINA
NEPAL
Kathmandu
INDIA
Shivapuri
N
(d)
(c)
Gokarna
Kathmandu
Nagarkot
Thimi
Patan
LEGEND
Ridge
River
Recent deposits
Talus cone deposits
Patan Fm.
Thimi Fm.
Gokarna Fm.
Boregaon Terrace deposits
Pynggaon Terrace ,,
Chapagaon Terrace ,,
Lukundol Fm.
Kathmandu Complex
Chapagaon
10 km
Bagmati R.
Lukundol
Nakhu K.
Phulchowki
b
Gangetic Plain
Siwalik Hill
Lesser Himalayas
Great Himalayas
pop up
N
MFT (HFF)
CCT (MDT)
MBT
MT (MCT)
piggy back basin
Kathmandu
CF
OST
OST
1 2 3 4
0
80 km

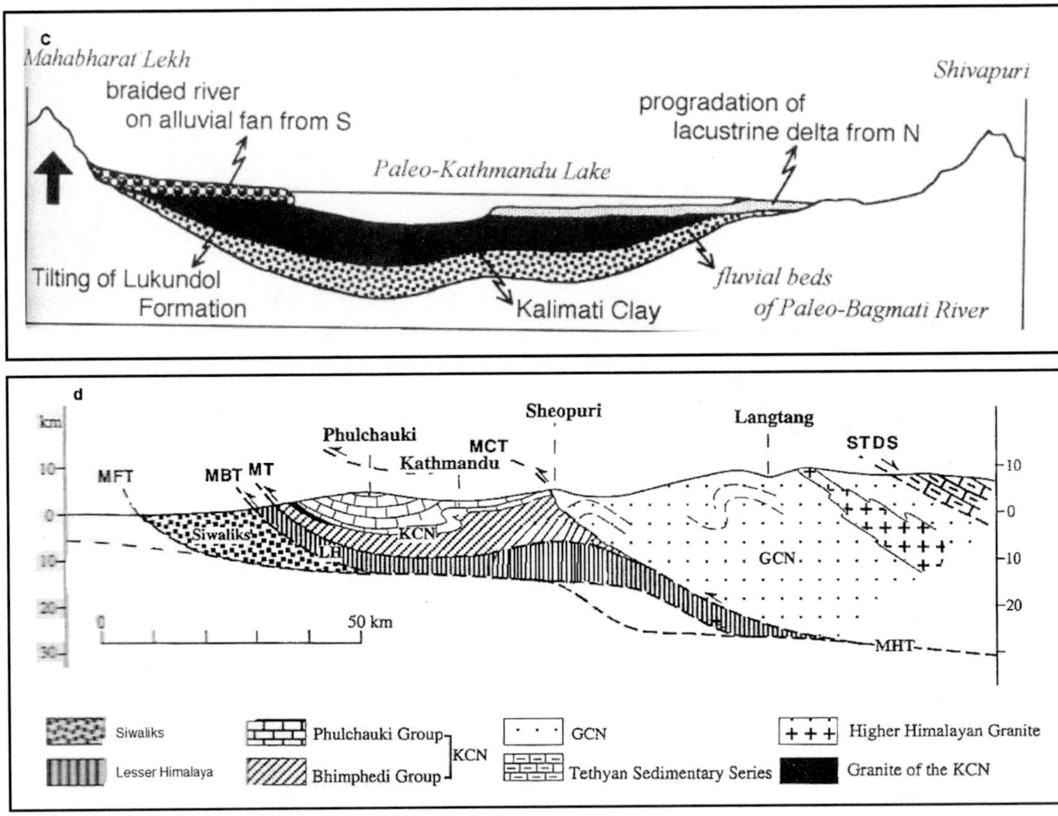

Figure 5.12 (a) Geological Map of Kathmandu Valley (after Gautam et al., 2001); (b) Geological cross section of area around Kathmandu Valley (after Sakai, 2001); (c) Geological cross section across Kathmandu Valley (after Sakai, 2001) and (d) Geological cross section of Central Nepal (after Rai, 2001)

5.11 Southern Tibet

In between the Himalaya and the Tanglha tectonic zone a mountainous terrain is present in the southern Tibet. The Gangdese Mountain Belt and the Nyainquentanglha Mountain Belt constitute this hilly region. The Indus Suture Zone decides the southern boundary of the Gangdese belt and the Bango Co-Dongquiao Nu River Fault (BCDNRF) demarcates and northern boundary of Nyainquentanglha belt from Tanglha tectonic zone. Though the granitoid rocks dominate area but fair space is also covered by the metamorphics of varying grade, sedimentary rocks and ophiolites. These rocks are Palaeozoic to Tertiary in age; however, high- grade metamorphics are debated to be Proterozoic or Palaeozoic in age (Bordet,1979; Sheng and Wei, 1983 Xuchang and Yanlin, 1984).

The Gangdese

Nyainquentanglha area shows complex tectonic evolution during the Mesozoic and Tertiary. Restricted presence Carboniferous-Permian rocks indicated affinity of this area with the Gondwana of south. In the beginning Mesozoic rifting occurred along the present

Indus Suture Zone and a sea started growing and this sea continued to spread till the beginning of Cretaceous and eventually closed during Eocene and Oligocene. During Triassic, clastics, pelites and carbonates are deposited in the southern area and intrusions of basic rocks also occurred. In Jurassic, shallow water sediments were deposited over large area. Since the beginning of Late Jurassic deep-water deposition commenced and this phase continued till Early Cretaceous. During this period over 5000 m thick strata were deposited and intense volcanic activity also occurred in island arc type environment. During Late Cretaceous, the area started experiencing compressive forces causing folding and faulting of the sediments and eventual closure of basin began. Till the Oligocene, by the time basin was completely uplifted. flysch and molasse were deposited in different parts. In this short history, granitoids were profusely intruded in this region. Three zones of granitoids namely the Nyainquentanglha (150-60 Ma), the Gangdese (120-20 Ma) and the zone of small rock bodies (50-10 Ma) are identified (Zhao and Morgan, 1985). These zones consist of diorite, quartzdiorite, granodiorite, quartz-monzonite, biotite granite, two-mica granite, microcline granite and tonalite. These granitoids belong to intermediate-acidic (I type) and acidic (S type) types. The magmas for I-type originated by different degree melting in the lower crust or upper mantle while for S-type in the upper crust through its partial melting.

DIMENSIONS OF HIMALAYAN GEOLOGY
A. K. Biyani (2006) : pp.161-176
Satish Serial Publishing House, New Delhi
E-mail: ssp@ndf.vsnl.net.in, hkjain1975@yahoo.com; website: www.satishserial.com

EtadyoniniBhutani arvanityupadharaya
Aham krtsnasya jagatah pranhavah pralayastatha (Geeta, 7.6)
I am the source of all creations and in Me, again it dissolves.

CHAPTER - 6

PALEONTOLOGY

The Himalayan stratigraphy covers a very large time range spanning between Late Archaean to Quaternary and have several time interruptions varying between region to region. Older tectonic divisions like of the Higher Himalaya and the Lesser Himalaya are as a rule unfossiliferous in nature except few formations of the Lesser Himalaya showing presence of stromatolite, brachiopods, trilobita, conodont and other fossils. Barring Higher Himalaya, presence of mega and micro-faunal and floral remains are recorded from all divisions of the Himalaya, many of the post-Proterozoic Himalayan formations such as of Tethyan areas are richly fossiliferous. The conspicuous presence of mega fossils in these areas caught the attention of geologists even in the eighteenth and early nineteenth centuries and their meticulous efforts established the stratigraphical sequences in many areas, which remained more or less, unchanged even now. The earliest fossil record is undoubtedly of stromatolites, which are present in the carbonate horizons of Mesoproterozoic age. Himalaya matched itself with the rest of the world in preserving the vastly diversified life. Foraminifera occurred since onset of Palaeozoic. Subsequent vicissitudes of many biotic pulses are also preserved in the Himalaya.

Paleontology played a very valuable role in establishing the geological history of the Himalaya. Not only fossils provided the geological age of sedimentary rocks but also helped in reconstructing the stratigraphical column in several areas and long-range correlation is also successfully attempted in the widely separated areas of Outer, Lesser and Tethyan divisions. Fossils successfully solved the age dispute of many stratigraphic formations, for example, age of Krol and Tal formations were earlier proposed between Late Proterozoic to Cretaceous by many workers but was authoritatively solved initially with the help of conodont fossils by assigning Late Proterozoic age to Krol Formation and Cambrian age to Tal Formation, this was later supported by identification of trilobita, brachiopods and other trace fossils. Paleontology based stratigraphy of Tethys and the Outer Himalaya stood firmly on the test of subsequent advancements. Some fossiliferous localities deserve the status of type horizons of stages/zones, for example Krol-Tal boundary could be a Precambrian-Cambrian type area. Punjabian (Saxonian) Stage of basal Late Permian system is named after Productus

Limestone bed of western Salt Range. In many areas formations are finely divided into biostratigraphic zones such as Subathu Group of Garhwal, Himachal, Jammu and Kashmir. This is divided on the basis of foraminifera and Spiti Shales of Jurassic-Early Cretaceous age is divided into nine biostratigraphic zones based on different ammonoid fossils assemblages respectively. Stromatolite, polynofossils, vertebrate fossils are also used for such exercises in different regions of the Himalaya. Fossils are also used for interpreting the palaeogeography, palaeoecology, palaeoclimate and tectonic and depositional environment of many sedimentary formations. Fossils can be used in quantitative deformation/strain measurements in Himalaya.

Himalayan fossil 'Saligram', preserved in the Tethyan rocks, is worshipped by devote Hindus either in their homes or in temples. Black, smooth, thickly cherty spherical to oval shaped stones 'Saligram' have ammonoids fossil in the core. Chert covered ammonoids resemble to famous Sudarshan Chakra (an ultimate weapon) of Lord Vishnu (the Sustainer). Silurian crinoidal limestone of Godawari area, Kathmandu is a famed building stone in Nepal.

6.1 Fossiliferous Horizons of Himalaya

Aerial coverage of sedimentary rocks in the Himalaya exceeds the combined surface occurrences of igneous and metamorphic rocks. A major portion of sedimentary rocks is abiotic in nature but sedimentary sequences of some areas are highly enriched in the biotic contents. More than 10 km thick, more or less uninterrupted mainly Palaeozoic-Mesozoic sequence of Tethyan Himalaya, developed from Kashmir in the west to Bhutan/ Arunachal Pradesh in the east is one of the best fossiliferous region of the world. Life record of more than 500 million years is preserved here. Some sections of Siwalik rocks of Pakistan and western India is a treasure house of Neogene vertebrate fossils. So is the case of Trans Himalayan sedimentary rocks of Ladakh or Late Cretaceous- Eocene marine transgression deposit named as Kakara- Subathu formations or for microfossils Krol-Tal boundary formations of Dehradun-Mussoorie area.

Many fossils species, orders, classes or phylum have long geological range such as gastropods, lamellibranchia or foraminifers, but their age wise continuous geological record is not reported from the combined stratigraphic column of all division of the Himalaya. These missing records can be attributed to many factors; it might be hidden under thick vegetation or soil cover or could not be traced due to topographical constraints or lack of investigations or concerned fauna or flora might have not deposited in that section and period of time. A concerted effort may reveal the presence of unrecorded fossils in future. Gupta (1971, 1973 and 1975) compiled the palaeontological records of the Palaeozoic fossils from the Himalayan region.

The important fossiliferous regions of Himalaya are as follows:

Precambrian: Stromatolite occur in the Lesser Himalayan region of Jammu and Kashmir Change, Himachal Pradesh, Uttaranchal, Nepal and NE India. They are reported from carbonate horizons namely Deoban, Mandhali, Krol of Uttaranchal. The Upper most Proterozoic time horizon named as Edicaran contains tracks and trails of animals are also reported from the Himalaya.

Cambrian: Cambrian metazoans occur in Baramula and Sarchu areas of Jammu and Kashmir; Keylong, Lahaul and Kinnaur of Himachal, Garbyang, Mussoorie area and Martoli Formation of Uttaranchal, Thakkhola of Nepal, Hazara Formation in Pakistan and in Bhutan.

Ordovician: Shamsh Abari Syncline, Hundwara, Lidar Valley, Naubug Valley, Baramula, Changru, Shiala of Kashmir; Lahaul and Spiti of Himachal Pradesh, West of Niti Pass in Uttaranchal and in Kali Valley and Thakkhola of Nepal as well as in Bhutan and Sikkim.

Silurian: Harpatnar Beds, Naubug Beds Ganga Nagar, Marhaum, Shamsahbari, Baramula, Nagmarg of Kashmir, Lahaul and Spiti valleys of Himachal Pradesh, Variegated series of Uttaranchal and Bhakia Sain Phyllite, Phulchowki of Nepal and in Bhutan and Sikkim.

Devonian: Tanwal in Kashmir, Muth Quartzite of Himachal Pradesh and Uttaranchal and Phulchowki of Nepal and Tanachu Series in Bhutan.

Carboniferous: Syringothyris Limestone, Passage Beds, Productus Limestone, Fenestella Slates, Agglomeratic Slates, Panjal Volcanics, Gangamopteris Beds, Zewan Formation, Kalhel Limestone, Linty in Kashmir, Lipak Series in Lahaul, Spiti and Kinnaur of Himachal Pradesh Kali and Sisne Formations in Nepal.

Permian: Permian fossiliferous rocks occur at Zewan in Vihi district and Sarcha Series in Ladakh; Kinnaur and Spiti in Himachal Pradesh, Kring Krang Series and Lower Bijni Unit of Doggada, Uttaranchal; Sarchu Series, Bichom Bhareli Yamune, Miri Formations and Abor Volcanic Formations of Arunachal Pradesh Gondwana Formation of Kosi in Nepal, Bhutan and Sikkim.

Triassic: Pir Panjal, Lidar and Sind Valleys, Wardwan, Hunmu Khunmu, Pastannah, Guryal ravine Kishtwar in Kashmir; Kalhel Limestone of Chamba, Dras, Jojila, Zanskar, Rupshu, Luch-lung-La in Ladakh; Yannam, Lilang and Linty in Spiti and Niti, Chocolate Series, Kalapani Limestone, Kuti Shales and Kioto Limestone in Uttaranchal.

Jurassic: Jurassic fossils occur in Ladakh, Animal in Jammu and Kashmir, Baltic, Joni La Tandi Synclinorium in Chandra Valley, Spiti and Zanskar valleys, Kioto Limestone, Lapel Beds, Oolite Beds in Himachal Pradesh, Kuti, Kali, Dharma Lissar valleys and Lapthal, Chidamu of Uttaranchal and Tatlung in Arunachal Pradesh.

Cretaceous: Indus Flysch, Dora and Khalsi in Ladakh and Kargil regions, Chambal Beds of Spiti, Guimal Sandstone, Chikkim Series of Spiti, Neelkanth Bed/Kakara Fm of Uttaranchal; Taltung Formation Nepal and the Amile Formation of Arunachal Pradesh.

Palaeocene: Pir Panjal Range in Jammu & Kashmir, Kakara Formation of Jammu, Himachal and Uttaranchal and Geku Formation of Arunachal.

Eocene: Subathu Formation, Bhainskoti of Nepal and Dalbuing Formation of Arunachal and Muree Group, Nummulitic Limestone and Ladakh in Jammu and Kashmir, Jarbutta Formation, Tosh Formation, Masem Formation and Bhainskoti Formation of Nepal.

Oligocene: Early Miocene – Dagshai/Muree/Dharamsala formations and Kasauli Formation in Jammu, Himachal Pradesh and Uttaranchal and Ladakh Molasse Group in Jammu & Kashmir.

Mid-Miocene: Neogene - Neogene fossils particularly vertebrate occur discontinuously in the Siwalik or equivalent formations, in the Outer Himalaya and also in the 'duns' (valleys) formed by Quaternary movement like Karewa, Kathmandu and others.

6.2 Fossil Museum of Himalaya

Sedimentary rocks of the Himalaya store varieties of animals (mega and micro) and plants (stromatolite, pollens, leaf and wood) fossils. In addition to these, ichno fossils (cast and mould), microbiota and fossils of doubtful affinities are also described in this brief account by mentioning the representative fossil names. In case of fossils of long geological range, their names are mentioned once when they appeared first time, though they may be also present in the rocks of younger ages. An alphabetically phylum, class/order wise occurrences of important animals and plants fossils are as follows:

Actinozoa (Anthozoa)

This class belongs to phylum Coelenterate, which commonly includes coral, hydrozoans and sea anemones. These animals occur either individually in deep water or in the colonial forms in shallow warm water. Actinozoan made first appearance in the Cambrian. The important forms recorded in the Himalaya are:

Ordovician: *Streptelasma, Heliolites, Monotrypa*

Silurian: *Favosites, Halysites, Cyathophyllum, Syringopora, Columnopora, Calostylis.*

Devonian: *Hallia, Lindostroemia, Endophyllum, Calceola.*

Carboniferous: *Laphophyllum, Dibunophyllum, Caninia, Namistium. Amplexus* and *Zaphrentis* in Permocarboniferous.

Triassic: *Columnophyllia*

Palaeocene - Eocene: *Feddenia.*

Ammonoidea

This order belongs to class Cephalopod of phylum Mollusca. Ammonoids are marine animals having bilateral symmetry. Some of the ammonoids such as *Baculites* are extinct. Fossils belonging to order Nautilodea and Dibranchia are also included in the list.

Ordovician: *Gonioceras, Crytoceras, Orthoceras, Trocholites.*

Silurian: *Orthoceras, Cornulites, Diaphorostoma, Kionoceras.*

Carboniferous: *Pseudomonotis, Nautilus.*

Permian: *Xenaspis, Papanoceras, Schizodus, Cyclo.*

Triassic: *Ceratites, Danubites, Meekoceras, Ophiceras, Otoceras Owenites.*

Jurassic: *Belemnites, Phylloceras, Hecticoceras, Hoplites, Perisphinctes.*

Cretaceous: *Anisoceras, Baculites, Acanthoceras, Acanthodiscus*

Brachiopoda

The marine brachiopod animals were more abundant in the Palaeozoic and Mesozoic. They prefer shallow water.

Cambrian: *Lingulella, Obolella, Discellomus, Obolus, Botslordia*

Ordovician: *Orthis, Plectambonites, Scenidium, Rafinesquina, Hindella, Triplecia.*

Silurian: *Doleorthis, Orthotetes, Pentameras, Chonetes, Eospirifer, Retzia.*

Devonian: *Atheyris, Schizophoria, Stropheodonta, Spirifer, Rhynchonella, Cyrtina, Atrypa, Meristina.*

Carboniferous: *Productus, Derbia, Syringothyris, Dielasma, Aulosteges Aviculopecten.*

Permian: *Marginifera, Ambikella, Hemiptychina, Neochonetes, Linoproductus*

Triassic: *Cyrtorostra, Metzelia, Etheripecten*, *Doinella.*

Jurassic: *Rhynchonella, Atrate, Ostrea, Protocardium.*

Cretaceous: *Cardium, Pseudomonotis, Gryphaea, Holestephnamus.*

Blastoidea

Extinct marine blastoids belong to class echinodermata. They show very restricted presence in Palaeozoic only. The reported genus is *Stephanocrinus.*

Branchiopoda

Animals of this sub class belong to class crustacea of phylum arthropoda. They are fresh water; marine and brackish water animals Genus *Estheria* is reported from Himalaya.

Bryozoa

This phylum is also known as Polyzoa. They are marine colonial animals living mostly in associations.

Cambrian: *Stephanocrimus, Eoscharophora* from Tal Formation.

Carboniferous: *Fenestella, Protoretepora.*

Coccolithidae

Belongs to class Mastigophora of phylum Protozoa. They are calcareous disc like animals. The reported genera from the Himalaya are *cycloplacolithella, Satpulispora, Desmochitina.*

Conodont

Conodonts are extinct, tooth like marine fossils having notable presence of phosphate in their composition. Their appearance is similar to scale conodonts (teeth of annelid worms) or fishes; but their exact nature is yet unknown.

Cambrian: *Amphigeisina, Cordylodus, Drepanodus, Furnishina, Hirsutodontus, Onetodus, Proacodus, Prascanodus, and Protohertzina Ulrichodina* reported from Tal Formation near Dehradun.

Devonian: *Ancyrodella, Icriodus, Palmatolepis, Polygnathus, Siphonodella, Bryantodus* from Phulchowki area of Nepal.

Triassic: *Anchignathodus, Neogondolella, Neospatheodus, Hindeodella, Gondolella, Ellisonia* from Guryal ravine of Kashmir.

Conularida

This group belongs to phylum Coelenterata. These were extinct marine animals with chitinous conical shell. *Conularia* genus of conularida is reported from Permian rocks of Arunachal Pradesh.

Crinoidea

This is a class of phylum Echinoderma whose present day representatives are sea lilies and feather star like marine fauna. Body is consisting of a stem, a calyx and moveable arms. They first appeared in Ordovician age.

Silurian: *Parapisocrinus, Pisocrinus, Crotalocrinus, Pyrocystites, Craterina and crinoids* stem.

Permian: Calceolispongia.

Triassic: *Calamophyllia, Spiriferina, Mormolatella.*

Cystoidea: Cystoids are rare Palaeozoic fossils belonging to Phylum Echinoderma.

Cambrian: *Eocystites* in Haimanta Series.

Ordovician: *Cadiacystis, Dendrocystis*

Dinoflagellata

They are Protozoans belonging to class Mastigophora. Some believe that they are algae. Important genera are *Spiniferites, Hystrichoklolporna, Homotryblium Cyclonephedium* are reported from Arunachal Pradesh.

Gonyaulacysta, Tenna, Hexagonifera and *Tenua* are reported from Palaeocene Eocene Subathu Formation.

Foraminifera

They are very common Protozoans belonging to class Gymnomyxa. Forams are marine in nature and are first appeared in the Cambrian.

Carboniferous: *Glomospira, Hyperammina, Lituotuba, Septammina Nodosinella.*

Permian: *Involutina, Triloculina, Nodosaria, Textularia, Fusulina, Schwagerina.*

Triassic: *Lingulina, Veginulina, Cytharina, Glomospira, Lituotuba, Ammobaculites.*

Jurassic: *Rotalia, Cristellaria, Truncatulina, Lagena*

Cretaceous: *Cristellaria, Textularia, Dentallina, Orbitolina, Palmula, Cibicides, Globigerina, Purgo.*

Palaeocene: *Ranikothalia, Lockhartia, Dictyoconoide, Daviesina.*

Eocene: *Nummulites, Assilina, Astora, Cantus, Lepidocyclina, Alveolina.*

Miocene: *Chiloguembelina. Miogypsina.*

Gastropoda

Gastropods are very well known fossils of phylum Mollusca. They are fresh water as well as marine animals. They made first appearance in the Cambrian.

Cambrian: *Lapworthella, Hyolithes.*

Ordovician: *Raphistoma, Bellerophon, Lophospira.*

Devonian: *Pleurotomaria.*

Carboniferous: *Eumphalus*

Permian: *Loxonema, Meekospira, Mourlonia, Neilsonia.*

Triassic: *Euomphalus, Worthenia.*

Jurassic: *Pleurotomaria, Cerithium*

Cretaceous –

Eocene-*Graphia, Trigonia*, *Seila*

Miocene: *Natica, Sinum, Mitra, Terebra, Conus.*

Pliocene: *Melanoides, Ferrissia, Guraulus, Viviparus, Tricula.*

Pleistocene: *Carychium, Gastrocopta Vertigo, Lymnea.*

Graptolitoidea

Graptolites are good index fossils of Lower Palaeozoic age. They belong to class Hydrozoa of phylum Coelenterata. Fossils are marine in nature.

Ordovician: *Climacograptus, Glassograpts*

Silurian: *Monograptus, Didymonograptus, Cheirocrinus.*

Hydrozoa

Hydrozoans are poorly fossilized, marine water individual or colonial coelenterates. The Silurian genus *stromatopora* and Rhaetic *Helerastidium* are reported from Kashmir.

Insecta

Class Insecta belongs to phylum Arthropoda. They occur in enormous number of forms and first appeared in Devonian age. Permian age *Kashmiroblatta* genus is reported from Kashmir.

Labigrinthodont

The genus *Archegosaurus* occurs in the Gangamopteris Beds of Carboniferous- Permian age in Kashmir.

Lamellibranchia

Lamellibranch belongs to phylum Mollusca. It is a well-known class of marine fossils which having wide occurrences.

Ordovician: *Pterinea, Ctonodonta*

Silurian: *Palaeoneilo*

Carboniferous: *Conocordium, Aviculopecten, Myalina.*

Permian: *Aviculopecten, Burmesia, Goniopora, Phestia*

Triassic: *Daonella, Halobia, Lima, Megalodon, Pecten, Avicula, Mytilus, Lorbis.*

Jurassic: *Aucella, Ostrea, Nucula, Leda, Arca, Trigonia, Pleuromya.*

Cretaceous: *Cordium, Gryphaea, Tellina, Opis, Hippurites.*

Eocene: *Sinodia, Parinomya, Venericardia, Cardium.*

Oligocene: *Ampullina, Melania, Potamides, Bittium*

Miocene: *Chlamys, Dosinia, Barbatina, Lucina, Drillia, Pitar.*

Pliocene: *Lamellidons, Indonaia, Pisidium.*

Pleistocene: *Parreysia, Gastropata, Philalanka, Pupilla, Gulella, Eulota.*

Microbiota

These are microscopic fossils. These tiny fossils belong to any phylum of plants and animals kingdom. Microbiota reported from the Lesser Himalaya are as follows: *Gunflint, Myxococcoides, Huroniospora, Eomycetopsis, Baltisphaeridium, Schimatispaeridium, Trachysphaeridium, Glenobotrydion, and Radiosutradialis.*

Other microbiota groups are

Acritarch: *Granomarginta, Vavasosphaeridium, Lithodinium.*

Cynophytes: *Satpulispora.*

Hystrichospherids: *Ascodium, Annedotubulata, Hystricho sphaeridium, Hystriochosphaera Gonyaulax, Pterocystidiopsis, Cyclonephedium, Tenua, Deflandrea.*

Hydrolithids (Protozoa): *Conotheca, Hyolithes, Holithes.*

Hyolithelminthids: *Hyolithellus, Psuedoothotheca, Spirellus*

Lapworthellids: *Lapworthella*

Poriferids: *Chancellonoids*

Hylitha: *Circotheca, Conotheca, Turcuttheca*

Nanofossil

These are microfossils of uncertain affinity and resemble with living coccolithophores. Following genera are reported from Krol Formation: *Tergestiella, Microhabdulus, Micula, Zygolithus, Tetralithus, Nannoconus, Disolithus, Mitrolithus*.

Oncolite

This microfossil group is reported from Krol Formation. The documented genera are *Vovatella, Ambigolamellatus, Vermiculites* and *Vericularties*.

Oncoids (Catagraph)

These are unattached organic calcareous concentric structure formed by the action of bacteria of blue green algae and move freely on the floor of the basins. Following genera are reported from Krol Formation *Volvatella, Ambigolamellatus* and *Vescicularites*.

Ostrocoda

Ostrocodes are small sized, lentil shaped bivalve animals living in all type of aquatic environment. Ostracoda sub-class belongs to class crustacea of phylum arthropoda. They are reported from many geological horizons of Himalaya.

Cambrian: *Olivooides, Hertzina, Loleoloides*.

Triassic: *Cyatherella, Healdia, Hungarella, Ledahia, Ogmoconcha*.

Palaeocene/Eocene: *Bairdia, Bathocypris, Hermanites, Cythirella, Botryococeus, Pediastrum, Paracypris, Ilyocypis*.

Pliocene: Pleistocene – *Dolerocypris, Eucypris, Cypretta, Candona, Darwinella* from Karewa Formation of Kashmir.

Pleistocene: *Cypris, Potamocypris, Strandesia, Candonopsis, imnocythese*.

Pteropoda

Pteropoda belong to class gastropod of phylum mollusca. They are minute pelagic, swimming animals and form calcareous ooze on ocean floor. Their established geological range is Cretaceous to Recent.

Sponge

They are simple highly porous soft body animals belong to phylum porifera. Their skeleton is composed of organic fibers or crystalline spicules Genus *Hegelian* and *Dictyospongia* are reported from Himalaya.

Trilobita

Trilobita is well-known marine index fossils of Palaeozoic age. Trilobita occurs in Krol Fm and in the Tethyans of Kashmir and Himachal Pradesh. Impressions of trilobita are reported from the Lesser Himalaya.

Cambrian– *Tonkinella, Anomocare, Conocorphe, Redlichia, Microdiscus, Dryctocephalus, Ptychoparia, Olenus, Chuangia. Pagetia, Triagnostus, Diplagnoyuas, Opsidiscus, Annamitia, Shahapsis, Hundwarella, Iranolessia* from the Middle Cambrian Puhru Group of Northwestern Kashmir (Parcha, 2005)

Ordovician: *Asaphys, Cheirusus, Prosobiscus, Calymene, Illaenus.*

Silurian: *Phacops, Acidaspis, Flexicalymene.*

Devonian: *Phacops.*

Carboniferous: *Phillipsia, Euphemus, Capulus.*

Vertebrates of Siwalik

Siwalik Supergroup of Neogene age is richly fossiliferous in some regions like Potwar plateau in Pakistan, Jammu, Pinjor (Chandigarh), Hardwar in India and in some parts of Nepal (Nanda, 1982, Corvinus, 1993) and Arunachal Pradesh. Besides vertebrate fossils large number invertebrate and plants fossils are reported from this group. However, Siwaliks are known for their vertebrate fossils. The important vertebrate groups and genera described from the Siwaliks are-

Primates: *Simia, Papio, Seminopithecus, Palaeopithecus, Dryopithecus, Ramapithecus, Sugrivapithecus, Cercopithecus, Macacus, Sivapithecus, Indraloris, Bramapithecus, Palaeosimia.*

Carnivores: *Hyaenarctos, Mellivora, Mustella, Vulpes, Crocuta, Felis, Viverra, Indarctos, Lutra, Amphicyon, Dissoprolis, Palhyaena, Vishnufelis, Martes, Canis, Sinictis.*

Elephants: *Mastodon, Stegodon, Elephas, Dinotherium, Prostegodon.*

Ungulates: *Rhinoceras, Equus, Sus, Hippopotamus, Camelus, Giraffa, Indivatherium, Buffelus, Bison, Bos, Hemibos, Moschus, Leptobos, Teleoceras, Aceratherium, Hipparion, Vishnutherium, Tragocerus, Hylotherium, Conohyus, Anthracotherium, Dorcabune, Sanitherium, Listriodon, Telmatodon, Progiraffa.*

Rodents: *Hystrix, Rhizomys, Nesokia*

Birds: *Phalacrocorax, Pelecanus, Struthio, Mergus.*

Reptiles: *Crocodylus, Gavialis, Rhamphosuchus* (Crocodiles), *Varanus* (Lizard)

Turtles

Snake: *Pythons*

Fishes: *Ophiocephalus, Chrysichthys, Rita, Arius.*

Pigs: *Palaeochoesus, Conohyus, Listriodon, Lophochoerus, Propotamochoerus, Dicoryphochoerus*

Murids (rats and mice): They originated 13.75 Ma ago in the Siwaliks.

Antemus, Progonomys, Mus, Karnimata, Parapelomys, Dilatomys, Millardia, Bandicota, Tatera, Nesokia

Rhinocerotids*: Paraceratherine* reported from Chakwal of Pakistan

Vertebrates: Non-Siwalik

Non-Siwalik vertebrate fossils are reported from Ladakh (Nanda and Sahni, 1990, Prasad et al., 2005) and Tethyan sequence. *Amblypterus* fish's teeth, scales, fins and mandible and hind limbs are reported from Kashmir. Fish, crocodiles, mammals, reptelia, turtle, marine mammals fossils are reported from Subathu Formation. Early Miocene mammals namely *Trilophodon, Dorcatherium* are reported from Pakistan. Fatehganj Zone of Kalakot area shows presence of *Kalakotia, Aulaxolophus, Paratritemnoden. Indohyus and Bunodentus* Lukundal Formation along Bagmati River in Kathmandu shows presence of remains of tooth of elephas, pig, *Artiodactyla,* pieces of horns, metacarpals of bovid etc. Cricetid rodent mammalia (*Democricetodon*) is reported from Ladakh Molasse of Early to Middle Miocene age.

Whales (Cetaceans) evolved, first time, in the Subathu and Kuldana formations about 53.5 Ma ago (*Himalayacetus subathuoensis* and *Diacodexis pakistanensis*) in landmass condition, *Pakicetus, Ambulocetus* and *Ichthyolestes* early species of fluvial origin were evolved around 49-48 ma later whales adopted marine habitat (Sahni, 1999).

Plant Fossils

Following types of plant fossils are reported from the Himalaya.

Stromatolites

Filamentous algae and cyanobacteria form laminated calcareous structure are described as stromatolites. The earliest reported stromatolite is 3800 Ma old, however their peak development has taken place in Riphean (Late Proterozoic) and they declined since the beginning of Palaeozoic. Some stromatolite like *Conophyton* is an index fossil. Columnar stromatolites are used for biostratigraphic horizonation. Himalayan stromatolites are reported from Jammu, Dharamkot, Aut (Kulu), Shali (Shimla), Deoban, Tundu Pathar, Sataun, Gangolihat, Kapkot, Krol and Baxa in Darjeeling and Dhading in Nepal.

Forms - *Colonfiella, Conophyton, Cylindricus, Baicalia baicalica, Mastoviella columnanis, Kussiella kussiensis, Omachtenia granensis, Platella talwarensis, Anabaria radiali, Thyssagetes, Parallel ophyton, Paniscallenia, Orthid, Parmites, Jurusania, Irregularia, Gymnosolen, Poludia, Tungussia, Ilicta, Minjaria, Septentrionalis, Rahaella, Yugmaphyton, Callumnoefacta* are present in the Lesser Himalaya. The Lesser Himalayan stromatolites show change in microstructure from streaky or banded form in Mesoproterozoic age to vermiform in Neoproterozoic to microstromatolites in the end of Neoproterozoic and patchy or grumous in the Early Cambrian.

Microstromatolites are calcareous algae. The recorded forms are *Renaleis, Ephiphyton* and *Oleckmia*.

Microbialite formed by benthic microbial community are reported Krol Belt sequence.

Pollens and Spores

Study of fossil pollens is called palynology. Pollens are reproductive part of plants present in the flower. Being microscopic in nature they are distributed widely and extensively used in palaeoenvironmental studies and correlation among formations. Pollens are reported from the Trans, Suture Zone, Tethyan, Lesser- and Outer Himalaya (Sharma, 1993).

Forms: *Trilete, Zonate, Monolete, Monosaccte, Non-Striate, Disaccate, Striale Disaccate, Monocalpale, Insertalsedis*.

Genus: *Sphaerochara, Tectochara, Chara, Couperipallis, Lakiapolis, Polycolpites, Triorites.*

Parasaccites, Plicatipollenites, Barakarites, Scheuringipollenites, Platysaccus, Striatites, Faunipollennites, Verrucosisporites from Manjir Fm of the Himachal Pradesh

Tricalpopollenites, Aquilapopollenites, Classopolis, Monololpotes, Acanthio-triletes from Tal Formation. *Ephedripites, Striainaperfarites, Scabratriporties, Pinuspollenites, Tetracolporites* from Eocene Dharamsala Formation.

Dalbuing Formation of Eocene Age in Arunachal Pradesh revealed wide varieties of pollens a few of them are. *Cyathidites, Stereisporites, Dandotiaspora* (Pteriophytic) *Parastriopollenites, Apiespollenite* (Gymnosperm) *Inaperturopollenites, Lilacidites, Spinainaperturites* (Angiosperm) *Mirchystridium, Hysteichosphyeridium* (Phytoplanktons), *Protoeacidites, Myricipites, Triorites* (Triporate).

From Palaeogene rocks of Cherra Formation: *Biretisporite, Palmaepollenites, Schizosporis, Retialetes, Polycalpites, Rhoipites, Nyssapollenites, Dandiotiaspora.*

Siwalik: *Cyathicidites, Alsophilidites, Todisporites, Leptolepidites, Klukisporites, Cedripites*.

Karewa: *Riccciaceae, Pinaceae, Batulaceae, Caprifoliaceae, Faguceae, Anacardiaceae*.

Chhongtash Formation: *Parasaccites, Pluicatipollenites, Divarisaccus, Caheniasaccites, Faunipollenites, Striatites* and *Ginkgocycadophytus*.

Leaf Impressions

Leaf impressions are mostly reported from the Himalayan Gondwana formations. Besides Gondwana, impressions are also reported from the Outer, Lesser and Tethyan sequences.

Ordovician: *Bythorphis.*

Devonian: *Psilophyton*

Carboniferous: *Rhacopteris, Sphenopteris*

Permian: *Schizoneura, Vertebraria, Glassopteris, Rotundocarpus, Cordaicarpus, Trizygia, Phyllotheca, Dictopteridium, Gangamopteris* from Arunachal Pradesh.

Jurassic: *Raphaelia, Piazopteris, Acrostichopteris, Taeniopteris Nilssonia, Cycadites, Pterophyllum* from the Ladakh and Nepal and *Elatocladus, Araucasioxylon* from the Bhutan.

Eocene: *Apocynophyllum, Gremiopsis, Sophera* from Arunachal Pradesh and also *Dipterocarpoxylon, Cynometroxylon.*

Siwalik: *Zizyphus, Dicotylophyllum, Ficus, Pinus* (Arunachal Pradesh).

Plio-Pleistocene: *Pinus, Cedrus, Picea, Abies, Betula, Caspinus, Corylus, Aluns, Juglans, Plantago, Ephedra, Salix, Maoutia, Ulmus, Aceracaesium, Aesculus, Berberis, Buxus, Pyrus* from Karewa. *Terminalia* from Upper Siwalik of Arunachal Pradesh

Fossil Wood: *Prunus* of Rosaceae family of Mio-Pliocene age is recorded from Ladakh (Guleria et al, 1983) *Gymnosperm* wood of Upper Devonian age from Bhikia Sain Phyllite.

Edicaran Fossils

Edicaran is upper most time horizon of Proterozoic era. Edicaran biota comprises of soft-bodied metazoans, which are represented by primitive coelenterates characterized by discoid and framboidal forms with few arthropods, echinoderms etc. and are present in horizontal to sub horizontal burrows. Fossils of this time horizon are reported from Krol-Tal of Nainital. The reported assemblage comprises of *Tiragina, Medusinite, Beltanella, Beltanelliformis, Pteridinium, Charinodiscus, Nimbia.*

Trace Fossils

Ichnofossils

These are also known as bioturbation structures developed by activities of various organisms such as resting, crawling, grazing, feeding and dwelling. Some of these structures are similar to tracks, trails or footprints. With these imprints, type or kind of life existing at the time of sediments deposition can be interpreted. Following ichno fossils are reported from Himalaya, *Skoliths, Planolites* from Miri Formation of Arunachal Pradesh. *Bergaueria, Cruziana, Didymaulichnus, Monomorphichnus, Neonereites, Palaeophycus, Plagiogmus, Planolites, Rusophycus, Skolithos, Treptichnus* of Early Cambrian (Upper Meishucunian to Lower Qiongzhusian*)* and *Skoliths* and *Aulichnites* impressions of colonial worm are reported from Tal Formation (Mathur and Srivastava, 2005). *Chondrites, Thalassinoides* have been reported from the Permotriassic Kalapani Limestone. *Planolites, Zoophycos, Frlanolites, Taphrhelminthopsis Monomorphichnus, Diplichnites* are the impressions of trilobita, coelenterata, mollusca and annelid. *Bergaueria, Cruziana, Chondrites, Diplichnites, Dimorphichnus, Planolites, Phycodes, Rusophycus, Taphrhelminthopsis* are reported from the early Cambrian Kunzum- La Formation of Spiti

(Parcha et al. 2004). Tangri et al. (2003) reported *Arenicolites, Didymaulichnus, Gordia, Helmithopsis, Monomorhichnus, Phycodes, and Planolites* of Late Precambrian-Middle Cambrian Tethyan Pele La Group of Bhutan.

Mould and Cast

Mould is the external impression of a fossil while cast is the filling in a cavity formed by solution activities. Moulds and casts of animals and plants fossils are described form several locations of the Himalaya for example tubes and burrows of worms are reported from Kashmir, cast of *Rusophycus* (trilobita) from Tal Formation.

6.3 Biozonation

Detailed biozonation on the basis biostratigraphy of fossiliferous formations are worked out in the several Himalayan regions of the Outer, Lesser, Tethyan Himalaya and of the suture zones.

On the basis of larger foraminifer, mollusca and bryozoans fossils assemblages' eight biozonations on regional scale of the Kakara- Subathu formations of Jammu, Shimla and Garhwal are reported by Mathur and Juyal (2000). These zones are as follows:

Diplocava nilkanthi Zone (Zone I)

Daviesina tenuis Lockhartia conditi Zone (Zone II)

Nummulites burdigalesis – N. subramondi Zone (Zone III)

Cordiopsis subathooensis- Turritella subathooensis Zone (Zone IV)

Nummulites rotularius- Steginoporella Zone (Zone V)

Assilina laxispira- A. placentulas grande Zone VI

Assilina spira abrardi Zone (Zone)

Muscles nuttalli- Parinomya blanfordiana Zone (Zone VII)

Zone I and the lower part of the Zone II belongs to Kakara Formation (Maastrichtian-Thanentian) and the upper part of Zone II to Zone VIII to the Subathu Formation (Ypresian-Early Lutetian). The biostratigraphic zonation study reveals not only precise age determination of the formation(s) but also assessment of marine and non-marine condition, tectonic and geomorphic setup of the region and palaeoecology and environmental conditions of the region.

6.4 Palaeontology in Tectonic Interpretations

Palaeontological findings in conjuncture with stratigraphy have been widely employed in tectonic interpretation in the Himalaya. Some fossils are very sensitive to change in depth, environment of deposition and of climate.most importantly their distribution reveals paleogeographic conditions of the region at the time of their burial. Fossils have been used to workout pre breakup position of present continents, timing of initiation of breaking process of mini cluster of continents as well as individual continent, palaeogeograhic locations of continents, nature and age of growing chasm between continents, cimatic belts traversed by moving continents, interface of one moving continent with other continents and subsequents relationship between interfacing continents. Time and duration of each event can be precisely constrained. Disappearing bridges confines the geographical distribution flora and fauna while evolving land bridges have paramount importance in migration and dispersal of flora and fauna. On the basis of palaeontological data, timing of plate break up and southward regression of Tethys Sea have also been sucesfully worked out in the Himalaya. Mammalian fossils in sedimentary Kakara and Subathu formations of India and Pakistan reveals transition from neritic to shallow marine to intertidal to lagoonal and finally fluvial setup describe the events related with closure of Tethys and initiation of evolution of Himalaya. The Lower Tertiary Indus Formation consists of Sumdha Gompa, Nummilitic, Jurutze and Gongmarula members (ascending order) Mathur & Juyal (1997. Late Ypresian Nummulitic Member contains marine fossils like *Nummulites, Assilina, Lockhartia* and other larger forams. Followed by other fossils like *Turritela,* Turritellids, *Diplodonta*, oysters, *Seila* of shallow marine to fresh water conditions. These assemblages suggest closure of Tethys in the beginning of early Late Ypresian from the Indus Suture Zone. The seawater regressed to south in Zanskar region in the late Ypresian; the topmost Kong Formation contains derived marine fossils like *Nummulites, Assilina, Ranikothalia, Miscellanea, Daviesina, Glomalveolina*, miliolids, rotalids. The Kong Formation is assigned post early Cusian (upper part of Ypresian) on the stratigraphic ground. After this no marine sediments are recorded from the Zanskar region. In Himachal Pradesh and Uttaranchal, Kakara and Subathu formations were deposited in a southwardly transgressive sea during the latest Maastrichtian and continue up to early Lutetian time. In top Subathu sequence brackish to fresh water fossils like *Seila, Physa, Bulinus*, oyster, *Neocyprideis* and *Ilyocipris* are found followed upward by the deposition of fresh to brackish water Dagshai and Kasauli formations that marks the complete withdrawal of the Tethys sea due to progressive upliftment of the area. The relatively thin cover of the Kakara and Subathu formations on the interior region of Lesser and Higher Himalaya would have been eroded away (Mathur, 1997). Palaeontology has been elegantly employed for working out the time of emplacement of ophiolitic melange of Ladakh in the post-early late Ypresian and Indian – Asia collision took place in early late Ypresian times. A continuous deep marine condition during Late Cretaceous: early middle Eocene, on the basis of marine fossils was identified in the Indus Suture and Zanskar Tethyan zones.

Fossil flora and fauna have been applied to plate movement, in Jurassic time the position of Indian landmass was such that plant and animal migration was readily possible across the Atlantic into the Gondwanaland. Before breakup of Gondwanaland in Kimmeridgian-Tithonian (Late Jurassic) two phytogeoprovinces namely *Cerebropollenites* in north and *Microcachryidites* in south comprising of Australia, Africa, Antarctica, South America and India were prevailing. The southern continent started breaking up during Tithonian-Neocomian and India started separating from Australia/Antarctica in Tithonian. During Early Cretaceous *Microcachryidites* flora prevailed and in Middle Cretaceous *Asteropollis* appeared while in Late Cretaceous Constantinisporis flora developed. Proxapertites and *Spinizonocolpites* along with triprojectate pollens e.g. *Aquilapollenites, Scollardia, Pulcheripollis* and *Cranwellia* became dominant during Maastrichtian-Palaeocene. The tropical Malaysian flora arrived in India during Oligocene- Miocene when Indian plate welded with Asian plate. Arcto-Tertiary flora became dominant in Miocene when land connection established with north after collision. The fossil pollens recorded from the Siwalik belong to Pinaceae, Ericaceae, Fagaceae, Rosaceae, Juglandaceae and Asteraceae families. After this new species in may genera developed in post-Miocene time in Himalaya. Land connections between different continents of Gondwanaland in different period of times have also been identified with the help plant fossils. Similarly palaeoclimate was also inferred during different geological periods e.g. in Early Permian to Late Permian cool climate was prevailing, in Early Triassic India was located between 23°S to 66°S latitude and climate was warm temperate that prolonged up to Lower Cretaceous (20°S to 30°S), in Campanian-Maastrichtian southern India was warm, central India arid and northern India experienced tropical climate. In Palaeocene India was situated in 0° to 23°S, the southern part experienced arid or warmer than today's climate During Early Eocene to Oligocene tropical climate, in Miocene to Pliocene time aridity was increased. Temperate dicotyledonous taxa replaced mixed deciduous forest along the Himalaya particularly eastward because of increase in aridity and temperature extremes (Mehrotra, Venkatachala and Kapoor, 2005).

6.5 Fossils in Palaeoecology and Palaeoclimate of Himalaya

Juyal (1997) worked out palaeoecology and palaeoclimate of Kakara-Subathu formations on the basis of rich fossils assemblages. Six faunal communities from bottom to top have been identified in these formations namely Maastrichtian *Diplocava* community (Late Maastrichtian) developed at the base Kakara Formation consisting of echinoid spines, calcareous algae and moluscan shells indicates normal salinity and clear shallow water conditions. This was followed by *Daviesina- Lockhartia* community (Thanentian) having presence of *Rotalia*, echinoid spines, *Venericardia* and miliolids developed in reducing environment at a depth range of 72-90 m or outer neritic depth, upper part of Kakara Formation. *Nummulites-Assilina* community developed in Subathu Formation at shallow part of neritic depth or sub tidal zone of tropical to subtropical waters; *Cordiopsis* community comprises *Bicorbula, Venericardia, Nautilus,* Turritellids (molluscs) and *Dentokrithe, Alocopocythere* and *Cushmanidea* (ostrocodes) developed in warm to temperate shallow (<60 m depth) condition or low intertidal zone; oyster community developed in high intertidal zone and lastly *Seila* community developed in fresh to brackish water condition in the different zones of Subathu Formations in the Ypresian to early Lutetian times.

DIMENSIONS OF HIMALAYAN GEOLOGY
A. K. Biyani (2006) : pp.177-201
Satish Serial Publishing House, New Delhi
E-mail: ssp@ndf.vsnl.net.in, hkjain1975@yahoo.com; website: www.satishserial.com

Prakrtistho hi bhunkte prakrtijangunan *(Geeta, 13.21)*
Nature is said to be the cause of all material activities and effects

CHAPTER - 7

PETROLOGY

Like any other orogenic belt, Himalaya do displays wide diversity covering the entire spectrum in its petrological make up, for example unconsolidated/semi consolidated sediments of recent to sub-recent origin to huge batholiths of several hundred-kilometers length to paired metamorphic belt. This section deals with salient characters of the three conventional groups of rocks namely igneous, sedimentary and metamorphic. Each group of rocks has its own significance in shaping the Himalaya and more importantly in tracing the evolutionary history of this young mountain range which embodying rocks as old as Late Archaean to as young as Recent sediments.

7.1 Igneous Rocks in Himalaya

Igneous rocks occur in all tectonic divisions. These rocks occupy fairly large area particularly in the Lesser, Higher and suture zones Himalaya and show significant diversity in their petrological compositions. However, granite and basalt dominate over other rock types as they are not only have wide aerial coverage but also formed in several episodes of magmatism. During Himalayan orogeny light colored leucogranites are formed mostly between 24-19 Ma duration as a result of anatexis in deep-seated condition.

7.1.1 Granites

Granites are usually referred by non-specific term granitoid, because granites display considerable variations in mineralogical composition (Pitcher, 1993). Granites occur in different forms and shapes, in different tectonic set up and are products of several magmatic episodes. All these factors contributed towards detailed study of granites in the Himalaya.

Sharma (1983) identified nine linear belts of granites in Trans Himalaya and Himalaya, namely the:

1. The northern most Karakoram belt of the Trans Himalayan region of 150 by 20 km dimension and granites are of 25 ± 2, 35 ± 15, and 209-216 Ma old.
2. Ladakh-Deosai belt is of 600 by 20-80 km dimension showing up to 5 km exposure thickness and is associated many differentiated phases of granites.

3. Granitoids of Tethyan Himalayan belt of 600 by 50 km dimension occur as independent intrusive masses situated south of suture zone e.g. Jaspa Kang Granite of 495±16 Ma old.
4. Vaikrita Thrust sheet granite belt of Higher Himalaya, these granites formed due to crustal anatexis during Himalayan orogeny. The younger granites are two mica tourmaline granite of 29-19 Ma age while pre-orogenic older granites are biotite granite formed in 675 to 495 Ma duration.
5. Jutogh-Manali thrust sheet granite belt of Higher Himalaya, granites are formed 1890-2336±193 Ma duration e.g. Chaur and Dalhousie granites.
6. Chail thrust sheet granite belt of Higher Himalaya are mylonitic in nature and petrogenetic dates vary between 1430 to 1170 Ma duration; a second phase of granitic activity characterized by two-mica and porphyritic occurred between 640 to 486 Ma duration e.g. Kaplas, Debguru, Manshera and Narayanthan granites.
7. The remaining three belts occur in the Lesser Himalaya with different country rocks associations and tectonic settings; one with slate, graywacke and siltstone e.g. Kulu, and Ramban granites, second associated with arenaceous rocks e.g. Amritpur Granite of district Nainital and third belt is associated with calcareous rocks and are situated near MBT.

Valdiya (1983) correlated characters of granite with epi, meso and kata grades of metamorphism. Rao (1983) used kinematic basis (pre-, syn- and post orogenic) for classification as each kinematic setting has its own distinguishing characters. Thakur (1983 b) recorded three phases of granites of pre-Himalayan granite magmatism (1800-2000, 1200-1400 and 500 Ma) and two Himalayan phases namely early Eocene and Late or post Miocene granites. The number of magmatic events changes substantially if later geochronological data are incorporated in analysis.

In Pakistan, Sham (1983 a) identified six well-defined belts of concordant granite namely the Nanga Parbat, Nauseri, Lahor in Kohistan, Lower Swat, Malakand and Ambela Granite Complex. Mitchell et al. in Shams (1993) reported eight types of setting for granites in Nepal namely the intrusive granites in the Lesser Himalaya, south of the MCT augen gneissic granites, klippen granites in Tethyan, calc-alkaline granodiorite plutons and granitoid of ring complexes of southern Tibet. Powar (1983) noted three periods of granite emplacement in eastern Kumaun namely the granitoids of Almora Nappe, Krol parautouchthon and Almora klippen and Central Crystalline Zone granites.

7.1.2 Important Granitic Bodies

Himalayan granites show huge variation in size, shape, thickness, and mode of emplacement. Sub-types of granites are identified on the basis of difference in chemical composition and in the mineralogy (Odell,1983). Same of these granites are also differentiated and are associated with economic minerals.

Ladakh Batholithic granite is the largest granite of Himalaya. It is 600 km long and has maximum width of 80 km and having more than 5 km exposed thickness. It is situated between Indus and Shyok Sutures. It is an arc like body having convexity towards north. In Ladakh, five magmatic cycles are recognized in this complex. The cycle was initiated through eruption of Shyok Volcanics consisting of basalt and andesite followed by Khardung Volcanics, non-marine lava flow, and emplacement of Ladakh Granite on regional level and finally porphyritic andesite. Ladakh Granite is dominated by granodiorite and quartz diorite. In Upshi and Chumathong areas granites occur in three varieties namely the quartz monzodiorite and granodiorite, leucogranite and porphyritic granite. It is also associated with ultrapotassic melanocratic minette dyke of post collision origin (Ravikant and Guha, 2002). Evidences of copper, lead, zinc, molybdenum and magnetite mineralizations are present in it.

Nanga Parbat-Haramosh massif is a huge diapir. This diapir is a part of Karakoram Batholith of 600x30 km dimension. It is situated north of Shyok Suture. It is 150x20 km size plutonic body trends in NW-SE direction. It is predominantly consisting of granite, tonalite and granodiorite and has minor presence of dykes of aplite and pegmatite. Large negative anomaly of over 500 milligal indicates its huge thickness. Kohistan Batholith exposed in the west of Nanga Parbat in the northwest Pakistan occupies nearly half of the Kohistan area (36000 sq km). It is made up of tonalite, granodiorite, diorite, aplite and pegmatite and is formed in five magmatic phases spanning between 102-24 Ma duration (Shams, 1983).

Outside of suture zones, granites occur as small to medium size batholiths and plutons. Leucogranites occupy the highest peaks of Higher Himalaya from Nanga Parbat in the west to Kedarnath, Gangotri, Badrinath, Manaslu, Shisha Pangma, Everest (Powar, 1983; Pognante and Benna, 1993), Makalu and in the high peaks (7000 m) in the north Bhutan. Upper Jurassic Tangmarg Granite of Kashmir was described as A- type granite (Rao et al., 1995). The Ordovician (489±20 Ma) Hant Granite of Baramula area of the Kashmir Himalaya occupies nearly 50 sq km in the north of Wular Lake. Granite is intruded in Salkhala and Dogra slates and shows peraluminous, S-type nature (Rao et al., 1990). The 1895 ± 54 Ma old Wangtu Granite in Himachal Pradesh occurs in two forms namely coarse porphyritic granite and fine grained granite. The crystallization age of the Chor granitoid is 823±5 Ma. These varieties are formed by fractional crystallization process. The coarse grained is of earlier origin and fine grained is of later origin (Singh, 2003). Sayanachatti mylonitic granite of Yamuna Valley, Uttaranchal is 6 km long body formed 1972 ±102 Ma ago (Singh et al., 1986); granite is mylonitised during Himalayan orogenic movement in ductile condition (more than 15 km depth) and later transported along with crystalline nappe to present position. It is formed by crustal anatexis process (Biyani, 1999). Six granite bodies namely Dadeldhura granite, Simchar (Le Fort et al., 1983), Palung, Ipa, Narayanthan, Sindhuli and Udayapur occur in Lesser Himalaya of Nepal. Dadeldhura granite occupies an area 1400 sq km and is extension of the Champawat Granodiorite of Kumaun and exhibit gneissic characters. Biotite, muscovite and tourmaline varieties of granites are recognized in the Dadeldhura Complex. These 470 ± 5.6 Ma old granites show S-type characters. In the Higher Himalaya of Nepal, granites are reported from Makalu, Manaslu, Mugu and Saipal areas (Kaphle, 1992; Einfalt et al., 1993). Zero Gneiss of Arunachal Pradesh batholithic size, shows two varieties of granite namely biotite granite (1914 ± 23 Ma) and tourmaline bearing leucogranite (1536 ± 60 Ma). Granite is widely distributed in the Arunachal Pradesh. Lohit Plutonic Complex is of batholithic dimension (250 km long and 100 km wide) composed of acidic-intermediate igneous rocks and high–grade metamorphic rocks. Diorite, tonalite, granodiorite, hornblende granite, leucogranite, tourmaline bearing pegmatite, biotite- garnet gneiss, banded gneiss, schist and marble are present in this complex (Misra and Singh, 2003).

In five areas of the Himalaya namely the Nanga Parbat, west of Badrinath, Kedarnath, Api, and Manaslu its continuation in Shisha Pangma, Everest and Makalu in northern Nepal and in southern Tibet and in Chomolhare and Kankapunzum and its northwest extension into Chung la, Chomolhare range, Gophu la and in eastern Lunana lake granites occur above 8000 m heights. Lunana granite occurs as laccolith.

7.1.3 Associated Rocks with Granites

At many place, granites show differentiation thereby producing many differentiated rocks. In West Pakistan alkali granite is associated with syenite, microgranite and carbonates, Ladakh granite is associated with tonalite (Pande and Rai, 1979), gabbro, metabasics, norite, hornblende pyroxene and nepheline syenite (Jan and Asif, 1983; Kempe, 1983).

7.1.4 Acid Volcanic Rocks

Acid volcanics are reported from several places e.g. andesite from Saltoro Range, rhyodacite ignimbrite along with gabbro, hornblendite peridotite and serpentinite in Shyok Volcanic Zone, Calc-alkaline volcanic rock and lava of dacite and andesite from Khardung, acid volcanics zone (Sharma and Gupta, 1983) and rhyolite, andesite and dacite from Hazara (Mohammad, 1983) and Kohistan Sequence.

7.1.5 Origin of Granites

Majority of the Himalayan granites are S-type formed due to anatexis of crustal material at deeper level inside the crust during Himalayan orogenesis i.e. in Tertiary (Biyani, 1998 b). Tangmarg in Kashmir Valley and the Trans Himalayan granites are I-type in nature.

7.1.6 Mineralizations in Granites

Mineralizations of metallic ores are reported from many locations but so far no workable deposits associated with Himalayan granites are reported. Dadeldhura Granite shows presence of tungsten; copper lead, fluorite in Ladakh Batholith and lead, zinc molybdenum and uranium occur in Lahor pegmatoid of northern Pakistan (Butt, 1983).

7.1.7 Basic Rocks

Basic rocks occur in the form of dykes, sills and in other intrusives forms and as lava flows in all tectonic divisions of the Himalaya. Major basic rocks occurrence have been studied in detailed particularly from geochemical and petrogenetic angles but individual minor dykes and sills have not received attention. In many instances, these rocks are metamorphosed to varying degree into chlorite schist and amphibolite. Geochronological database of basic rocks is very small and with the existing data no definite assessment on their time preference of emplacement or intrusions can be attempted; however, the oldest basic magmatic rock is of Mandi-Darla lava flows (2510 Ma) of Sundernagar Formation of Himachal Pradesh (Ahmad and Bhat, 1987).

In the Higher Himalayan Yamuna Valley, three large size basic sills occur at Wazri, Hanumanchatti and in the Paligad Valley and a lava flow (?) overlie the Lesser Himalayan footwall. All these rocks are quartz tholeiite type and magma for them was generated in the mantle of heterogeneous composition. The core part represents primitive basalt with Mg number varying from 49 to 73. The Garhwal Volcanics occur in Garhwal Group as lava flows and sills and dykes. These rocks contain clinopyroxene (augite) and plagioclase (andesine and labrodorite) as main phases and in minor amounts andesine, epidote, calcite, zeolite and quartz occur (Ahmad. and Tarney, 1991). Basic volcanics are also extensively developed in Bhimtal-Bhavali area of Kumaun. Magmas for these basic rocks were originated from non-pyrolitic heterogeneous upper mantle. The liquidus temperature of magmas were in the 1050°-1250°C range and before emplacement, magma underwent considerable fractionations. Spilite occur in the Adibadri area of the Garhwal (Srivastava, 1979). In Mandi-Darla volcanics of Himachal Pradesh plagioclase is andesine to oligoclase composition and pyroxene is mainly augite of aphyric nature and exhibits quenched texture suggesting rapid cooling of the lava (Taron, 1979). Basic volcanism has very limited distribution in the Lesser Himalaya of Nepal. However, in the Higher Nepal Himalaya, these rocks occur as minor amphibolitic bodies (e.g. Ganesh Himal) and perhaps associated with acid magmatism. Lesser Himalayan basic rocks occur in the Kuncha and Tansen Groups (Bashyal et al., 1989). Abor Volcanics of Arunachal Pradesh occur as highly jointed and fractured rocks, the lava flow contains microphenocrysts of augite and plagioclase (An_{56-42}). The volcanics are extensively altered.

These volcanics are emplaced in various tectonic setting for example Panjal Traps represents rift environment, Dras Volcanics in island arc, ultra basics in the back are basins and basic rocks across the melange zone in the mid oceanic ridge setting.

7.1.8 Ophiolites

Ophiolite represents the ancient oceanic floor that has come to the surface due to disruption in normal subduction process. This process of upthrusting is called obduction. Ophiolite is standard sequence comprising of peridotite, dunite, pyroxinite, serpentinite, gabbro, sheeted dykes, pillow lava and pelagic sedimentary rocks; if this standard sequence gets disrupted due to tectonic overhauling then intermixing of the rocks occur and this intermixed condition is described as ophiolitic mélange. Ophiolites and ophiolitic mélange occur in the Indus and Shyok suture zones. Melange in Indus suture occurs at Sumdo village and to the east of Kyon Tso and an ophiolite sequence spreading over 5 sq km occurs south of the Nidar village.

7.1.9 Ultrabasic Rocks

The ultrabasic rocks are mainly reported from suture zones. These rocks are supposed to be obducted slices of upper mantle and few of them have been subsequently serpentinized. They show autochthonous and allochthonous nature. The allochthonous occur as nappe and klippen. Some of their outcrops contain considerable area e.g. Jungbwa Nappe and Spontang klippen in Zanskar.

The important locations of ultramafic rocks are margin of Ladakh Granitoid (hornblende pyroxenite), Indus Suture Zone (pyroxinites and norite), Kohistan (pyroxinites and hypersthene diorite), Kargil (lamprophyres), Dras Volcanics (peridotite, dunite, hornblendite, hypersthene gabbro) Shyok Volcanics (peridotite and serpentinite), Chushul (diorite), Darjeeling (mica lamprophyre), Jungbwa in Nepal peridotite, nepheline syenite and alkaline gneisses in Ampipal and Gorkha in Kuncha Group of Nepal (Dhital, 1995) and metapyroxinite/peridotite and serpentinite in Khitarbari and Tenga formations in Arunachal Pradesh.

7.1.10 Carbonatites

Carbonatite dykes are reported from Koga and Jhambil in Swat, Tarbela and Khungi in Mardan and in Sillai Patti carbonatite complex of North Pakistan. Anastomosing dykes occur in metasedimentaries or at the contact of metasedimentaries and granite gneiss, and contain calcite, siderite, biotite, hornblende and magnesio-arfvedsonite. These dykes are situated between MBT and MCT and are 15 cm to 15 m wide (Butt et al., 1989).

7.2 Geochemistry of Igneous Rocks

At present majority of research contributions on hard rock petrology are centered on geochemistry. Recent advances in instrumentation for geochemical analysis like atomic absorption spectrophotometer, X-ray fluorescence, induced couple plasma, electron microprobe have made analysis of sample comparatively easy affair which was once highly tedious and time consuming job. Major, minor and trace elements and rare earth elements and microprobe analysis of igneous rocks are now widely employed for identification of minerals, in inferring prevailing conditions at the time of magma formation, evolution of magma, change in composition of magma during upward movement, surface and subsurface tectonic conditions at time of emplacement and in case of solidified rocks identification of minerals, mineral chemistry, classification of rocks, temperature and pressure condition at the time of emplacement and fractionation history during crystallization (Ahmad and Bhat, 1987).

Each of these aspects are rigorous evaluated and in turn geochemistry helps in solving stratigraphic and tectonic problems as well as also provide better understanding of the rocks.

Thousands of geochemical analysis of majority of igneous and metamorphic are available in journals and books and are being published regularly by the researchers and the professional organizations. Representative major elements analyses of selected rocks are given in the Table 7.1 and trace elements analysis is given in the Table 7.2.

Table 7.1. Major elements geochemical analysis of Himalayan igneous rocks

	1	2	3	4	5	6	7	8	9	10
SiO_2	57.93	75.05	50.89	47.13	50.04	38.97	60.35	49.45	58.93	43.96
TiO_2	1.02	0.67	1.00	0.57	1.64	2.06	0.78	1.03	0.67	3.77
Al_2O_3	18.97	11.07	14.75	13.69	14.39	13.66	12.14	15.1	16.2	10.66
Fe_2O_3	8.09	4.90	2.17	2.30	4.82	11.08	0.83	8.88	5.71	8.42
FeO										5.98

MnO	0.01	0.07	0.13	0.19	0.16	0.15	0.01	0.29	0.12	0.18
MgO	4.94	1.46	8.92	14.05	0.65	5.37	10.78	9.46	2.44	10.29
CaO	1.07	1.16	5.40	8.63	9.79	10.14	0.48	6.86	6.37	10.27
Na_2O	1.63	2.82	4.64	1.19	1.98	0.8	0.78	4.82	3.63	2.37
K_2O	4.86	1.75	1.14	0.49	0.62	5.58	9.94	0.25	2.06	0.86
P_2O_5	0.07	0.19	0.36	0.37	0.15	0.26	0.10	0.22	0.22	1.20
H_2O										2.40
Sum	98.68	99.14	98.15	97.22	84.24	88.07	96.19	96.36	96.35	99.80

	11	12	13	14	15	16	17	18	19	20
SiO_2	48.54	40.30	40.70	68.76	72.80	47.31	71.58	53.0	47.16	52.17
TiO_2	0.87	0.00	0.00	0.20	0.29	16.53	0.40	1.32	1.25	0.97
Al_2O_3	13.31	0.76	0.64	17.62	10.65		15.03	14.4	16.12	15.27
Fe_2O_3	2.90	1.63	1.55	0.20	4.75	13.48	1.26	13.09		8.6
FeO	7.45	13.46	12.71	0.30	-	0.00			14.31	
MnO	0.13				0.14	0.11	0.05	0.14	0.16	0.14
MgO	8.03	40.12	38.30	0.42	0.10	6.98	0.21	5.4	8.77	8.24
CaO	11.12	2.55	5.32	3.81	0.26	4.32	0.06	6.26	2.54	6.84
Na_2O	2.73	0.04	0.19	6.04	4.70	3.42	3.31	4.91	1.85	3.43
K_2O	0.31	0.01	0.00	0.25	0.02	1.54	5.10	0.70	1.26	1.92
P_2O_5	0.06	0.00	0.22	0.01	0.02	0.25	0.12	0.17	0.09	0.28
H_2O	2.31									
Sum	97.76	98.87	99.63	97.61	93.73	93.94	97.12	99.39	93.51	97.86

	21	22	23	24	25	26	27	28	29	30
SiO_2	54.35	46.47	49.7	68.2	70.11	69.79	56.09	68.38	72.63	50.96
TiO_2	1.40	2.64	1.4	0.4	0.53	0.34	1.07	00.53	0.29	2.50
Al_2O_3	13.72	15.28	13.6	14.1	13.57	11.25	17.48	16.02	13.83	10.21
Fe_2O_3			12.7	3.7	5.98		2.15	00.42	3.28	
FeO	11.15	13.54				6.34	3.78	02.94		17.94
MnO	0.15	0.19	0.2	0.1	0.11	0.36	0.06	00.10	0.03	0.14
MgO	11.83	8.48	7.0	1.7	0.98	1.63	4.43	01.05	0.82	3.74
CaO	7.15	8.87	11.1	2.6	1.32	3.64	4.78	00.94	1.25	7.41
Na_2O	0.02	1.46	2.0	2.4	2.69	2.60	4.66	02.72	2.57	0.99
K_2O	0.05	0.65	0.4	4.7	4.34	0.12	3.80	04.48	5.05	1.15
P_2O_5	0.18	0.43	0.1	0.2	0.24	0.13	0.32	00.09	0.25	0.17
H_2O							0.49	02.13		
Sum	100.0	98.01	98.2	98.1	99.87	96.2	99.11	99.8	100	95.21

	31	32	33	34	35	36	37	38	39	40
SiO_2	50.55	65.24	66.66	70.18	54.23	67.31	54.72	74.30	53.94	19.12
TiO_2	2.68	0.002	0.51	0.30	0.83	0.37	0.79	0.12	0.90	0.74
Al_2O_3	10.33	17.926	14.68	17.99	15.55	15.29	16.39	13.43	14.43	7.32
Fe_2O_3			4.62	0.05		3.77		2.06	9.90	5.91
FeO	18.21	0.073		1.08	8.98		9.28			4.0
MnO	0.15		0.07	0.09	0.14	0.09	0.14	0.04	0.13	0.34

MgO	4.09	0.021	0.91	1.27	8.08	1.59	6.73	0.15	4.42	3.44
CaO	7.22	0.018	2.31	0.10	5.45	2.41	5.97	0.48	8.52	25.50
Na_2O	1.62	0.189	3.10	4.11	3.96	4.29	4.99	2.87	2.85	1.05
K_2O	0.60	14.846	5.68	4.19	2.16	4.67	0.45	4.91	1.60	3.85
P_2O_5	0.12	0.002	0.16	0.09	0.16	0.21	0.03	0.18	0.19	
CO_2										21.99
Sum	95.57	98.317	98.7	99.45	99.54	100.0	99.49	98.54	96.88	93.26

1,2- Rohtang 3,4- Volcanics of Rohtang area, 5- Siwalik Vulcanite, 6- Dowar Khola, 7- Khewra Gorge, 8- Dir Metavolcanic Andesite, 9- Basaltic Andesite, 10-Shewa Shabaz Garhi Meta Gabbro, 11- Mahak Amphibolite, 12-Toratiggu Dunite, 13-Toratiggu Peridotite, 14- Toratiggu Plagio-granite, 15– A- granite in Warasak Khyber area, Pakistan 16- Dargan Ophiolite 17- Hant Granite, 18-Garhwal Volcanics, 19-Shyok Volcanics, 20- Shyok Volcanics Sati Bridge, 21, 22 -Abor Volcanics, 23 - Sainj Metabasalt, 24-Kistwar Granite, 25-Tangmarg Granite (A-Type), Kashmir, 26- E - Ladakh, Nidar Ophiolite, 27- Ladakh Batholith Chumatang, 28- Mandi Granite, 29- Sayanachatti Mylonitic Granite (Uttarkashi), 30- Hanumanchatti Metabasalt, 31- Wazri Metabasalt, 32- Amritpur Granite, 33- Wagntu Granitoid H.P., 34- Doda Granitoid J&K (Guha and Gupta, 1988), 35- Bhimtal Volcanics, 36- Jaspa Granite, 37- Mandi Darla Volcanics, 38- Dadeldhura Granite, Nepal, 39- Jutogh Amphibolite W. Himalaya, 40- Sillaipatti Carbonatite.

Table 7.2. Trace elements analysis of some rock samples of the Himalaya (in ppm)

	1	2	3	4	5
Ni	72	80	77	84	65
Ba	172	310	nd	nd	124.6
Rb	121	32	42	62	48
Sr	1228	772	90	248	553
Y	49	48	23	22	23
Zr	647	117	83	154	129
Nb	200	57	5	8	6
V	175	107	-	-	-
Cr	91	450	156	209	252
Sc	nd	12	-	-	-

1- Volcanics around Shergol village, Ladakh 2- metavolcanics(blue-schists), Baltikhar village, Ladakh, 3- Shukur village, Karakoram, 4- Sati bridge, eastern Karakoram, 5- Tegar village, eastern Karakoram.

7.3 Metamorphic Rocks

Metamorphism is the response of imposition of temperature, pressure and fluids on the preexisting rocks. These agents of metamorphism usually act combinedly but they may also act also individually and finally they change the texture and mineral composition of subjected rocks. Nearly 40 percent surface area of Himalaya is covered with metamorphic rocks and interestingly it is largest inverted metamorphic region of the world. Isograds are mapped in metamorphic terrains (Bhattacharaya, 1981).

7.3.1 Distribution of Metamorphic rocks in Himalaya

The entire Higher Himalaya, barring some locations occupied by sedimentary klippen, is a metamorphic terrain. The Lesser Himalaya is extensively invaded by Higher Himalayan crystalline nappes in many regions and many of the older formations of this division are mildly metamorphosed but major part of it is of non-metamorphic rocks. The basal formations of the Tethyan Himalaya are mildly metamorphic in nature but impact of metamorphism is not uniform everywhere. The suture zones are partly metamorphic and partly non-metamorphic. The Outer Himalaya is exclusively non-metamorphic in nature.

7.3.2 Types of Himalayan Metamorphism

Several types of metamorphism like regional, static, contact, cataclastic etc. are observed on the basis of relative influence metamorphic agents (Winkler, 1976). In Himalaya, such types of metamorphism are recognized in different regions. Most dominating and extensively distributed is regional metamorphism accompanied with deforming processes (Windley, 1983; Barnicoat and Treloar, 1989). Next in importance is cataclastic metamorphism; it is localized along major shear planes like MCT and MBT, Tethyan Thrust and also along thrusts and faults of lesser dimensions. All along the MCT, a strip of 8-15 km width of dynamic or cataclastic metamorphism occurs in which rocks are thoroughly mylonitised. In MBT it is just 5-200 m wide zone of brittle to brittle–ductile deformations. Retrograde metamorphism is recorded as waning event in most metamorphic rocks. Plutonic or deep-seated metamorphism is reported from suture zones. Contact metamorphism caused by igneous intrusions is also recorded in many locations but to great extent its characters are camouflaged by imposed regional metamorphism. High-pressure blue schist facies metamorphism is recognized in the 12 km long and 1 km wide Shergol ophiolitic mélange of the Indus Suture Zone, Shangla and Allai in Kohistan, east of Kargil, western Ladakh, Sumedo in the eastern Ladakh and in the Darjeeling Hills.

The regional metamorphic rocks show three patterns namely normal i.e. high grade rocks at lower topographical level and low grade at higher, reverse or inverted or hot side up in which high grade occurs at higher topographical or structural level while low grade rocks at lower level and domal metamorphism in which isograds form concentric rings.

Almost everywhere regional metamorphism is of Barrovian type with the development different isograds in succession (Das, 1978). In Alaknanda Valley M 2 as Buchan type event was recognized. Porphyroblast- matrix relationship study carried out in the Yamuna and Satluj valleys reveal all six stages of the textural evolution (Biyani, 2000 c, Gururajan, 1994).

7.3.3 Polyphase Metamorphism

The Himalayan metamorphic rocks show mono and polyphase characters. Mono or one event metamorphism is restricted particularly in younger Phanerozoic rocks and occur in the outer Lesser Himalaya, Tethyan Himalaya and also in suture zones. However, many petrologists favor polyphase nature of Himalayan metamorphic rocks (Acharyya 1979; Staubli, 1989). Permian age metamorphism is proposed for eastern Himalayan rocks. A weak event around 500 Ma ago is also suggested. Metamorphic dates are also related with Caledonian and Hercynian orogenic movements. Many workers have suggested that signatures of the pre- Himalayan metamorphic event(s) are obliterated by the powerful Tertiary metamorphism. Tertiary Metamorphism is dated at various dates such as 65 Ma (at time of collision) and punctuated events between 50-20 Ma, 30 Ma and 8 Ma or even in Pleistocene.

7.3.4 Relationship between Metamorphism and Deformation

Himalayan metamorphism is correlated with deformational events occurred during orogenic movement (Ghosh, 1981;Pecher, 1989). In majority of cases regional metamorphism outlasted the deformational events i.e. it is syn-to post-kinematic in nature. One or more events of metamorphism are recorded on the basis of textural study. Two to four events of metamorphism are recorded in Proterozoic rocks. Among these multi events, one event is invariably retrogressive in nature, this may be mild to extensive one (e.g. in suture zone) and one event may be of cataclastic nature, if such type of deformation occurs in the region. Usually one to three events of progressive metamorphism of reported. In these progressive events the Miocene or main Himalayan metamorphism is most strong as well as widely distributed one. Peak period of Himalayan metamorphism does not coincides with peak period of deformation this is explained either by plate thickening or plate breaks off in earth' s interior.

Metamorphism is also related with igneous activities in many parts of the Himalaya (Das Gupta et al., 1979). In case of contact metamorphism date of metamorphism and emplacement of intrusive mass are synchronous in nature. The Miocene event is associated with early basic and late acidic magmatism metamorphism is confined in the Proterozoic rocks (Vanney and Hodges (1996). Cataclastic metamorphism is correlatable with thrust/ fault movement. During waning phase of imposed P-T conditions, retrogressive metamorphic effects set in the rocks.

7.3.5 Timing of Metamorphism

This is debatable issue. Himalayan metamorphism refers to metamorphism of rocks during Himalayan orogeny. The main event, as stated earlier, strongest one happened in Miocene period. But many workers also propose metamorphic events of prior to and post-Miocene time (Sinha Ray, 1975,1982 a). It is believed that earliest metamorphic occurred in the Proterozoic time or even at the boundary Archaean and Proterozoic period (Verma, 1989; Joshi, 1995). If Himalaya is considered as rejuvenated extension of peninsular India, then certainly possibility of Proterozoic metamorphism cannot be ruled out. Another metamorphic event of 1800 Ma has also been proposed. A weak event around 500 Ma ago is also suggested. Metamorphism dates are also related with Caledonian and Hercynian orogenic movements (Gururajan and Virdi, 1984). Permian age metamorphism is proposed for eastern Himalayan rocks. In Nepal, 600-500, 80-60, 40-30 and 20-10 Ma metamorphic episodes in the Higher Himalaya are recorded. Five pre- Himalayan metamorphic events

>2000 Ma, >1950±3 Ma, 1865±3 to 1887±5 Ma, 650±2 Ma and >466±2 Ma in NW Pakistan Himalaya are reported on the basis of ^{40}Ar-^{39}Ar dating (Baig et al., 1989).

Many workers have suggested that signatures of the pre- Himalayan metamorphic event(s) are obliterated by the powerful Tertiary metamorphism. Tertiary Metamorphism is dated at various dates such as 65 Ma (at time of collision) and punctuated events between 50-20 Ma, 30 Ma and 8 Ma or even in Pleistocene. In Zanskar region, monazite reveals peak metamorphism occurred between 32-30 Ma and Sm- Nd method revealed 31-28 Ma, the duration of metamorphism was 35-25 Ma while in Garhwal metamorphic duration was 44-26 Ma. In Kaghan Valley peak metamorphic condition around 40 Ma ago. This date of peak metamorphism is significantly different from other Higher Himalayan regions.

7.3.6 Extents of Metamorphism

The autochthonous or paraautochthon of Lesser Himalaya and Tethyan Himalayan rocks are metamorphosed in lower to middle part of green schist facies P-T condition. In suture zones, static metamorphism of blue schist facies condition (high pressure and low

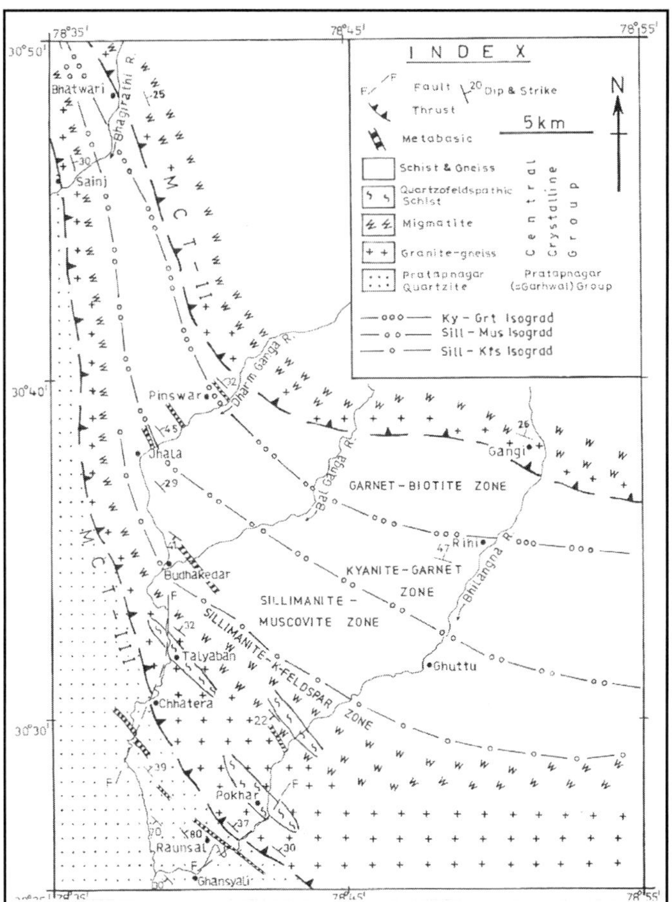

Figure 7.1. An example of zonal mapping in the Higher Himalaya of Garhwal region (after singh et al., 1996).

temperature) is sporadically recorded in exposures of small size. Haimanta Formation in Satluj Valley is reported to contain kyanite–talc association; white schist, an unusual combination formed at extreme depth. The Higher Himalayan region shows widely distributed metamorphism peaked to upper amphibolite facies. Granulites are reported, in isolated patches from Karakoram and eastern Himalaya, however, calc-granulites are present in many areas.

7.3.7 P-T Conditions

P-T condition of metamorphism is estimated through petrogenetic grid, fluid inclusion and by electron probe methods using exchange reactions between mineral pairs. Vaikrita rocks of Kedarnath area has yielded 600°-800°C/6-8 kb, 450°-550°C/5-6 kb and 400°-300°C/4 kb for M1, M2 and M3 events respectively (Metcalfe, 1993). In Garhwal Nappe 664°C/5.9 kb, Dudatoli-Almora 712°C/6 kb and Central Crystallines of Almora 759°C/6.2 kb, white schist of Haimanta Formation >10 kb, in the eastern Nepal/Tibet area 550°-680°C/8-10 kb, 650°-750°C/4-7 kb and 600°-700°/2-4 kb for M1, M2 and M3 events respectively (Le Fort et al., 1986) and in Zanskar HHC yields peak temperature 780°C, 10.5 to 11.5 kb pressure and in MCT zone peak P-T condition is 770°C/6.2 kb by EPMA technique. In Bhagirathi Valley crystallines, fluid inclusion study reveals 495°-540°C/4.9-5.7 kb P/T condition. Between Pali-Yamunotri area of Higher Himalayan Yamuna Valley is 500°C/4.5-5.2 kb, P-T condition is estimated by grid method (Biyani, 1997 c; Biyani, 2000 c; Hodges et al., 1988, Hubbard, 1989). In Mabja Dome, rocks of sillimanite and chloritoid zones of core and periphery yielded 705±65° C and 820 ± 100 MPa and ~500° C and ~150-450 MPa temperature and pressure respectively,

Blue schist facies formed in 350°- 450°C and 9-11 kb pressure range. In Allai melange zone blue schist was formed in 380°-450°C and 7-8 kb pressure range, its retrogression occurred at 300° C and pressure 3-6 kb. Eclogite lenses are formed around 850° C and between 26-29 kb pressure condition in Karakoram region. Eclogite at Besal in the upper Kaghan valley is formed 650°± 50°C and 13-18 kb (~60 km depth) pressure conditions.

7.3.8 Metamorphic Belts in Himalaya

Metamorphic belts (Miyashiro, 1992) are reported from Ladakh, Karakoram and Kohistan area of western Himalaya. Virdi (1981) identified three belts namely the Pongong-Baltit belt in the Karakoram, Shyok-Gilgit low to medium pressure belt and Indus-Kohistan high- pressure belt in this area. In Pongong-Baltit belt, Karakoram intrusive granite is present in the schist, gneisses and marble. Barrovian type greenschist-amphibolite metamorphism (Jurassic?) is super imposed by low-pressure andalusite-sillimanite type and restricted contact metamorphism (95 ± 6 Ma). In Shyok-Gilgit belt, metamorphism is of low to medium pressure and medium to high temperature type and shows development of greenschist-amphibolite facies series in Ladakh greenschist–amphibolite–pyroxene granulite facies series in Kohistan. The metamorphism took place in magmatic arc and back arc basin environment under 300° to 370° C/km geothermal gradients. Indus- Kohistan belt a paired metamorphic belt. In this belt blue schist-green schist facies series rock occur in the Ladakh (high pressure/low temperature) and blue schist – amphibolite – garnet granulite facies series rocks (moderate to high temperature and pressure) occur in Kohistan. Geothermal gradient for high-pressure metamorphism varies between 15±18° C/km (Sychanthavong, 1990 b).

7.3.9 Contact Metamorphism in Himalaya

Contact metamorphism is reported from many places of the Himalaya. Generally contact metamorphism is superimposed on the earlier or later regional metamorphism. It is found around vicinity of granitic intrusions or in the xenoliths and margins of large basic bodies e.g. areas around Champawat Granodiorite in Kumaun, metamorphism of Lesser Himalayan footwall quartzite in Yamuna Valley. In Lakhwar dam site in Dehradun district, chlorite schist and quartzite xenoliths present in a basic dyke intruded in Chandpur Formation of Garhwal Lesser Himalaya. Zoned skarn is reported from the Chaur Hills of Himachal Pradesh (Das, Okrusch and Olesch, 1988). The Baltoro Batholith intruded around 21 Ma ago and was responsible for producing andalusite-bearing hornfels in high temperature low-pressure contact metamorphic condition. As a result of contact metamorphism small knots of fibrolite and biotite replace cordierite and muscovite.

7.3.10 Inverted Metamorphism in Himalaya

As stated earlier reverse or inverted metamorphism or inversion of isograds refers to a condition in which high-grade rocks occur on higher structural or topographical levels in the present day crust. Louis de Loczy observed such condition as early as in 1878 in Mohan et al. (1989) and by Ray (1947) and Sinha Ray (1982 b) in Darjeeling and Sikkim Himalaya. Himalayan inversion is explained by two principal mechanisms one says that inversion is a real or, primary or syn-metamorphic while other view believes that it is apparent, or secondary or post-metamorphic in nature (Arita, 1982). The former view is most favored one and is explained by hot iron model. This model proposes that inversion is result of movement of hot crystalline nappe over colder footwall. The crustal scale underthrusting of the Lesser Himalaya is responsible for transient (short duration) inversion of down buckled S-shaped isotherms. This model is linked up with evolution of MCT, and believes MCT is cause of inversion. The MCT is responsible for movement of hot crystalline nappe. The latter model believes in that the overturned or recumbent folding of the crystalline rocks in which entire sequence of overturned limb is exposed on the surface causes inversion.

Inversion is also explained by other mechanisms like thrust stacking, contact metamorphic effect of Higher Himalayan leucogranites (Harris et al., 1993), shear or frictional heating along the MCT (Graham and England, 1976; Sharma and Kizaki, 1989), distributive ductile shearing and preferential flow of fluids during metamorphism. All these basic mechanisms have been scrutinized thoroughly and weaknesses in them were highlighted, the recent trend is to combine two/three mechanisms together according to geological setting of the area and to provide reasonably logical mechanisms of inversion (Biyani, 1998 a).

7.3.11 Petrographic Characters of Himalayan Metamorphites

Many varieties of metamorphic rocks ranging from the prehnite- pumpellyite facies to granulite and blue schist facies are recognized in the Himalaya (Varadarajan, 1974). Their general mineralogical assemblages of common metamorphic rocks are as follows:

Slates: Slates occur in the Outer and Lesser and in Tethyan Himalaya.

Common mineralogical assemblages of slates are quartz-sericite-chlorite; quartz-chlorite-graphite.

Phyllite: Phyllite is widely present in the Lesser Himalaya and Tethyan Himalaya and also occurs in Higher Himalaya and Indus Suture Zone.

Assemblages

Quartz-biotite-muscovite-chlorite; quartz-muscovite-chlorite; quartz-chlorite-sericite.

Blue Schist

Blue schist is reported from many locations of western Himalaya particularly in the suture zones e.g. Shin-Kamer and Marin area of Allai Melange Kohistan and Ladakh in India. The reported assemblages are-

Quartz-albite-muscovite-glaucophane-garnet ± rutile+epidote–chlorite;quartz-glaucophane-muscovite-epidote-rutile; glaucophane-actinolite-garnet-sphene.

Phyllonite:occurs in fault planes.

Assemblages

Sericite-chlorite-quartz; chlorite-sericite; carbonaceous chlorite.

Schist- Schists of varying mineralogical composition are present in the Lesser, Higher and in Suture zones of the Himalaya.

Assemblage

Quartz-biotite-muscovite; biotite-muscovite, garnet-biotite, garnet-biotite-staurolite; garnet-biotite-kyanite; garnet-muscovite-biotite-feldspar; feldspar-muscovite-garnet, graphite-biotite-garnet; garnet-plagioclase-mica; garnet-staurolite-kyanite; garnet-mica-plagioclase-garnet-staurolite, quartz-plagioclase-biotite, garnet-sillimanite-orthoclase-muscovite; and -biotite-garnet-garnet-fibrolite; andalusite-biotite-garnet-graphite-fibrolite, andalusite-biotite-staurolite -muscovite-kyanite-fibrolite-garnet.

Migmatites and Gneisses

Restricted presence of migmatite is reported mainly from the Higher Himalaya and its nappe and klippen in the Lesser Himalaya. Gneisses are mainly present rocks of Vaikrita nappe (Kano, 1984).

Granulites

Granulites are reported from the Kohistan region in Pakistan and from the northeastern India. In Kohistan, pyroxene granulite occurs in a 30 km wide belt between the Nanga Parbat and Dir. It is formed around 800° C at 7-8 kb pressure. It is composed of plagioclase, orthopyroxene, clinopyroxene with subordinate amount of quartz, opaque, amphiboles and biotite. The pyroxene granulite was earlier reported as metanorite. Garnet granulite is reported from the Jijal Complex in Kohistan.

Assemblages

Quartz–feldspar-muscovite; garnet-feldspar-biotite, garnet-biotite-muscovite-orthoclase; quartz-biotite-muscovite-K-feldspar-epidote;garnet-biotite-muscovite-plagioclase-garnet;garnet-biotite-muscovite-plagioclase-sillimanite-kyanite-garnet-cordierite.

Eclogite

Eclogite lenses are reported from the Kaghan valley, Tso Morari Crystalline, and Ladakh. Fresh eclogite contains garnet, phengite, clinopyroxene, omphacite, rutile and carbonates and in matrix amphibole, phengite quartz. Retrograde eclogite shows presence of amphibole, biotitic, epidote, paragonite quartz and zoisite. Eclogitization took place 55±17 Ma ago. Omphacite-garnet-quartz-rutile±phengite are reported from eclogite of the upper Kaghan valley

Ultra High Pressure Metamorphic Rock

Ultra high-pressure metamorphic rocks are those rocks, which formed at pressure greater than 28 kb, and temperature may be around 700° C. These rocks are reported from Tso Morari area and contain garnet having inclusions of quartz, coesite, magnesite, clinopyroxene, and rutile; zircon, dolomite and magnesite; clinopyroxene (omphacite- jadeite), amphiboles (glaucophane, barrosite, edinite, actinolite and tremolite), phengite, calcite, dolomite, aragonite, magnesite, talc, kyanite, paragonite, epidote- zoisite/ clinozoisite (Sachan et al., 2005).

Mylonitic Gneiss

Mylonitic gneisses are present in the MCT zone or in the nappes or klippen of crystallines rocks of Higher Himalaya occurring over the Lesser Himalaya. Mylonitisation is readily visible in the quartzites and granites.

Mineral Assemblages of Common Metamophic Rocks of Himalaya

Quartz-plagioclase-zoisite-clinozoisite-chlorite-biotite-muscovite-sphene-iron oxide; garnet-chlorite-biotite-sericite-plagioclase-epidote-ilmenite.

Quartzites

Pure and impure quartzites occur as regionally metamorphosed, mylonitised and thermally metamorphosed rocks in all divisions of the Himalaya. Quartzites show interesting variations in texture and in mineralogical compositions.

Assemblages

Quartz; quartz – paragonite; quartz – muscovite-biotite; quartz – sericite-biotite-zircon, quartz – muscovite-biotite-garnet.

Marble: Pure and impure marble occur in the Lesser and Higher Himalaya (Biyani, 2000 b).

Assemblages

Calcite: calcite – phlogopite, calcite – paragonite, calcite-dolomite-quartz-muscovite-zircon-iron ore.

Calc-Schist: Mostly confined in the Higher Himalaya and are reported from few areas e.g. Badrinath or Ganesh Himal area of Nepal.

Assemblages

Quartz-calcite-phlogopite-muscovite-chlorite, calcite-diopside-clinozoisite, orthoclase-plagioclase-phlogopite.

Metabasics: Occurs widely in the Lesser, Higher and Tethyan Himalaya as sills and dykes of varying ages.

Assemblages

Hornblende-biotite-andesine-garnet-quartz; hornblende-biotite-andesine-quartz; hornblende-biotite-chlorite; hornblende -augite-olivine-biotite- chlorite; hornblende - chlorite -clinozoisite-zoisite-pistacite-biotite.

7.4 Sedimentary Rocks

Sedimentary rocks cover a major part of the crustal surface. These rocks widely vary in their mineralogical and chemical compositions and are formed in different depositional environments. The Outer Himalayan terrain is mainly made up of sedimentary rocks of Tertiary age while in the Lesser, Tethyan, suture zones and in the Trans - Himalayan areas sedimentary rocks dominate and few sedimentary klippen occur over the Higher Himalaya. The oldest unmetamorphosed sedimentary rocks, believed to be of Middle Proterozoic age, are present in the Lesser Himalaya and the youngest one is of Holocene age, for example well-compacted/cemented Doon Gravels in Doon Valley of Uttaranchal

A variety of sedimentary rocks occur in the Himalaya and their brief description is as follows:

7.4.1 Rudaceous Rocks

Conglomerate and breccia are main constituents of this category. Conglomerate comprising of rounded pebbles and is reported from many stratigraphic formations of the Himalaya. The best-known horizons are Boulder Conglomerate of the Upper Siwalik and Blaini Conglomerate of the Kumaun and Himachal Lesser Himalaya. It also occurs in Dirang, Miri, Nahan, Krol, Tal and Indus formations as well as in Ladakh and Tethyan sequences. On the basis pebble and matrix content, oligomictic (Dirang Formation in Arunachal) and dimictite (Miri and Bichom formations) are recognized. Dhital (1992) identified three horizons of dictite varieties of conglomerates from Barahakshetra Eastern Nepal, Tansen Western Nepal and Sallyan area of Mid-western Nepal. Intraformational (Nahan Formation) and psuedoconglomerate (Kamlial Formation) are also reported. Conglomerate also serves as indicator of unconformity e.g. base of Kakara Formation.

Angular fragments or pebbles constitute breccia. These angular fragments are formed by many processes such as weathering, landslide, cataclasis, collapse of overlying rocks volcanic fragments on, pyroclastic etc. Breccia is relatively less distributed in the Himalaya. Epiclastic or sedimentary breccia is reported from Subathu Formation while tectonic breccia is present in many fault zones e.g. Main Boundary Thrust and Main Central Thrust.

7.4.2 Arenaceous Rocks

Arenaceous rocks are composed of sand size (2.0-0.0625 mm) particles. Sandstone is a common rock in this group. Sandstone is classified into many subdivisions on the basis of proportion of quartz, feldspar, rock fragments and type of matrix as well as on the basis of cementing material or matrix. Relatively pure sandstone (>95% quartz) is designated as orthoquartzite or quartz arenite. Himalayan sandstones show considerable variations in their texture, color, mineralogical compositions, degree of compaction and mode of origin. Fine to coarse-grained occasionally gritty sandstones are composed of perfectly rounded to sub angular (Berinag), sand particles matrix (e.g. in Middle Siwalik) and cement (e.g. Tal) bind the quartz grains together. Sandstone occurs in white (Muth and Tal), gray (Nagthat), black (Neelkanth), grayish white (Bichom), green (Siwalik), brown (Damuda), red (Kamlial) and yellow in color. Siliceous, feldspathic, micaceous, calcareous, micaceous, clayey, glauconitic and ferruginous varieties of cement/matrix are reported from the Himalayan sandstones. Arenite, arkose, lithic and feldspathic greywacke, and lithicarenite are common types of sandstones. Quartzite variety is reported from Miri, Muth, Berinag, Nagthat, Tanwal and, Tal formations.

7.4.3 Argillaceous Rocks

This category includes siltstone (0.0625-0.004 mm), shale (>0.004 mm), mudstone, clays, marls etc. Siltstones of varying colors are reported from Krol, Tal, Siwalik, Bhareli, Yinkiong, ophiolitic melange zone and several other formations. Siltstones contain clay minerals, quartz, carbonates and other minerals.

Himalayan shales are commonly black, gray, grayish white, white, brown, red, purple and green in color and occur in finely laminated to massive forms Some shale horizons are fossiliferous like Fenestella Shale, foraminiferal shales and Siwalik shales show impression of leaves at many places. Compositionally shales are described as ferruginous, siliceous, sandy, silty and novaculites types. Mudstone is also reported from some places. Shales are extensively present in all divisions except the Higher Himalaya.

7.4.4 Carbonate Rocks

Limestone, dolomite and magnesite are common rock types of this category. Excluding the Outer and Higher Himalaya, sedimentary carbonates occupy considerable space in the remaining divisions of Himalaya. Hard and massive to bedded dolomite in different shades of color are present in the Krol, Chillepam, Calc Zone, Kioto, Hilang Suit, Kerabari and Dhading formations, and in the Tethyan Zone.

Limestone also varies considerably in color, texture and mineralogical compositions. Pure limestone occurs at Hathipoan in Dehradun, Jammu, Chamba and other areas. Limestone of Neelkanth area near Rishikesh is of crystalline blue variety. Other types include sandy, cherty, oolitic, siliceous, clayey and feldspathic. The important limestone horizons are Krol, Kioto, Tandi, Kalapani, Subathu, Yinkiang, Larjung, Tilochopass Chekha and Tangcher Formations. In Bhutan and Nepal, limestones are reported from Syaldanda, Kanchi Tarsikot, Barbati - Chobar, Chandragiri, Hathisundh and other areas. Calcareous tufaceous deposits are reported from Kailash ridge in Nepal.

Magnesite ($MgCO_3$) is present in Jammu, Chamba, Almora, Khari Dunga, Ranighat, Phulping and Dhunchekhola areas. Being of economic significance, magnesite is described in detailed in the chapter of mineral resources.

7.4.5 Flysch and Molasse

The Swiss geologists for Alpine orogenic sediments first use the term flysch and molasse. Flysch is thick marine turbidite sequence of graywackes like sandstone (31%) interbedded in a rhythmic fashion with indigenous shale (64% by volume). It is mainly clastic in nature shows notable absence of carbonates and occasional presence of volcanics; however, chert and radiolarites may rarely present. Flysch sediment is syntectonic in nature. Molasse is coarse clastic marine to continental sequence of immature sandstone (lithic and sublitharenite), shale (silty gray and red in color), mudstone, clay stone and thick conglomerate. The molassic sediment is produced by denudation of uplifted mountain i.e. post-tectonic in nature. Several flysch or flyschoidal formation have been recognized from the Lesser Himalaya such as Sundernagar Formation, Shimla Slates, Chakrata Formation, Chandpur, Daling formation in India and the Tansen-Rukum-Himchuli Belt, Kuncha Formation, Nourpul Formation and Ranighat formations in Nepal.

The Siwalik Super Group, Indo-Gangetic Alluvium and Quaternary sediments deposited in Duns like Dehradun or Kathmandu valleys represent Younger Molassic sediments.

7.5.6 Minor Sedimentary Rocks

Tuff and Agglomerate

These rocks are formed by consolidation of pyroclastic materials ejected during volcanic eruptions. Dividing line between tuff and agglomerate is placed at 2 cm. Tuff (<2 cm) occurs in Panjal Trap in Kashmir and welded tuff (ignimbrite) in Yinkiong Formation of Arunachal Pradesh, agglomerate (>2.0 cm) is reported from Shergol ophiolitic melange at Khangrol village in Ladakh.

Lahar

Lahar reported from Ladakh is composed of rounded pebbles and boulders of acidic volcanic clasts and red jasper embedded in matrix.

Chert and Jasper

Chert and Jasper occur in two forms namely as nodules and bands. Chert beds are present at the contact of Krol and Tal Formations, in Indus Ophiolites and jasper bed is reported from the ophiolitic melange zone e.g. at Shergol.

Intertrappean Bed

Conglomerate, shale, calcareous sandstone, black shaley quartzite, carbonaceous sand occurs as intertrappean beds in Abor Volcanics of Arunachal Pradesh.

Salt Deposit

Salt beds containing halite, sylvite, carnallite, anhydrite and other minerals of evaporite origin are present in Mandi area of Himachal Pradesh and Salt Bed in Pakistan. Thin encrustations of salt layers are deposited by waters of mineral springs and are reported from Tentang, Cheligaon, Kumsumkhola and Lamjung area of Nepal.

Phosphorite

Sedimentary-bedded gray to black, granular thin to thickly bedded phosphorite (P_2O_5 >19.5%) is present in Dehradun-Mussoorie area in Uttaranchal, at Hanley in Zildat Ophiolitic melange in India and Dharan and Phulchowki area of Nepal. Phosphorite also occurs as nodules in Bichom Formation.

Coal and peat

Gondwana and Tertiary coal occur in bedded and ball forms at many places in the Lesser Himalaya and in the Siwalik. Peat occurs in Karewa of Kashmir and at many places in the Siwaliks and carbonized wood in Siwalik.

Calcrete (Kankar, Nari/Caliche)

Calcretes are widely distributed in 20 million sq km area on the globe and are formed in region of 40-60 cm annual rainfall. Calcretes are near surface terrestrial secondary calcium carbonate rich accumulation in soils and bedrocks formed in marginal areas of lakes during periods of low strand, in the marginal sea present in arid, semi- arid, sub humid, temperate and even in polar climate receiving Quaternary sediments. In Himalaya calcretes are reported from the Muree, Dagshai, Siwaliks, Karewa and from the Indo Gangetic plain. Calcretes are useful in the palaeoclimatic analysis particularly carbon dioxide level in the palaeoatmosphere (Singh and Tandon, 2002).

7.4.7 Quaternary Sediments in Himalaya

Exposures of Quaternary (Pleistocene and Holocene) sediments occur in almost all parts of the Himalaya. These sediments are deposited in the palaeolake, river valleys and slopes of the mountain or even on the hill tops by many processes such as lacustrine, glacial, fluviatile eoline, landslides etc. Compact to partially consolidated lacustrine silt, clay, granules, boulders etc. sediments are reported from many palaeolakes for example upper reaches of the Satluj basin, Skardu, Hunza, Chilas, Attargo, Li, Dankar, Hanse-Kioto, Tsokar, Lamayuru and others. Partly lacustrine and partly fluviatile sediments such as gravel, sand, silt clay, pebbles and partly consolidated sediments occur in many tectonic valleys called *duns*; the notable examples of *duns* are Kashmir Valley, Pinjor Valley, Dehradun Valley in India and Thankkhola Dang Dun, Katmandu Valley and Pokhra Valley in Nepal. Thickness of sediments in duns are several hundred meters for example nearly 380 m in Kathmandu Valley and more than 600m in Pokhra Valley of Nepal and nearly 500m in Dehradun Valley. Fluviatile sediments occur in river valleys either on the floor of valleys or on the valley sides in the form of river terraces. In Kaligandaki valley of Nepal, highest level of river terraces is 300 m above the riverbed. Three to five terrace levels are common in the Himalaya. Fluviatile sediment is mostly composed of boulders, pebbles, silt, sands and clays. Marshy sandy, silty and clayey sediment are common on either slopes of the Outer Himalaya. Rivers are also making major fans on the Outer and Lesser Himalayan slopes e.g. three major fans are present on the northern slopes of Dehradun Valley while southern slope of this valley is formed by coalescing fans. Talus deposits are very common all along the river or mountain slope and are formed by normal weathering or mass wasting actions. Glacial sediments in the form of moraines occur at high altitudes of interior regions. These fresh angular sediments are highly heterogeneous in size. Moraines are reported from Kashmir, Himachal Pradesh, Uttaranchal, and Nepal. Glacial hills are reported from Karewa. Fine, yellowish brown sandy loess of eoline origin is present in the Karewa Formation of Kashmir (Singh, 1982). Present Sedimentological study of Neogene and Quaternary sediments are being attempted from the sediment lithofacies identification such as clast supported massive gravel (Gcm), matrix supported gravel (Gmm), graded-matrix supported gravel (Gmg), massive sandstone (Sm), planar cross bedded sand (Sp), horizontally stratified sand (Sh), sand-mud alteration, silt-mud (Fsm) with lithologs preparation and their depositional environment point of view.

7.4.8 Environments of Deposition

Himalayan sediments are deposited in variety of geological environment namely deep water e.g. in suture zone to shallow water marine or even in continental conditions. Chakrata flysch is deep-water deposit, Krol and Tal are tidal flat deposits, Karewa is lacustrine and Upper Siwalik is fluviatile. Other depositional environments are glacial, lagoonal, fluvioglacial, swampy and marshy, deltaic, arc-trench and continental. Deep-water rocks are comparatively less abundant in the Himalaya. Most of the sedimentary rocks are formed in shallow marine condition.

7.4.9 Cycles of Sedimentation

The long geological history of the Himalaya is responsible for repeated evolution and termination of sedimentary basins of varying sizes. Several cycle of sedimentation have been recognized in the Himalaya. In the western Lesser Himalaya of Himachal Pradesh and Uttaranchal four cycles of sedimentation Sundernagar-Damtha (dominantly siliciclastics with bimodal volcanics and granites, 2500-1600 Ma), Shali-Deoban-Larji (carbonates with minor clastics, 1400-900), Shimla-Jaunsar (siliciclastic, argillite with carbonate, volcanics, diamictites and evaporites in upper part, 900-700) and Blaini-Infrakrol-Tal (diamictites, clastics and carbonates, 700-550 Ma) have been recognized. Each cycle is followed by a time gap of varying duration. The prominent unconformities are 1600-1500, 1000-900, 725-700 and 500 Ma old. In Siwalik, nine cycles of sedimentation have been recognized.

Size of Basins of Deposition

The size of basins varies in time and space. There are examples of large basins covering the entire length of the Himalaya, for examples in a foreland basin Siwalik sediments were deposited or in a Tethyan basin in Tethyan rocks were deposited, medium size basin is represented by Krol Basin and Quaternary basins were of small size.

Basins of Deposition

The present form of Himalaya is result of interaction of variety of processes and the main body fabric of this mountain range is shaped by the cumulative effects tectonism and of sediments deposition, which have taken place many times in independent basins. The size of basins varies considerably in time and space. The oldest Himalayan basin probably existed around 3000 Ma ago in which present day metamorphosed Higher Himalayan sediments were deposited. The size and depth of basin is a matter of speculation and estimate becomes complicated if we consider the Higher Himalayan is extension of peninsular cratonic mass of south and it is not yet clear how much part of it is consumed by subduction of Indian plate. However sediments of this basin subsided more than 25-30 km depth as upper amphibolite facies metamorphism takes place around such at this depth. In case of Lesser Himalaya, four megacycle of sedimentation namely Sundernagar - Damta, Deoban-Shali, Shimla-Jaunsar and Krol Cycles have been recognized in Proterozoic and early Palaeozoic time in the Western Himalaya. Mega Krol Cycle is followed by three minor cycles of later age namely in Permian (Fenestella beds) and in Cretaceous (Kakara/Shell limestone) and finally again a medium but wide cycle in Eocene age (Subathu Formation). The basin size for each of these cycles of sedimentation is considerably large having lengths in hundreds of kilometers and width tens to few hundred kilometers (Valdiya, 1995). For example, present day dimension (350x120 km) of Krol Basin is probably much smaller than its original dimension as basin was subjected to severe folding and faulting. However, an extremely large basin existed in Palaeozoic-Mesozoic time in which Tethyan sequence deposited. In Neogene time Siwaliks sediments were deposited in a foreland basin in south. These basins were having more than two thousands kilometer length and several tens to hundreds of kilometers width. Indus Suture Zone also formed a very long but narrow basin parallel to edge of Indian plate. This basin has complex sedimentation, magmatism and metamorphic history starting from Palaeocene to end of Tertiary. The Quaternary basins, mostly lakes, are of comparatively small size. Several such basins for example Lamayuru, Kashmir Valley, Doon Valley, Kathmandu Valley are reported from different parts of the Himalaya. The Kashmir Basin was 140 km long and 60 km wide depression and has maximum depth 1300 m. Further to the west of Kashmir Basin, Campbellpore and Peshawar basins of tectonic origin were formed in Plio-Pleistocene time. South of Outer Himalaya, Indo Gangetic Plain, a foredeep, is situated. This basin covers nearly 0.8 M sq km area. It is perhaps largest basin of the Himalaya system.

7.4.10 Present Day Sediment Evolution

Himalaya is a region of active chemical weathering as it falls in mid-latitude region. As a result of chemical weathering enormous amount of sediments are generated. The three

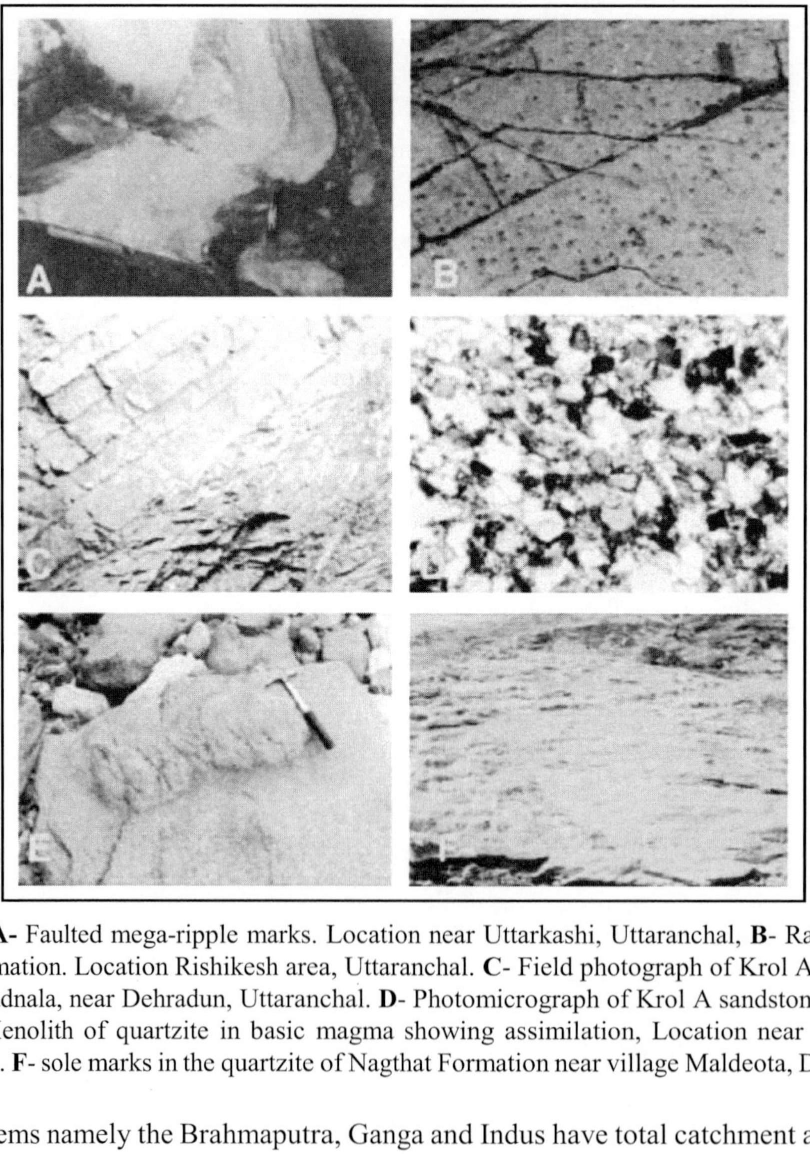

Plate 7.1. A- Faulted mega-ripple marks. Location near Uttarkashi, Uttaranchal, **B-** Rain prints in Kakara formation. Location Rishikesh area, Uttaranchal. **C-** Field photograph of Krol A sandstone. Location Judnala, near Dehradun, Uttaranchal. **D-** Photomicrograph of Krol A sandstone, Location same. **E-** Xenolith of quartzite in basic magma showing assimilation, Location near Uttarkashi, Uttaranchal. **F-** sole marks in the quartzite of Nagthat Formation near village Maldeota, Dehradun.

rivers systems namely the Brahmaputra, Ganga and Indus have total catchment area nearly 1 M sq km in the Himalaya with sediment generation of 1500 million tons. The Ganga and Brahmaputra rivers have 1.656 M sq km catchment area including peninsular India and this catchment yields nearly 1000 Mt sediment every year. This sediment is either consumed in enlarging the Bengal Delta are in developing Ganga Submarine Fan (GSF). The GSF is world's largest fan started evolving 16 Ma ago and contains 1.13×10^{16} tons of sediments with annual accumulation rate of 665 Mt. In the history of past seven million years, the sediment accumulation rate abruptly increased by 4.5 times in last one million year.

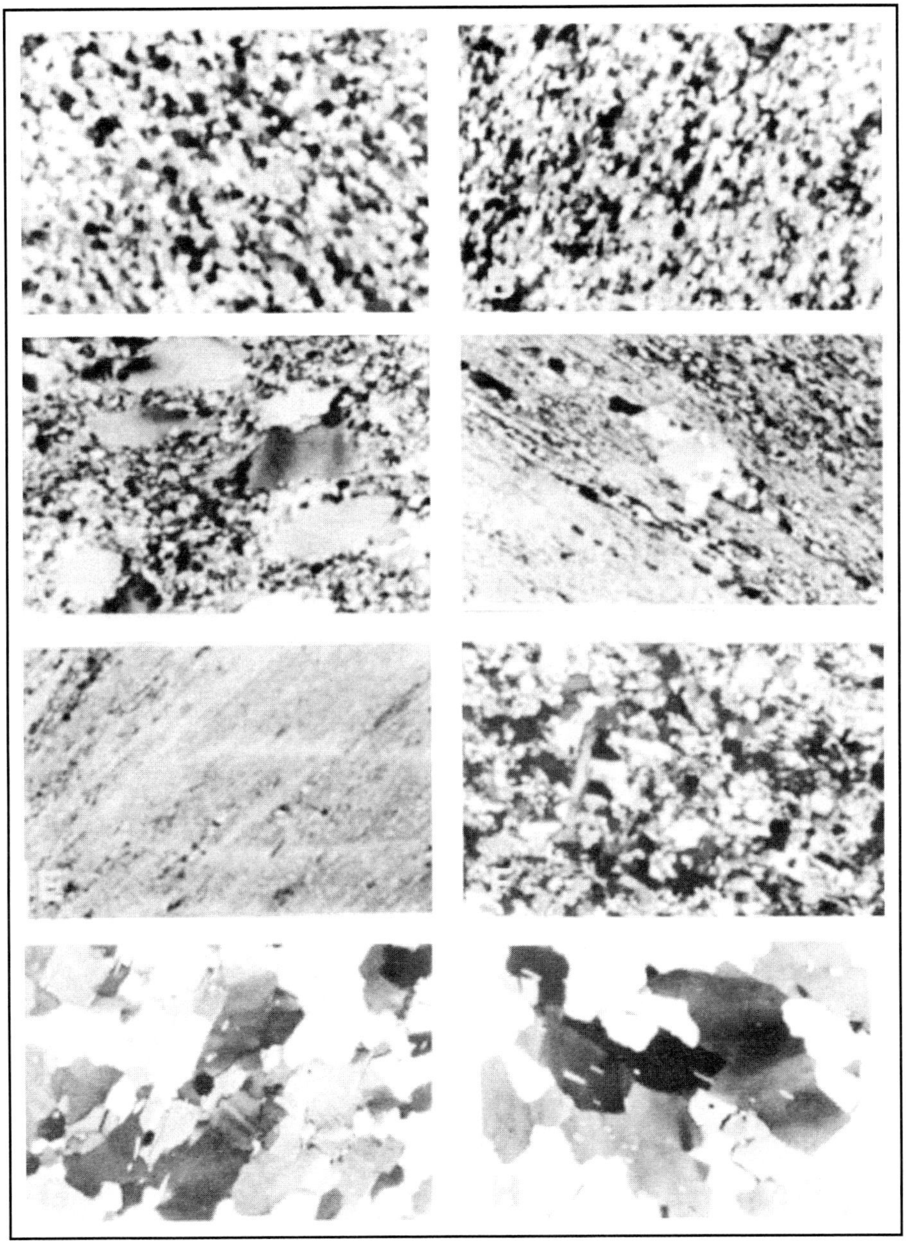

Plate 7.2. Evolution of Quartzite in Higher Himalayan Yamuna valley, Uttaranchal **A**: Paligarh quartzarenite of Lesser Himalaya half km from MCT showing weak preferred orientation, **B**- Increase in grain size by metamorphism caused by overriding higher Himalayan thrust sheet, **C**- Quartzarenite just near the contact of MCT showing sporadic increase in grain size, **D** and **E**- Pyrophyllitic quartzites present on the top of MCT zone, **F**- Coarsening of grain size due to regional metamorphism near Banas **G**- Janakichatti quartzite showing further increase in grain size and triple point, **H**- Yamunotri quartzite showing maximum increase in grain size. Note: Crystallization of paragonite in quartz grains in G and H. Straight boundary indicates absence of deformation and metamorphism at high temperature, granoblastic polygonal structure is indicative of high degree stability

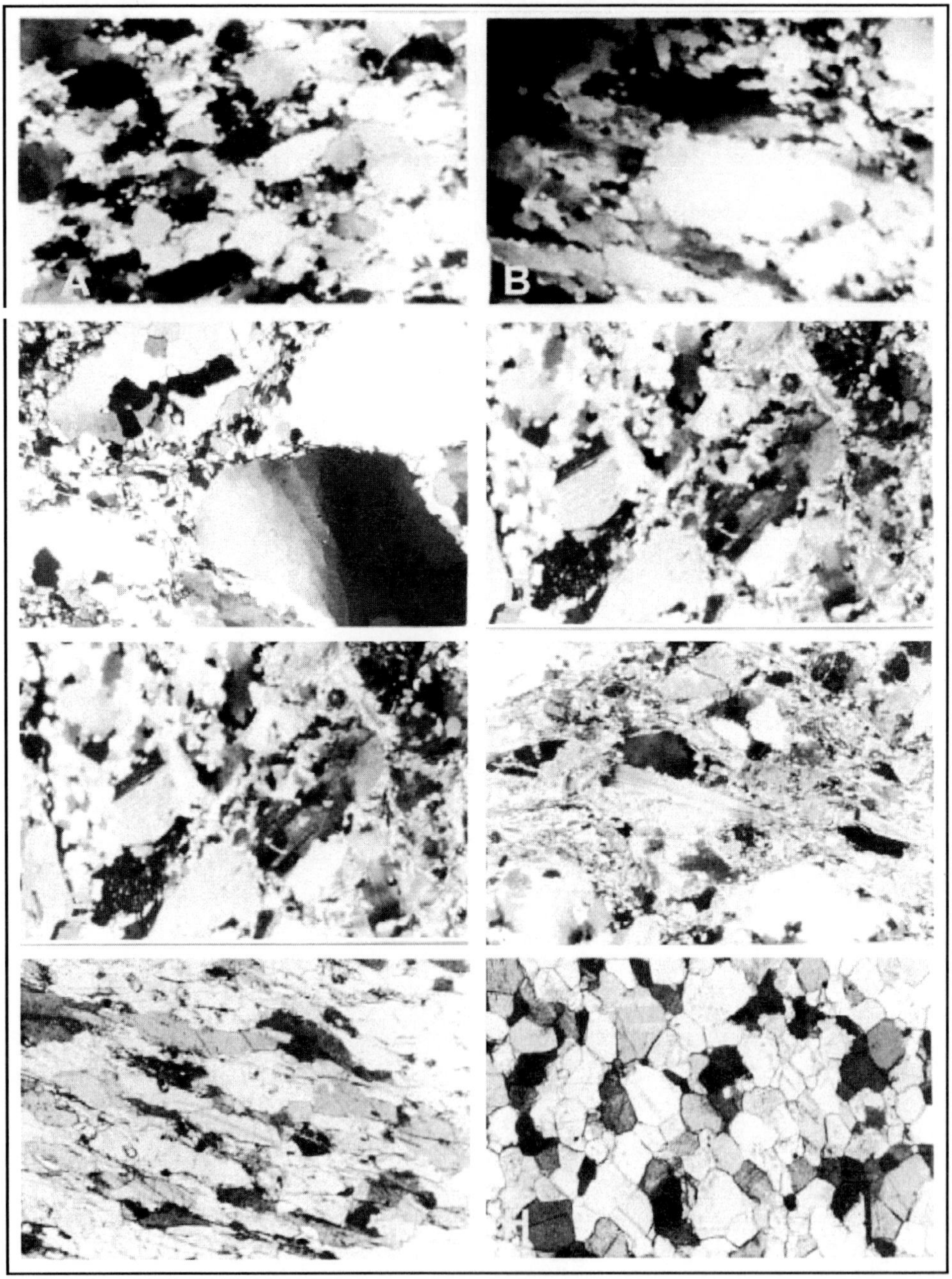

Plate 7.3. Development of mylonitic texture in Yamuna Mylonite Zone, Uttarkashi, Uttaranchal.
A, B and C- Development of mortar texture and progressive coarsening of grain size in ductile deformation condition. Note: in B quartz grains flattens due to high stress. D, E and F- Mylonitisation of Sayanachatti mylonitic granite showing mortar texture, crushed grain boundary and partial conversion of orthoclase into microcline and frayed mica grains. G- Flattening of calcite grains (approx. 1:8) due to high stress created by thrust sheet in marble of Phulchatti, H- Crystallization of limestone into marble at Janakichatti, triple grain boundary indicating crystallization in relatively static conditions

In Ganga river, out of 794 Mt sediments the contribution of Tethyan Himalaya is <10%, Higher Himalaya 80±10%, Lesser Himalaya 20±10%, Outer Himalaya <19% Indo-Gangetic Plain <10% and of peninsular India is <10%. Of this sediment amount 729 Mt passes through Farrakha in the west Bengal and the difference of 65 Mt is consumed in the construction of on land flood plains. Of 729 Mt, 328 Mt passes through Hooghly River and rest goes in Bangladesh. The Ganga and Brahmaputra rivers combinedly carry 980 Mt sediments, in Bangladesh out of which 550 Mt deposited in the flood and delta plains of the Bangladesh and rest i.e. 430 Mt goes in the sub aqueous delta and Ganga fan. The bed load in Ganga and Brahmaputra is estimated between 600-2500 Mt (Wasson, 2003).

DIMENSIONS OF HIMALAYAN GEOLOGY
A. K. Biyani (2006) : pp.203-207
Satish Serial Publishing House, New Delhi
E-mail: ssp@ndf.vsnl.net.in, hkjain1975@yahoo.com; website: www.satishserial.com

Sahasrayyugaparyantamaharyadbrahmano viduh
Rtrim yugasahasrantam te horatravido janah (Geeta, 8.17)
Those who know from realization Brahma's day as extending to a thousand Mahayuga
and Brahma's night as extending to another thousand mahayugas,
they alone know the reality about time.

CHAPTER - 8

GEOCHRONOLOGY

Geochronology deals with measurement and interpretation of time. Geological time is measured by two methods namely non-radioactive or relative and by radioactive or absolute techniques. The non-radioactive methods are based on fossils or study of comparative field and laboratory studies. These methods provide a relative idea of time, for example statements like "Himalaya is formed by Tertiary orogenic movements" implies that mountain formation has taken place between 65-1.8 Ma duration from today or "… on the basis of fossils record Cambrian age is assigned to Haimanta Formation," means deposition of Haimanta sediments had taken place between 542-500 Ma ago i.e. in duration of 42 Ma. These statements do not tell whether the process has actually taken place in the entire duration as stated or in a part of described duration. However, with the better documentation of fossils record time limit can be further restricted but certain amount of ambiguities would still remain. In case of absolute time, measured mostly by radioactive methods one can get precise information after a careful scrutiny of geological setting, when a particular event happened, for example, Sayanachatti mylonitic granite in the Yamuna Valley of Uttaranchal is 1972±102 Ma old, this means we have a reliable date of crystallization of granite that vary in a narrow range. Relative methods of age determination are rapid and cheap while absolute methods are not only very expensive but also time consuming. Because of cost and time factors total number of radiometric dates reported from the Himalaya may not exceed more than 1000 today. These dates are dealing with many geological activities and are published in various journals and many of these dates are quoted in the books.

Himalayan radiometric dates are related with age of formation of igneous and sedimentary rocks as well as rate and duration of sedimentation, time and duration of metamorphism(s) and deformation(s), date of initiation of thrusting, date and duration of upliftment, time of plate formation or separation, rate of plate movement and time of collision, rate of decompression and exhumation, an impressive number of dates are dealing with neotectonic activities, geomagnetic anomaly, palaeoseismicity, dating of topographic features, glacial activities and palaeoclimatic interpretation. From combined interpretations based on absolute dates and field and laboratory observations one can infer time wise relative contributions of various geological processes involved in the Himalayan development and finally a reliable evolutionary history of the Himalaya can be synthesized. Thus arrived sound geological reasoning makes the geology subject stronger. Despite availability of a large data bank, a final statement in above mentioned aspects in still far because continuous refining of dates due to progressive advancement in instrumentation techniques and interpretations open up fresh scope of adjustment, however, the final story may not be much different of what is available now.

The older events preserved in the Himalayan rocks have been dated by U-Pb, Pb ratio, ^{39}K-Ar, Ar-Ar, Rb-Sr, Sm-Nd, ion microprobe and fission track methods. The younger geological events are dated by ^{14}C, thermoluminisence, and other methods. D. Krummenacher gives probably first radiometric age from the Himalaya in 1961.

This discussion based on compilation of 632 data reveals that most of data, nearly sixty percents, are related with acidic and basic rocks. Nearly eight five percents dates are related with acid magmatism. The analysis of data further suggests that magmatic processes remained fairly active in major part of at least 2000 Ma old history of Himalaya. This is supported by widespread development and occurrences of acidic (75%) and basic rocks (25%). The time duration in million year of important phases of acidic magmatism are 2120-1765, 1220-1200, 675-601, 581-545, 516-460, 365-315, 265-230, 115-94, 65-55, 48-40, 21-19, 16-11 and less 7-3 Ma. These thirteen phases can be grouped into phase of intense (1220-1200 Ma, 516-460, 115-94 and 65 to end of Tertiary), active (2120-1765 or 675-601 etc.) sporadic (365-315 etc.) and diffused magmatism (in this category nearly 20% data population fall and are not part of above mentioned thirteen phases). On an average one acidic magmatic activity has taken place in 20 Ma years and particularly the Tertiary magmatism is most dominant one (Bhandari and Singh, 1979). There are many periods, of wide duration, of occasional acidic or absolutely no magmatism periods such as in 1800-1500 (i.e. 300 Ma), 1200-675 (525 Ma), 460-365 (95 Ma), 315-265 (30 Ma), 235-115 (120 Ma) Ma duration. In the above 1080 Ma duration, only four acid magmatic events occurred.

The major phases acid magmatism can be correlated with local and global tectonic disturbances for example magmatism during 115-95 Ma in related with plates formation from the fragmentation of Pangaea or Tertiary magmatism in general is an integral part of Himalayan orogenesis and five phases of Tertiary magmatism (65-55, 40-40, 21-19, 16-11 and 7-3 Ma) are related with collision or thrust initiation or regional upheaval events.

As stated earlier radiometric data dealing with basic magmatism (24) are comparatively less. This could be attributed to occurrences of basic rocks in innumerable small or minor size bodies such as dykes or sills. The large size plutons or batholiths are exceptional while the well-defined flow covering large surface area are few e.g. Panjal Traps, Rampur metavolcanics, Mandi-Darla volcanics or Abor volcanics (Bhat and Ahmad, 1990; Bhat and Lefort, 1992). In addition to size factor, most of the basic rocks are metamorphosed to varying degree, thereby upsetting the radiometric clocks. The phases of basic magmatism can be roughly grouped into 110-74 (5), 60-40 (3) and 30-10 (9) Ma and rest eight events are sporadically distributed in the stratigraphic column of Himalaya. Barring few dates, by and large the basic magmatic events show high degree correlation with acidic events. Bhat (1998) identified four basic magmatic events of mantle origin in the western Himalaya. The earliest event of the Late Archaean-Early Proterozoic age is followed by the events of Middle Proterozoic, Late Proterozoic and of the Late Mesozoic age. These events characteristically become younger in the northward and in the westward directions.

The age of different events of metamorphism is a matter of intense debate among petrologists. Empirically main event(s) of metamorphism occurred in Tertiary age and subsidiary or weak events in pre-Tertiary age and in both cases precise timing of peak metamorphic condition are interpreted differently. The age of metamorphism is determined by dating a newly crystallized mineral formed in that particular metamorphic episode. Variations in dates of such minerals suggest that time of peak metamorphism vary from region to region in Himalaya for example in Annapurna region of Nepal Himalaya two metamorphic peaks had arrived around 35 Ma (high temperature) and 15-13 Ma (less temperature) ago. In Everest region, hornblende gives 21 Ma age of peak metamorphism while in Bhagirathi Valley peak condition occurred around 19.8±2.6 Ma ago. A metamorphic event belonging to early Riphean age (1491.8 ± 59.7 Ma) is also reported 2-3 Km above the Munsiari Thrust in Bhagirathi Valley. The muscovite ages related with initiation of metamorphism are 17-18 Ma in Nepal and south Tibet while 21-22 Ma in Garhwal Himalaya and in the same region youngest age is 5.7 Ma.

The age data are also used to determine the rate of cooling in the Himalaya. It is found that cooling rates are not uniform in all the regions but varies from place to place and time to time. In MCT zone of Nepal rate of cooling was 380°C/ Ma in later phase while in early phase it was just 0.10°C/ Ma. In Garhwal region rate was 300° C/ Ma and 90° C Ma respectively. In case of leucogranite cooling rate may be as fast as 1000° C/Ma (Searle, 1995 and Searle et al., 1993).

The age of initiation of thrust activity and duration of thrust movement are also inferred with the help of radiometric data using the minerals that are grown during thrusting. It is believed that main Himalayan thrusts live MCT, MBT and MFT evolved in a sequential geochronological order. The MCT is the oldest and the MFT is youngest one. The MCT (Vaikrita) is believed to be activated around 29 Ma ago but MCT activation has been also proposed anywhere between 35 to 20 Ma in different region. It is proposed thrusting of VaikritaThrust continued up to 15 Ma ago while in Munsiari Thrust up to 6 Ma. The MBT and MFT were activated around 10 and 0.5 Ma ago respectively and they are still active one. In South Tibetan Detachment Zone brittle movement initiated between 16.37 and 16.67 Ma in central Nepal the higher reported age in this zone is 17.3 Ma and continued for at least 10 Ma. Similarly geochronological data are used for calculation of stress and strain rate and rate of upliftment. The rate of upliftment varies from sector to sector and has been discussed in the section of neotectonics. Rate of sedimentation has been calculated on the basis of radiometric data for example it is 14 m/Ma in Krol Belt sequence of Lesser Himalaya, 20 m/ Ma in Tethys sequence while 270-710 m/Ma in Siwaliks. In many areas of Siwalik Formation e.g. Paramandal-Uttarbeni, Jammu-Nagrota, Balli, Samba-Mansar and Jabbar Khas sections in India a sound magnetostratigraphy is established with the help of radiometric and palaeontological data. In the similar way magnetostratigraphy of Siwalik rocks are established in many parts of Nepal. Geochronological data revealed the dichronous nature of the lithofacies boundaries between Middle-Upper Siwalik i.e. lithofacies boundary is demarcated at 5.26 Ma ago in Jabbar Khas area while 3.90 Ma ago in Nagrota section in Jammu region. The lower boundary of the Boulder Conglomerate formation is also dichronous in character it is as young as 0.6 Ma in Parmandal region and as older 1.72 Ma in Nagrota area of Jammu, this led to interpretation that foreland propagating thrust and ancestral drainage system played the role in deposition of thick conglomerate beds.

Entire domain of plate tectonic process right from fragmentation of Pangaea, separation of India Australia and Antarctica, northward drift of India, rotation of Indian land mass, collision of Indian plate with Eurasian, closure of Tethys sea and finally growth of Himalayan mountain have been dated rigorously and dates are being refined time to time in the light of new observations.

Geochronological data provided better understanding of many Pleistocene and Holocene activities such as palaeoseismicity, rate of neotectonic upliftment and sedimentation, rate of faulting and thrusting, evolution of life, variation in palaeoclimatic characters, rate of subsurface water flow, glacial activities i.e. advancement and retreat, migration of animals, human beings and plants and rate of various geomorphic processes. Thus it can be stated interpretations based on geochronological and other data helped in better understanding of evolution of the Himalaya and many concepts prevalent in recent pasts have been modified in the light of new observations and interpretations.

8.1 Dates of Acidic Magmatism in Himalaya

The important granitic bodies with ages in Ma (Islam et al., 2005; Singh, 2005) of the Himalaya are -

1. Early Proterozoic

Iskere Gneiss (1852 ± 14); Kotla orthogneiss (1836 ± 1); Shang Granodiorite (1864 ± 4); Munsiari (1890 ± 155); Namik (1910 ± 88); Tawaghat (1906 ± 220); Dhakuri (2315 ± 135); Chaili, Bhilangana valley (2121 ± 60); Rihee – Gangi, Bhilangana valley (1841 ± 86); Ghuttu, Bhilangana (1763 ± 116); Chirpatiya, Bhilangana valley (1708 ± 131); Amritpur granite near Nainital (1585 ± 192); Joshimath-Guptkashi (1950 ± 200); Naitwar, Tons valley (1811 ± 133); Hanumanchatti, Yamuna valley (1972 ± 102); Bhatwari, Bhagirathi valley (2047 ± 119); Wangtu Satluj valley (2025 ± 86); Wangtu (1895 ± 64); Wangtu (1866 ± 10); Magladgad (2068 ± 5); JWGC, Wangtu (1860); JWGC, Sainj (1840); JWGC, Kishtwar (1840); Bajaura (1840); Luhri (1840).

2. Isotopic data of 1400-1200 Ma granites of NW Himalaya

Bandal (1220 ± 100); Nirath-Baragaon (1430 ± 150); Manikaran (1230 ± 120); Baijnath (1130 ± 110, 1320 ± 370); Gwaldam (1300 ± 80); Brijranigad, Bhilangana (1276 ± 12);

Ingedinala, Bhilangana (1139 ± 46); Rampur-Padampuri (1238 ± 128); Amritpur (1110 ± 131).

3. Isotopic data of Early Palaeozoic Granites of NW Himalaya

Almora Kumaun (506 ± 20); Ranikhet, Kumaun (485 ± 55); Chaur, Himachal Pradesh (530 ± 40); Akpa, Rakcham, Himachal Pradesh (477 ± 29); Kaplas, H.P. (552 ± 2); Mandi, H.P. (507 ± 100); Dalhousie, H.P. (472 ± 50); Manikaran H.P. (467 ± 45); Jaspa, H.P. (512 ± 16); Kulu, H.P. (518 ± 8); Mandi, H.P. (545 ± 12); Rohtang H.P. (601 ± 19); Kangan J&K (470 ± 11); Polokongka La, J&K (487 ± 25); Korzok, J&K (487 ± 14); Nyimaling J&K (460 ± 8); Rupshu J&K (483 ± 1); Tso Morari Gneiss J&K (472 ± 2); Hant J&K (489 ± 20); Nyimaling J&K (456 ± 16); Polokongka La Jammu & Kashmir (487 ± 23); Rupshu J&K (475 ± 23); Tso Morari Gneiss (499 ± 8); Dadeldhura, Nepal (470 ± 6); Simchar, Nepal (466 ± 40); Palung Nepal(470 ± 4); Formation III, Augen Gneiss Nepal (513 ± 30); Kangmar, Tibet (562 ± 4, 485 ± 6, 484 ± 7); Nanga Parbat, Pakistan (400 – 500); Manshera, Pakistan (516 ± 16).

4. Ages from Karakoram axial Batholith of NW Himalaya

Darkot plutonic unit, Karakoram (109 ± 4, 111 ± 6); Hunza plutonic unit, Karakoram (97 ± 17, 105.7 ± 0.5); K2 Gneiss and Mustang Tower Gneiss (120-85); Kuk (63 ± 2); Sarbzea (43 ± 6); Baltoro (21 ± 0.5); Chilas Igneous complex, Kohistan, north Pakistan (111 ± 24); Gabug, South Tibet (43 ± 3); Lag hoi Kangri south Tibet (15.1 ± 0.5); Nyalam south Tibet (16.8 ± 0.6); Maitia, south Tibet (9.8 ± 0.2); Tangtse, Leucogranite (18 ± 0.6); Ladakh, India Panamic body (83 ± 9).

5. Isotopic data of Trans Himalayan granites

Gaik (235 ± 13), Dras volcanics (264), Kargil intrusives (103 ± 3 and 101 ± 2), Leh (63 ± 5, 49.8 ± 0.8, 58.4 ± 1, 61.5 ± 2, 60 ± 10), Shey granite (60 ± 10), Kohistan arc (102 ± 12, 54 ± 4, 49 ± 11, 26 ± 1), Kailash intrusives and volcanics (38.8 ± 1.3) and Gangdese (120-90 and 70-40).

6. Isotopic data of Higher Himalayan Leucogranite

Nanga Parbat Haramosh Massif (2.3, 5, 7); Mango Gusar Granite (37 ± 0.8); Dassu Gneisses (6.8 ± 0.2); Jutial, Karakoram Nanga Parbat (10-5.3); Sumayar Leucogranite (9.2±0.5, 8.6); Baltoro, Karakoram Batholith (21 ± 0.5); Leucogranite, Karakoram Batholith (21.4 ± 0.3); Chichi granite (22-16); Safat Leucogranite (26); Zanskar Leucogranite (20.8 ± 0.3); Ghumber Ranjung Leucogranite Zanskar (21.4 ± 0.1); Gangotri Leucogranite (21.1 ± 0.9) Shivling Leucogranite Gangotri Leucogranite (21.9 ± 0.5, 23 ± 0.2); Manaslu Leucogranite (29±1); Makalu Leucogranite (24±0.2); Everest Leucogranite (20.6); Annapurna Leucogranite (22 ± 1); Shisma Pangma Leucogranite South Tibet (17.3 ± 0.2).

DIMENSIONS OF HIMALAYAN GEOLOGY
A. K. Biyani (2006) : pp.209-219
Satish Serial Publishing House, New Delhi
E-mail: ssp@ndf.vsnl.net.in, hkjain1975@yahoo.com; website: www.satishserial.com

Prakrtim svamavstabhya visrjami punah punah
Bhutagramamimam krtsnamavasam prakrtervasat (Geeta, 9.8)
This whole multitude of beings subject to the influence of their own nature

CHAPTER - 9

GEOPHYSICAL STUDIES

Initial geological studies in Himalaya were mainly field oriented subsequently data and samples collected in the field were processed in the laboratories. In later stage of investigations various geophysical techniques were employed particularly for probing the subsurface phenomenon.

9.1 Interior of Himalaya

Deeper probe of earth's interior is reliably carried out by various natural and synthetic seismic techniques, in addition to this low cost gravity and magnetic techniques also provide useful information while the shallow probe is normally carried out through electrical, geological and drilling techniques. Three fundamental layers constitute earth are the crust, mantle and core. Mohorovicic Discontinuity separates the crust from mantle and the Gutenberg Discontinuity separates mantle and core. Crustal thickness is estimated by the surface, body and earthquake waves, DSS profile, gravity and isostasy methods. Thickness of the crust varies according to the tectonic setting of the region; it is just 5-6 km thick in oceanic region, (below abyssal plain), 20-25 km in thickness in ocean-continental transitional area (in continental slope and rise regions) and crust has average 35 km thickness in the continent regions. Again in continent, it is comparatively less thick in the cratonic/shield area and is thicker particularly in young folds mountain areas. Further crust is divided into two layers, the upper crust and the lower crust of granitic and basaltic composition respectively. These two sub-crustal layers are separated by a second order discontinuity called Conrad Discontinuity. The average thickness of crust in the peninsular India is 38.7 km (Kaila and Krishna, 1992).

The thickness of the crust in Indo Gangetic Plain varies between 40-45 km, however, the average thickness is 43 km in the plain and the upper crust is 20 km thick and lower crust is 23 km of which top 5 km (average) is constituted by Quaternary alluvium, though the maximum thickness of alluvium is reported to be of 10 km near sub Himalayan foothills. The average density of sediments near surface is 2.0 gm/cm^3. P and S waves velocities in the alluvium are 3.40 km/s and 2.0 km/s respectively. Below alluvium, between 3.5 to 20.0 km depths, the density increases to 2.60 gm/cm^3 the P and S waves velocity to 6.15 and 3.55 km/sec respectively. The 23 km thick (20-43 km depth) lower crust below Indo Gangetic Plain is inferred to have density 3.30 gm/cm^3. The P and S waves velocities in this zone are 6.58 and 3.80 km/s. The upper mantle beneath the Plain shows shield like characters and the velocities of P and S waves are 8.19 and 4.6 km/s respectively. The Low Velocity Zone (LVZ) situated at 100 km depth and the mantle between Moho and LVZ has density 3.30 gm/cm^3 and P and S waves velocity are 8.19 and 4.60 km/s respectively (Mahadevan, 1994). In Garhwal region of Uttaranchal the range of P and S waves velocities are 4.72 km/s and 5.89 km/s and 2.90 km /s and 3.39 km/s in the upper and lower crusts respectively.

In Himalaya, crustal thickness gradually increases in north direction and finally attains the distinction of world's maximum crustal thickness, 78 km, in the Great Himalaya region between Kanzalwan and Nanga Parbat (Hirn et al., 1984) and at Hanle, Indus suture zone, Moho is recorded at 85-90 km depth The average crustal thickness in the Himalaya is 55-60 km. The upper crust is 30 km thick while lower crust is 25-30 km in thickness. The crustal thickness estimated by gravity method in the Mount Everest is 55 km. In northeastern Himalayan region Moho is situated at 46.3 km depth and thickness of upper and lower crusts are 24.8 and 21.5 km respectively while another analysis yielded 41.5 and 22.2 and 19.3 km respectively. The P and S waves velocity in the granitic and basaltic layers of the Garhwal Lesser Himalaya is 5.2 and 2.97 km/s and 6 and 3.43 km/s. In Assam and Arunachal Pradesh, based on Pg, P* and Pn velocities of 143 earthquakes recorded during 1985- 1997, thickness of the crust is 45.3+1.4 km, 48.9± 1.2 and 51.8± 1.4 km below Brahmaputra Valley and at MBT and MCT respectively. The approximate thickness of the upper and lower crusts in these areas is 21.5 and 23.8, 25.1 and 23.8 and 28 and 23.8 km (Sitaram et al., 2001). The S wave velocity just below the crust 4.4 km/s in Garhwal Himalaya and LVZ is situated between 100-110 km depth. Within LVZ, S wave velocity drops to 4.0 km/s. Crustal thickness near MBT is in Jammu area is 44 km and 57 km in Pir Panjal range (Malinconico, 1989).

In Hindukush and Himalayan regions seismic data of 51 deep earthquakes yielded valuable the P-wave velocity and depth relationship across the Moho (33 km) the P wave velocity changes from 6.50 to 7.90 km/s and across 113 km depth 7.90 to 8.13 km/sec. In Himalaya, at 354, 404, 434, 825 km depths the velocity of P wave are 8.13, 9.55, 9.55 and 11.34 km/s respectively. In Hindukush region the P wave velocity at 362, 412, 545, 644, 675 and 822 km depths are 8.13, 9.67, 967, 10.84, 10.84 and 11.12 km/s respectively. It is inferred from these data that in Hindukush region 400 and 650 km velocity discontinuities are relatively elevated by 80-150 km and 30-80 km respectively. With the help of deep seismic sounding technique, 18 km thick sedimentary sequence is reported from Jammu and Srinagar synclines.

The dip of Moho is 3° in IGP while 15°-25° in Kanzalwan location of Ladakh. Moho it self is faulted at many places and 6 km throw is inferred in south of the Nanga Parbat. The top and bottom of asthenosphere i.e. upper Mantle in Kashmir is inferred to be situated at 150-180 and 340-365 km depth respectively.

9.2 Gravity Anomaly

Gravity anomaly is the difference between an observed value of gravity reduced to geoid (or reference equipotential surface) and a theoretical or normal value of gravity at a point on the ellipsoid (Rapp, 1989). From the basic data set several types of gravity anomalies can be computed for different interpretation purposes. The commonly used gravity anomaly types are Bouguer anomaly, free-air anomaly and isostatic gravity anomaly. Bouguer anomaly, which is negative in land area, suits very well for interpretation of earth structure and its interior. Bouguer anomaly shows regional correlation with altitude on the earth surface while free-air anomaly is opposite to it, means higher the regional elevation more negative is Bouguer anomaly. Gravity anomaly is measured in milligal (mgal) unit.

In Indo-Gangetic Plain, the Bouguer anomaly is –50 mgals near the northern edge peninsular India or southern boundary of the plain and this decreases to –220 mgals in Punjab or –250 mgal in Sarda depression situated near the northern boundary of the plain or –210 mgals in Gandak Depression. The anomaly contours show WNW-ESE trend in Punjab plain while E-W trend in Ganga plain. Anomaly value also depends on average density of rock formations; the Siwaliks are being of light density (2.1-2.3 gm/cm^3) in comparison to high density (2.55-2.60 gm cm^3) Vindhyan rocks so on gravity anomaly basis more than 10 km thickness of sediment is inferred in the south of Himalayan front. This indicates asymmetric slopes of foreland basin and basin is deepest near the northern boundary of the plain.

In Pakistan, Bouguer anomaly value is 50 mgals near the edge of shield and its value gradually increases to –400 to –450 mgals in Hindukush, -450 mgals in the Haramosh area and reaches to peak value –550 mgals in the Great Karakoram range. In a section across Central Indian shield, the value of Bouguer anomaly is –50 meals at the edge of shield, -350 mgals in the Higher Himalaya and –500 to –550 mgals in Tibetan plateau. The Kohistan area shows peculiar feature of positive and negative anomaly. In Myanmar, eastern subduction, the Bouguer anomaly value is 0 to –25 mgal in Bengal Basin, -175 mgal in the east of Sawsgang Fault, -120 mgals in Myanmar plain and +50 mgals over Shan Plateau.

The isostatic anomaly in the south and north of the plain is varying between –10 to –30 and –90 to –100 mgal respectively, when 30 km thickness is taken for computation. This is due to low density of sediment and bending of lithosphere. The same anomaly shows positive value in the range +80 to +100 mgals in Higher Himalaya. The isostatic anomaly in general show positive trend between HFF and to the north of suture zone (Main Karakoram Thrust) because of thickening of crust and inferred presence of higher density rocks in deeper part (Verma, 1991).

The effective elastic thickness for the Himalayan region is estimated as 50 km and depth of compensation 70 km by merging satellite and field gravity data. The effective elastic thickness suggests brittle nature of the crust for compensation of the topographic load (Rajesh andma, 2003).

9.3 Heat Flow

The solid earth is a giant heat engine sustaining all types of tectonic activities since its origin. The hot visible evidence interior is time to time manifested through the outpouring of lava by hundreds of active volcanoes distributed all over the globe and continuous discharge of hot water and steam from springs. The heat of interior is transmitted to cold outer surface mainly through conduction, convection or advection and also rarely by radiation mechanisms. The heat flow is measured in various units like milli watt per square meter (mWm^{-2}) or calorie per square centimeter per year ($Cal/cm^{-2}/yr$) or watt per year. The average global outflow of heat is annually estimated at $4x10^{13}$ W or 80 mWm^{-2}.

Limited numbers of heat flow measurements by conventional or chemical methods are carried out in the Himalaya. The synthesis of data reveals that Himalayan terrain is comparatively hotter than the peninsular India. The heat flow value in Himalayan region varies between 40 to more than 400 mWm^{-2} in different divisions of the Himalaya. On the basis of heat flows value the Himalayan terrain is divided into four zones (Ravishanker 1989, 1993).

Zone - I

The heat flow value in this zone is greater than 180 mWm^{-2}. This zone correspondence to Trans Himalayan region and is characterized by comparatively younger magmatic and volcanic activities. The youngest granite in this region is just 3.5 Ma old. At Chumathong 300±120 mWm^{-2} heat flow value is recorded and at Puga in Ladakh, the geothermal gradient near surface is estimated to more than 1000° C/km.

Zone - II

This zone comprising of the Tethyan and Higher Himalaya, is situated on the convex side of Zone I and heat flow value ranges between 180-100 mWm^{-2} with an average of 130±30 mWm^{-2}. The estimated geothermal temperature gradient in this region is 61°±20°C/km.

Zone - III

This zone corresponds to Lesser Himalaya and heat flow varies between 100-70 m Wm^{-2}.

Zone - IV

It is foothill zone and is characterized by subnormal heat flow value (varies between 70-40 mWm^{-2}) and geothermal gradient (17.5° C/km).

9.4 Earthquakes and Seismicity

The Himalayan arc is seismically one of the very active intracontinental regions in the world. Every year thousands of earthquakes of varying intensity/magnitude are recorded in the Himalaya. Earthquakes of higher magnitudes causing extensive damage to property and life and they are vigorously shaking the fragile rocks thus making slopes vulnerable to geocatastrophic processes.

Dr. T. Oldham initiated study of Himalayan seismicity, after great Cochar earthquake of 1869; he compiled the seismic record of Himalaya since historic time to 1885. His illustrious son R. Oldham continued the work further and made discovery of P and S waves by analyzing the June 12, 1897 earthquake of Assam. Since then Himalayan seismicity is being continuously monitored and several Indian seismologists who worked in Himalaya, earned respectable position in the comity of seismologists.

Great majority Himalayan earthquakes are of tectonic origin and caused by the sudden disruption of sub-surface strata (Gupta, 1992). The breaking up of rocks quickly releases huge amount of stored strain energy in the form of P, S and L waves. The earthquakes are recorded on instrument called seismographs hosted in seismological observatories. Varieties of seismographs like Wood Anderson, electromagnetic, analogue, digital are in use. Damage is qualitatively measured on 12 points Modified Mercalli Scale or by amplitude of waves or acceleration of vibrations caused by sudden rupture of strata on Richter Scale; in thc form of simple arithmetic numbers like 4.6, 7.8, 6.7 etc. These numbers described the magnitude (M) of earthquakes. The magnitude scale is further refined into surface wave magnitude (Ms), body wave magnitude (Mb), long wave magnitude (ML) and coda magnitude i.e. train of diminishing wave after earthquake. Seismic moment scale based on displacement of rock masses is yet to become popular. A chain of observatories between Srinagar in west to Kohima in the east is situated in Himalayan terrain and also many observatories are situated in the plains of Indus, Ganga and Brahmaputra. These observatories continuously monitor the seismicity of not only Himalaya but also of the adjoining regions and analyze the seismic data as soon as any seismic event happens.

9.4.1 Seismicity in Himalaya

Indian Standard Institute (ISI) has prepared a seismic zonation map of the India and divided the country into five seismic zones (Krishna, 1992); this is now revised into four zones beginning with zone II as there is no zone I. The entire Himalaya and adjoining plain fall under category IV and V i.e. an area of maximum earthquake incidence. Narula (1991) in another attempt identified five seismotectonic domains namely – Main Himalayan Seismic Zone, Kashmir Syntaxial Seismic Zone, High Plateau Seismic Zone, High Himalaya Seismic zone and Foot Hill Seismic Zone in Western Himalaya. In another classification Gupta and Srivastava (1992) divided Himalayan seismicity into six regions namely: Hindukush, Kashmir – Himachal Pradesh, Uttaranchal and Western Nepal, Eastern Nepal – Sikkim, NE India and Andaman – Nicobar region based on earthquakes characteristics like depth of origin, magnitude, active tectonic elements, (thrust, faults etc.) and related factors.

Many efforts have been made to compile the seismic records of the Himalaya and neighboring areas in recent years Chandra (1992) catalogued 391 events occurred during 1505 to 1963 and 399 events between 1964-1986 of surface wave magnitude (Ms> 5.5) and body wave magnitude (Mb> 5.0) in the area lying between 25°-36°N, 69°-83°E; 20°-30°N, 83°-100°E of Himalaya and vicinity. Several great earthquakes of highest intensity and magnitude such as Mandra (1669; Io = X), Kashmir (1828; Io = X), Kashmir (1885; Io = X), Kangra (1905, M 8.6) Nepal (1833; Io = X), Nepal (1869; Io = X) Bihar-Nepal (1934, M-8.4) Assam (1897; Io = XII or M – 8.7), Assam (1950, M-8.7) and in south of Himalaya namely in Delhi (1720) and Calcutta (1737) claming nearly 3 lakh lives each are recorded.

1905 Kangra (Himachal Pradesh) and 1934 Bihar earthquakes killed nearly 20000 and 11000 people respectively. The effect of seismic waves by 1897 Assam earthquake is felt in 4.5 M sq km area. The reevaluation of magnitudes of the Assam (1950) and Kangra (1905) earthquakes suggests Ms 8.0±0.1 and 7.8±0.05 respectively. The latest earthquake of October 8, 2005 Muzaffarabad in Pakistan was having focal depth 10 km and is of 7.6 M. The NE area is highly seismic in nature because of presence of three plates (Indian, Burmese and Eurasian) and since 1897, ten earthquakes M >7.5 struck in this region. 10 to 15 earthquakes of low magnitude are recorded daily in the northeastern region while 19 damaging earthquakes (M = 5.2-6.8) are recorded between 1980-1992 (Gupta, 1993). The figure of damaging earthquake for entire Himalaya is more than 650 (M = >5) between 1890-1992.

In entire Uttaranchal 384 seismic events of <2 to 3 M were recorded during the period between January, 1993 January, 1995. 230 events were originated from <10 km depth and 58 events had focal depth >15 km. Eastern Uttaranchal is seismically very active region of the Himalaya, 106 seismic events 0.5-4.2 ML are recorded during May, 1999 to November 2000. The range of seismic moment of these events are $5.92\times10^{(e10)}$) N-m to $11.43\times10^{(e10)}$ N-m and stress drop ranges between 0.01 to 9.5 bars for the earthquake 2.0-3.4 M/events indicating that the crust of the region is unable to sustain higher stresses (Paul et al., 2004). Another very active seismic region is northwestern Himalaya- Hindukush area, possibly due to plate subduction at considerable depth. The intensity of seismicity progressively decreases in Badakshan massif, Pamir Foredeep, Kohistan region, Aksai Chin zone and in South Kashmir region.

9.4.2 Tectonic Control

Underthrusting of Indian plate is responsible for deformation of rocks and that subsequently pushing the rocks in general in south direction and also piling up them one over the other. The endogenetic turmoiling is manifested in formation of large number of faults of different nature and many of these are still active. Time to time slips along these faults release stored strain energy in the form of generation and propagation of elastic waves in the earth (Narula and Shome, 1992). In northwestern Himalaya seismicity is identified along Indus and Kohistan/Shyok Suture zones. The seismic plane here dips in NE and extends up to 70 km depth. Riasi Thrust situated 60 km south of suture zone and in further south Hazara Thrust System (a set of faults parallel or perpendicular to suture zone with reverse and strike slip fault plane solutions) mainly controls the seismicity in this area. In western Himalaya seismicity is mainly confined along the MBT (its equivalent Panjal Thrust in Kashmir; Jawalamukhi, Nahan and Krol Thrust in Himachal Pradesh). In Garhwal Seismicity is confined to the north of MBT; comparatively shallow earthquakes (6-8 km depth) show strike slip solution and modest (15-20 km) earthquakes show thrust solution (Khatri et al., 1989). In Central Nepal, most of the earthquake epicenters lies north of the MBT and are confined till 20 km depth from the surface. The Satlitta Fault shows normal fault solution. 1991 Uttarkashi earthquake is attributed to movement in the sub-surface Delhi-Hardwar ridge (Sati and Nautiyal, 1994). A couple of earthquake originated in Ganga Valley have shown normal fault solutions. The Nepal-Bihar earthquake of 1934 was originated close to Lesser Himalaya and has shown thrust-cum-strike slip movement in Monghyar -Saharsa ridge at 57 km depth deciphered by seismic analysis. In the NE region, deepest earthquake is recorded from 91 km depth; again seismicity is high and is close to MBT. The 1897 Shillong earthquake had shown trend of seismicity transverse to regional structure. Tista, Gangatok, Kotrili, Bomdila, Padma, Madhupur and Sylhat are prominent seismically active transverse lineaments of NE region. Most of the seismicity in Tibetan Plateau is confined to shallow depth (5-10 km) although an isolated deepest event was originated as much as 85 km depth. The Tibetan area is in general state of tension, however in the northern Tibet fault plane solutions indicates both normal and strike-slip faulting. The minimum stress axis is oriented oblique to Himalayan range.

MCT is comparatively less active than MBT and along the surface trace of MCT earthquakes occur in 10-30 km wide zone and are originating from the depth less than 10 km; however, one reported earthquake is originated 32 km below ground surface (Pande, 1985). The lack of seismicity north of MCT is attributed to inactive nature of rigid Tethyan slab. The HFF is seismically less active fault. The 10-15 km depth of most of the earthquake is caused by presence of thick gouge in existing faults and due to transition from brittle to ductile environment caused by increase in temperature. Thus the unstable coseismic frictional slip is expected to occur in the above depth range. Great earthquakes are caused by slip over gently dipping plane, called detachment fault, which is separating the overlying sedimentary wedge shaped formations from the subducting Indian plate. 1897 Shillong earthquake indicates 16 ± 5m reverse slip on a 50° SSW dipping ESE trending intracrustal fault of 100-120 km length while in the 1905 Kangra earthquake the slip was of less than 5 m along a rupture of 110 km length that earlier was believed to be of 250 km length. The Kangra earthquake triggered an earthquake of >7 M in Dehradun within a few minutes at a depth of 30 km (Hough et al., 2005).

Recent seismicity also shows cyclic pattern Two periods of high seismic activities namely 1897-1916 and 1934-1951 are recorded while intervening period is also present one extending till now is of low activity (Chander, 1988). The average recurrence time of great earthquake in the Himalaya ranges between 200-270 years and for earthquakes of magnitude greater than 6 is 15 years for earthquake a magnitude is 2.5 to 3 years. Bilham and Ambraseys (2005) doubted the idea of recurrence interval of earthquakes on basis of comparison between present GPS driven convergence rate, 18 mm/yr, and evaluated convergence rate, 6 mm/yr, from seismic records of last 500 years. A cumulative seismic moment release in Himalaya is estimated of 1.3×10^{29} dyne-cm with <5 cm/yr slip rate in last 500 years and suggested that 500 years is insufficient time to record a complete earthquake cycle in Himalaya. The average release of energy by all earthquakes in the Himalaya is estimated at 3.7×10^{21} ergs for 100 km depth and 1.8×10^{22} ergs for 50 km depth and ultimate stress are 0.11×10^{-3} and 0.26×10^{-3}. The slip rates of great earthquakes are calculated around 17±2 mm/yr; however, it decreases from the east 18 mm/yr in Nepal to 10 mm/yr in Pakistan. The rate of upliftment on the strike parallel faults in the MBT zone in Nepal is calculated 3-4 mm/yr in the 400000 to 50000 years.

In general, earthquakes of shallow depth origin are more common and are widely scattered. This phenomenon is attributed to presence of many deep fractures and these fractures are in the process of adjustment. The estimated average ruptures length for an earthquake of M 8.4 in Himalaya is 270 km. Four great earthquakes of 1897, 1905, 1934 and of 1950 had produced ruptures of 500-550, 280-300, 300 and 300-450 km length and if 1803 and 1833 earthquakes are also taken into calculation then entire length of Himalaya (2400 Km) is ruptured by major earthquakes in 147 years of 19th and 20th century. Still in Himalaya some seismic gaps are observed e.g. between Kangra and Bihar earthquakes the gap is situated between 82°-87° E. In future major earthquakes would eventually fill this gap (Khatri, 1993,1999).

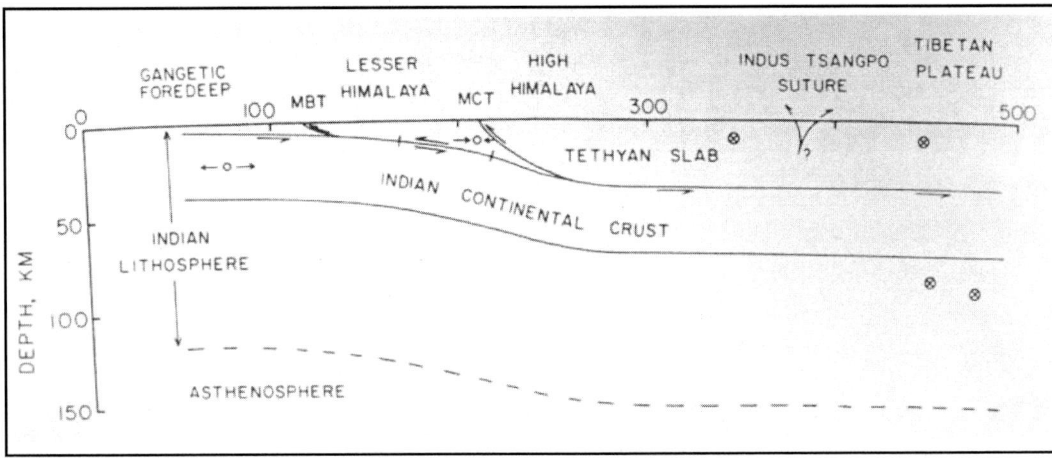

Figure 9.1. Evolutionary model that supposes all thrust earthquakes generate at the detachment surface (after Ni and Barazangi, 1984).

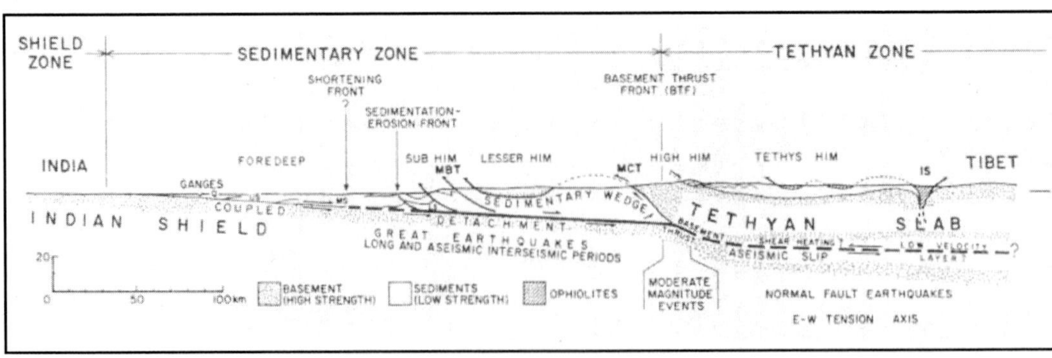

Figure 9.2. Steady state model of Seeber et al., 1981 that says that active Himalayan thrust coincides with the subducting slab. Detachment denotes that portion of fault situated between subducting slab and sedimentary wedge and basement thrust is portion of interacting slabs.

Ultimate cause of fault controlled Himalayan seismicity is explained by two models namely the evolutionary model proposed by Ni and Barazangi (1984), this model envisages that zone of plate convergence is progressively shifting in the south direction by formation of intracrustal thrusts in the Indian shield the process started from Indus Suture Zone and progressive migrated in the south along the Tethyan Thrust, MCT, MBT to MFT from Eocene time onward. In this model all thrust earthquakes define a part of the detachment between the underthrusting Indian plate and upper Himalayan block (Fig.No. 9.1). The other model is designated as steady state model of Seeber et al. (1981) which proposes presence of a detachment surface between under thrusting Indian plate (below) and overriding rocks of nappe, thrust, sedimentary wedge etc. (above). The separation surface is also called basement thrust fault, and earthquakes are generated whenever movement in the basement thrust takes place and earthquakes of modest magnitude occur in a narrow zone down dip from basement thrust fault (Fig. No.9.2).

9.4.3 Mid–Crustal Ramp

On the basis of microseismicity, seismic reflection analysis, gravity anomaly, geodetic measurement, uplift rate and geological observations a mid-crustal ramp at 15-25 km depth, a region of relatively steep dip or mid-crustal ramp (15°-20°), in the sub-surface Main Himalayan Detachment is inferred in the Lesser and Higher Himalayan transition zone (Gahalaut and Kalpana, 2001).

9.4.4 Reservoir Induced Seismicity

Reservoir Induced Seismicity (RIS) term is used for those quakes, which are generated due to accumulation of water in the dam. Many large dams (>1 cubic km reservoir volume) are constructed in Himalaya and many more are to be constructed in future. It is feared that large dams would enhance seismic risk but Gupta and Rajendran (1992) have shown that in case of 11 large dams the seismicity instead of increasing has actually decreased. This is attributed to thrust fault environment of Himalaya. In case of thrust faults, seismicity decreases due to impoundment of reservoirs; scientists are still skeptical about the role of percolating pore fluid under the reservoirs and dams.

9.4.5 Palaeoseismology

Study of historical seismicity or palaeoseismicity falls in palaeoseismology or active tectonics. In the quantitative parts approximate time and magnitude of the palaeoearthquakes are determined and in qualitative parts evidences of seismicity through earthquake borne structures or seismites are investigated. In the Himalaya terrain such studies are in the incipient stage A few palaeoseismic analysis of the Outer Himalaya e.g. Nadoh area of the Haryana is carried out and fault segments of few to couple of tens of km have been identified and their magnitudes as well as probable timing of events are determined. Soft sediments deformational structures are reported from the Pliocene- Quaternary fluvio-lacustrine sediments (5 Ma) of Khalsar and Tirit area of the Ladakh and Karakoram regions. Seismites are also reported from Sumdo area and Baspa Valley of the Himachal Pradesh.

9.5 Geomagnetic Deep Sounding (Gds), Magnetovariational (Mv) and Magnetotelluric (Mt) Investigations in Himalaya

These are geophysical methods of explorations for minerals and petroleum deposits. However, they can be used for study of large-scale subsurface tectonic features of the earth. Electromagnetic (EM) investigations record diffusion of electromagnetic waves in the earth and measure the vertical and lateral electrical resistance of the rock formations. Variations in the geomagnetic field of the earth are recorded through electromagnetic waves in the permanent as well as in the field based temporary observatories. In these observatories, electrical configuration of mega structures like regional folds and thrusts are recorded and are interpreted through electrical discontinuities. In the Outer and Lesser Himalaya, GDS study recorded a conducting layer dipping towards MBT and conformed the presence of a midcrustal conductor, a subsurface zone of higher electrical conductivity, probably formed by release of fluids during metamorphism. This mid-crustal layer correspondence with the seismogenic subsurface detachment plain present in the Himalaya. GDS study in a section from northwestern peninsular part of India to Himalaya revealed the presence of a major elongated Trans Himalayan Conductor bounded by Delhi-Hardwar ridge and Moradabad Fault that coincides with a localized zone of high seismicity. This conductor furrowed in three conductive zones in the west (Arora, 1993, Arora, 2000, Arora and Reddy, 2000).

Magnetotelluric study of the earth is based on response of the earth to the naturally occurring electromagnetic waves. In this study time variations of electric and magnetic fields of the naturally occurring electromagnetic fields are measured on the surface. Magnetotelluric studies in the 150 km long Mohand- Ramnagar profile have shown presence of a four- layered structure in the region. The top 50m thick layer comprised of alluvium, of 6-20 U-m resistivity, this is followed by a 700-1000 m thick Upper Siwalik layer of more resistivity (30-200 U-m), the third 3-4 km thick layer of the Middle and Lower Siwalik is of 9-15 U-m. The basement of high resistivity called Trans Himalayan Conductor identified by strong conductive anomaly is present at 5-8 km depth. This is a subsurface extension of 50-60 km wide Aravalli Mountain and is located 8-12 km depth. In the Pang- Phobang Profile of Ladakh- Karakoram area, Tso Morari Dome of metamorphic rocks is of very high resistivity (>1000 Ohm-m), this domal mass in south is followed by a vertical conductivity feature that correspondences to the thrust boundary of the Tethyan sedimentary rocks and in the north another the vertical conductor, Indus suture, separates the domal mass with granitic batholith of the Ladakh of >1000 Ohm-m and in the further north a 12 km deep fault separates the low grade metamorphic rocks with the high grade metamorphic rocks (Gupta et. al., 2005).

9.6 Global Positioning System (Gps) Based Study in Himalaya

The Navigation Satellite Timing and Ranging Global Positioning System (NAVSTAR GPS) system was first developed by the United States of America in 1978 for high accuracy satellite- based navigation and surveying system. Initially access to data was restricted and was only used for military purposes; however, later in 1993 use of GPS data for civil purposes were also allowed. Civil access to GPS data provided tremendous boost to current geodynamic research. To offset monopoly of the USA, European countries launched first satellite of Galileo system in December 2005 with some additional information system. Galileo will become fully operational in 2008. GPS is primarily being used for identifying the geographical location and for monitoring the rate movement on any scale. GPS data can be used in calculations of convergence, displacement and upliftment and strain computation and seismic hazard assessment in orogenic belts as well as in the continents it is also useful in the study of landslides. In India as well as world over very large number of scientific and technical institutions are measuring and interpreting the GPS data on regular basis that might be daily to periodic of variable time duration. GPS technique in crustal deformation and seismic monitoring is based on periodic measurement of precisely located baselines that are capable of detecting even the movement of a few mm/yr. In the field GPS is fixed on a concrete pillar and after a lapse four months time instrument is installed on the pillar and data are collected after an interval of time varying between 4 to 72 hours. Data are either recorded manually or goes directly into computers for further processing.

GPS based setup started functioning in India in 1994 for crustal deformation study as it can detect even very small movements, thus it has become an indispensable tool in current crustal deformation studies of Himalaya as GPS quantitatively measures the velocity of the plates and upliftment and subsidence of the land. On a 530 km long base line in the southernmost part of the south India, 54±3 mm/yr velocity of the Indian plate is determined with an angular rotation rate 0.44±0.03/Ma. The study further reveals that peninsular part does not behave as single rigid block but in between Delhi and Banglore convergence is taking place with 3±2 mm/yr rate which might have been accommodated in the Narmada –Son Lineament or with in the shield area by forming a bulge.

In India, eighteen GPS stations are installed in and around Himalaya. In Himalaya, 18±2 mm/yr or 1.8 m/centaury convergence rate is calculated on the basis of current GPS data and most of the Himalayan deformation is taking place between the MCT and MFT and the zone of maximum deformation coincides with the zone of maximum micro seismic activities. However, this rate is 14±1 in western Himalaya, 10-20 mm in the Uttaranchal, and 10-mm/ yr in the Darjeeling- Sikkim Himalayas. 11±4/ yr slip in the Karakoram Fault adjusted in east-west extension in the southern Tibet and westward movement NW Himalaya toward the Nanga Parbat, therefore, restricting the eastward extrusion or movement of the Tibet. GPS has not only detected slip movements in the Lesser Himalayan thrusts indicating building up of the strain that might be released by the seismic activities in future. This is further supported by the model inversion data that MFT is locked over ~100 km width and this zone is building up a slip deficit at a rate of 14±1 mm/yr and so far nearly 8-10 m of potential slip accumulated in Garhwal Himalaya (Banerjee and Burgmann, 2002, Mukul, 2004). In right lateral Karakoram strike-slip fault 11 mm/yr movements has been recorded. However, on the basis of beryllium concentration in the boulders of the moraine provided 4 mm/yr slip rate in the Karakoram Fault area (Brown et al, 2002).

The salient GPS findings are that velocity and deformation rates in the Himalaya is substantially different in different sectors implying different mechanisms of deformations are operating in different regions, the extension vector between the Himalayan GPS stations and Lhasa (Tibet) is consistent with the east-west extension of southern Tibetan Plateau, significant lateral variations in the convergence rates in the eastern Himalaya (10-12 mm/yr in the Sikkim–Darjeeling and Himalaya), identification and measurement of movements in the active faults and landslides and in estimation of current strain accumulation rates.

DIMENSIONS OF HIMALAYAN GEOLOGY
A. K. Biyani (2006) : pp.221-240
Satish Serial Publishing House, New Delhi
E-mail: ssp@ndf.vsnl.net.in, hkjain1975@yahoo.com; website: www.satishserial.com

Dvamimau purusau loke ksarascaksara eva ca
Ksara sarvani Bhutani kutasthoksara ucyate (Geeta, 15.16)
In this world there ar two kinds of entities perishable and imperishable; the bodies of all beings are perishable and the soul within is said to be imperishable

CHAPTER – 10

ENVIRONMENTAL GEOLOGY

Man is greatly dependent of the geologic system and to make life easier he is relentlessly modifying this geological system. The modifications are aimed at to provide congenial living conditions. Many of these modifications are detrimental in short duration and some are certainly catastrophic in longer one. The growth in population and subsequently rise in requirements are putting accelerating pressure on the geologic system; whether the area is habitated or not. Not only man, nature does modify or destroy herself either violently or slowly. Difference between man and nature lies in intensity or speed of their actions what nature does, for example, denudation over a very extended period of time, same results man accomplishes in much shorter duration i.e. in days or weeks by mining with the help of mighty technology at his command. Actions of man and nature hardly left any area untouched, however at same places influences of their actions are more visible than the other. In this context, Himalayan geologic environment is more vulnerable or sensitive to pressure of these actions because of inbuilt weakness in Himalayan terrain because of presence of partly consolidated sedimentary or tectonically crushed rocks spreading over a large part, close spacing of intricate network of active faults and joints, over steepened slopes, thunderous falls and turbulent flow of rushing waters, continuously sustained geomorphic inequilibrium conditions due to time to time jerking, feeble to violent, of land by earthquakes in addition to these presence of dense exploitative population preying on natural and forest resources, modifying the slopes and damming the rivers.

Combinations of slow and convulsive actions and reactions of nature and human activities in Himalayan terrain damaging the geologic system beyond repair in some regions while in general the entire terrain show sign of wilting of varying orders. Attempts to reverse the damage are exerting further pressure on the system. The certain environment damaging actions like sudden heavy down pour or cloud burst or global warming or loss of forest cover or faunal actions are non-geologic in nature but they cannot be excluded from the scope of environmental geology because of their overlapping influence on geologic processes.

Source wise important geologic problems faced by the Himalayan regions are as follows:

Problems Created by Nature

1. Seismic
2. Soil and Land Erosion
3. Snow avalanche
4. Silting up of rivers and lakes.
5. Change in course of rivers.
6. Glacial melting and Global Warming.
7. Desertification, Cloudburst, Heavy Rainfall
8. Global Warming

Problems Created by Man

1. Slope modification for civil constructions and agriculture.
2. Pollution of rivers and lakes.
3. Minerals exploitation
4. Reservoir induced seismicity.
5. Loss of vegetation cover, faunal actions and elimination of fauna

Problems Created by Nature and Man

1. Landslide and mass wasting.
2. Forest fire

In environmental geology terms like hazard, risk, disaster and ecology are frequently used. The term hazard is a potential threat to human and their welfare and is often employed for harmful action of any activity to human beings for example; a landslide is hazard if it is likely to kill human being, destroys dwelling or road, tunnel etc. It is always present. But if it occurs in remote region and go unnoticed then it is not a hazard. So if is the case with earthquake, submarine earthquake is not a hazard if it does not harm the properties of human being. Risk is synonym to hazard but risk has additional implication of the chance of a particular hazard actually happening i.e. probability of hazard occurrence. A comprehensive definition of disaster is given in the Indian 'The Disaster Management Bill, 2005' that defines 'disaster as a catastrophe, mishap, calamity or grave occurrences affecting any area, arising from natural or manmade causes, or by accident or negligence which result in substantial loss of life or human suffering, or damage to, or destruction of, environment, and is if such a nature or magnitude as to beyond the coping capacity of the community of the affected area' hence disaster is a realization of hazard. Ecology deals with correlation between all organisms living together in one and the same locality and their adoptions to the surrounding or in simple words relation of organisms to their surrounding living as well as non-living.

10.1 Problems Created by Nature

10.1.1 Seismic Hazard

Geological aspects of Himalayan seismicity have been discussed in the chapter of Geophysical studies. Himalaya is one of the highly earthquake prone intracontinental regions in the world. Earthquakes of feeble intensities or low magnitude frequently rock the Himalaya terrain, however, the devastating events occur infrequently. The impacts of strong events are horrendous and have created very deep psychological impact. Effect of 6.5 M Patan earthquake of December 28, 1974, created by activation of 30 km long fault, felt in Swat, Hazara, and Kohistan regions of Pakistan. This earthquake was originated between 10-15 km depth and killed nearly 1500 persons and destroyed 6000 houses and also badly damaged 130 km long stretch of Karakoram Highway by landslides. The recent example is Uttarkashi earthquakes of 20th October 1991. This event through seismically was not a powerful one (M=6.5) but has killed nearly 3000 persons and destroyed properties worth of thousands of millions rupees. The 1950 Assam earthquakes killed more than 1500 persons. Similarly 1905 Kangra earthquakes killed more than 20000 persons in one stroke. October 8, 2005 Muzaffarabad earthquake of 7.6 magnitude was most devastating earthquake in the Himalayan region that killed nearly 73256 persons in Pakistan Occupied Kashmir till November 1, 2005 and more than 1500 persons in India by destroying more than one lakh houses including 30000 houses in Jammu and Kashmir in India and also damaged the roads, bridges and triggered landslides on massive scale. The damage is estimated around $10 billion.

Internal dynamics of Himalaya is responsible for all earthquakes; therefore, menace cannot be averted. However, taking suitable area specific precautions, impacts of earthquakes could be minimized. We know today fairly well about area wise probability of earthquakes and about the expected magnitudes of likely striking earthquakes and characteristic earthquake in a particular area. Seismic gap in Uttaranchal and in the northeastern Himalayan region are likely to be potential areas for next devastating events. Due to continuous research in Himalaya geologists know very well which faults are active and palaeo-seismic investigation can tell us past seismic histories of these faults and repeat time of next events. In the civil constructions due weightage should be given to seismic risk and necessary precautionary measures can be incorporated during construction stage. A proper disaster management approach or plan is must to tackle the problem.

10.1.2 Soil and Land Erosion

The annual global output of sediment is estimated at 15.0 billion tons per year this includes nearly 1.2 billion tons or 7% contribution from the Himalayan rivers draining nearly 2.3 million sq km area or 1.4 percent of total continental area. Since the beginning of Neogene nearly 6 million cubic kilometer sediment is supplied by the Himalaya. It is enough to wrap the globe with 12 m thick blanket of sediment.

The enormous generation of sediments in Himalayan region hardly allows developing a thick and fertile soil cover over its sloping surfaces thus causing poor productivity of land in Himalaya. Heavy rainfall, steep slope, fractured nature of rocks and subtropical to temperate climate in most part of Himalaya are main factors for land erosion and sediments generation. These sediments likely to fill reservoirs at a faster rate thus reducing the water storage capacity but in lower reaches during flood add a layer of fertile soil every year in the plain, forms the deltas and sediments also choke the port like Kolkata where constant dredging is required to maintain the working condition in the port.

10.1.3 Snow Avalanche

Snow avalanche is sudden downward motion of snow mass, which may dislocate rocks, soil, ice, tree etc. coming in its way. In extreme cases the snow mass may attain speed up to 300 km/h and exerts pressure up to 50 tons per sq m. This phenomenon occurs on inclined slopes of higher altitudes of the Himalayan terrain, therefore, thus confined in poorly habitated remote regions. With rise in population snow avalanches are emerging as important geohazard. Avalanches every year kill large number of persons and destroy civil properties coming in the way as well as initiate soil erosion, rock fall, damming of rivers and deforestation in the Himalayan region. Avalanches may occur in loose snow and snow slab forms. Tunda village in Ladakh was totally destroyed in 1838 by avalanche. Kalpa was the headquarter of Kinnaur district of Himachal Pradesh but due threat posed by the avalanches, the headquarter was shifted safer village Peo. The other important tragedy caused by snow avalanche are death of 258 persons in Lahaul in 1979, 207 persons in Zozila Pass in 1988, 40 army person were killed in Kargil in 1994, 70 persons died on National Highway-1A of Jammu and Kashmir in 1995, 20 persons were killed in 1998 in Sikkim and again nearly 500 persons were killed in Jammu & Kashmir in 2005. 40 ft thick snow avalanches in March, 2004 in Bhagirathi Valley blocked the flow of the river Bhagirathi for 48 hours and as a result of avalanche wide spread damage in the form destruction of houses, roads, electric and telephone lines occurred and killed few persons and hundreds of wild animals between Gangotri and Gaumukh. Snow avalanches in February-March, 2005 did extensive damage in the Bhagirathi, Alaknanda and Yamunotri valleys.

Snow avalanches are frequent in the months of January to March. Avalanches usually occur during or immediately after the storms. Border vigilance, refuge in hills, winter sports, maintenance of civil facilities, cultivation, tourism, mountain adventure, research and trade are identified as reasons for avalanche related casualties in the Indian Himalaya. Avalanche hazard can be mitigated by active and passive methods. In active methods snow avalanche is controlled by structural methods like construction of snow bridges, snow rakes, snow nets, diversion dams, snow gallery, wedge mounds, catch dams, afforestation, artificial triggering, snow fence, jet-roof and wind baffles. Passive methods aims at avoiding the interaction with avalanches by creating awareness, preparation and publications of hazard maps and atlas, training, publicity through media and by advance forecasting of avalanches (Singh et al., 2004, Chaudhary and Chauhan, 2004)

10.1.4 Silting up of Reservoirs and Lakes

Reservoir silting in Himalayan region is a problem of main concern to executing and maintaining authorities of reservoirs. This is not a new problem and had been studied even in 1870. The silt production is estimated between 36000-40000 cubic meters per year from an area of 100 sq km. In Bhakra Dam, period between 1965-1973, annual rates of silt deposition was 2948.3-hectare meter while it was 3920.0-hectare meters during 1959-65. Adequate precautions are usually taken during planning stage of dams and in all major dams consideration for sufficient silt storage space is taken. Developing proper floral cover can reduce the supply of silt.

Lakes are, in general, temporary depression of variable size on the earth's surface. The life of Himalayan lakes are generally short as they get filled up with silt and clay more quickly than the lakes of peninsular India, Many examples of palaeolakes have been cited in the section 3.3. The present Himalayan lakes are also shrinking rapidly for example area Dal Lake in Kashmir Valley is reduced from 17 sq km to 10 sq km in last 50 years as Telbal Nala deposits 8000 tons silt every year and 40000 to 50000 tons of organic waste is produced by aquatic weeds. Wular Lake in normal time has spread of 32.5 km while in flood the area spreads to 56 sq km; it is shrinking with 6.4 m/yr rate because of heavy siltation from the watersheds of the Jhelum and Puhru rivers and reclamation of land for cultivation, settlements, nurseries etc. The water of the lake increasingly polluted by discharge of waste, sewage, affluent water and fertilizer rich sediments causing eutrophication and decomposition zones hence depletion in biological and chemical demands of oxygen. Nainital Lake is filling up with a rate 0.022 cm/year. Renuka Lake in Himachal Pradesh is also shrinking rapidly. Life of Himalayan lakes can be increased by removing aquatic weeds and plants, by making check dams on rivers and nala draining into lakes, controlling the urban waste, oil spillage from boats, animal waste and by initiating programme of periodic cleaning.

10.1.5 Changes in Course of Rivers

In southern foothill regions and in the Indo-Gangetic Plain shifting migrating channels of Himalayan rivers are either eroding the land or creating new usable surfaces from erstwhile channels. This is causing boundary problems between the states and also between the neighboring countries and on local level new land is added or taken away from existing agricultural fields. Thereby causing many legal, social and political problems. Within the mountainous terrain many disproportionately wide valleys are also carved out by shifting of channels. Tectonic activity, river bank erosion, landslides, change in slope conditions, temporary blocking of rivers due to landslides and snow avalanches and human activities are the main responsible factors for shift or change in the riverbanks. The Yamuna and Ganga rivers have shifted its course to the east by 10-40 km in last 3000 years, the Kosi river has shifted its course by 112 km in the west direction between 1738 and 1964 and the river Gandak has shifted in the east by 10.5 km over its mega fan in the period 1935-1975 due to tectonic activities. In Indian Himalaya shift is observed in the Alaknanda, Giri, Satluj, Baspa, Saryu, Goriganga, and Brahmaputra etc. and in Nepal, Kosi, Kanfai, Ratu, Bagmati, Tinall, Rapti Babai and Karnali rivers are presently shifting their channels.

10.1.6 Glacial Melting

Earth has under gone several periods of glaciations in its past, even in the Proterozoic era (2500-542 Ma) at least five ice ages were recorded and in the Late Neoproterozoic the entire earth was covered with the glaciers and it was described as a 'snowball earth' (Sankaran, 2003). The present Himalayan glaciers are remnant of fourth or last glacial age (130-10 ka). Recession of glaciers started about 14000-15000 years ago and due to continuous recession formerly 2 percent area of Himalaya under permanent snow cover is now reduced to 0.5 percent. The recession is attributed to periodic climatic changes caused by global warming. Warming of environment in the Himalaya was also recorded within a glacial period, for example four recent warming events recorded around 32, 20, 15.8 and 10 Ka in Ladakh. The Tibetan ice sheet of more than 1 km thickness melted completely between last glacial maximum (18 ka) and at beginning of Holocene (10 ka) with an average rate melting >10 cm/year. A study in Himachal Pradesh has revealed that 19.1 cubic km was stored in the 19 big glaciers in 1962 this has reduced to 14.71 cubic km in 2001, a loss of 23%. At present 20-30 percent waters in the Himalayan Rivers is contributed by glacial melting. Outside Himalaya in Schirmacher Oasis in the Antarctica, glaciers recession range between 2.25-10.79 m /yr have been recorded by the Indian expeditions of 2003-04.

The melting of glaciers added further youthfulness to rivers because increase in flowing volume of water and sediments load in the rivers, and the load is further contributed by emergence of loose moraines and erratics, which were initially entrapped in the glacial ice. If present trend of recession continues in future, then it should not be disappointing to accept the reality that permanent glaciers might vanish from the earth.

The lateral and the receding glacial moraines border a number of slowly expanding lakes. Receding moraines are converting into terminal moraines. These glacial lakes might burst when water exceeds the holding strength of the morainic materials and thus create a condition of the flash flood or lakes may also bursts by sudden fall of huge volume of avalanching snow into the lakes that cause and initiate failure of the morainic barriers. Flash floods in Bhutan were generated by such failures in lake Tarina in 1959, lake Lunana in 1960 and 1968 and lake Luggy (0.9 sq km area) bursted in October 1994. Bursting of lake Luggy resulted into discharge of 23 million cubic meters of water in 25 hours time with a peak discharge 1654 cumecs. 107 m deep Raphstreng Lake at 4350 msl in Bhutan was a small pond in 1940 and gradually grew to 1.5 by 1 km in size in 1978 and in 2001 its size was reported to be 1.94 by 1.13 km, the morainic barrier holding water of this lake is 175 to 400 m wide and reservoir holds 66 million cubic meters water. This lake was on the brink of failure, however timely remedial measure like excavation of spillway channel prevented the untoward incidence (Joshi, 2001). Pareechu river, a tributary of river Spiti belonging to Satluj river system, is situated in the Tibet, it was blocked by glacial debris in 2003 and slowly in the river Pareechu water level in the lake started rising ultimately water breached the obstacle in 2005 and the ensuing flood in the downstream region did tremendous damage. Government of Himachal Pradesh has identified 34 such potentially hazardous lakes in the Tibet.

Accelerating adventure tourism and glacial expeditions and military presence in the glaciated regions in last some decades have now become a serious cause of concern as these activities not only modifying the glacial surfaces but also leaving behind the varieties of garbage, waste and unused materials. With time this material is also incorporated within the glacial mass, some of such materials particularly toxic pollutants might be having potential of endangering glacial ecosystem. In Siachen glacier, the highest military field of the world, nearly 1000 kg waste material is generated by the Indian Army every day, this waste is packed in metal drums and dumped into crevasses and same might be situation in Pakistan occupied southern part of the glacier. However, an erroneous impression is being created that because of military presence, restricted up to 6000 m, the glacier might disappear by 2050. International Commission on Snow and Ice has observed that Himalayan glaciers are receding faster and if present rate of recession continues then glaciers might disappear by 2035. Measures for environmental safeguards are adopted by the Indian military but tourists, pilgrims; adventure expeditions hardly care for environmental safe guards as a result easily accessible glaciers like Gangotri, Pindari, Mount Everest and many others are highly polluted as well as dirty.

Chemical composition of glacial melt water

The chemical composition of glacial melt water of the Pindari Glacier, 1 km down stream from the snout, is reported by Pandey and Others (2001) pH of water ranges from 7.1 to 8.2; electrical conductivity 69 to 179 (micro/cm-1); HCO_3 508-1095; SO_4 444-1194; Cl 1-52; Ca 789-1616; Mg 148-543; Na 13-48; K 13-91; H_4SiO_4 5-53; TDS (mg/l) 76-161; TSM (mg/l) 53-4073.

Hydrogeochemistry of Thermal waters

Geochemical characteristic of thermal waters is given in Table 10.1

Table 10.1: Hydrogeochemistry of hot springs water of Himalayan region (in ppm)

	1	2	3	4	5	6	7	8	9	10
Temp. °C	90	55	81	84	50	40	44	44	52	38
pH	7.9	6.8	7.65	8.6	7.3	7.65	7.8	7.3	7.2	7.1
Sp.C	1045	2630	2730	2835	892	469	5309	1115	816	1801
HCO_3	276	112			258	264	1011	339	435	269
Cl	170	1485	370	122	148	13	1080	165	18	119
SO_4	33	22	123	241	85	10	34	24	28	370
Th	-	230							113	362
Ca	52	70	17	7.0	11	44	13	82	27	127
Mg	12	13	5		1	18	3	16	11	11
Na	155	490	530	360	190	19	1040	110	133	150
K	27	37	72	18	13	10	70	6	10	17
F	3.0	1.6	80	10	0	0.3	6.5	0.4	2.1	1.05
B	10	19	116	16			17	<1		
SiO_2	83	115	120	100	100	28	70	46	80	60
TDS	874	1630	2020	1247	702	279	2834	580	565	1017

1-Thermal water springs in Uttaranchal Yamunotri, 2- Badrinath, Alaknanda valley, 3- Tunga Valley, Ladakh, 4- Chhumathang, Ladakh, 5- Bashist Beas Valley, Himachal Pradesh, 6- Tatwna Gajkhad Himachal Pradesh., 7- Parvati Valley, Himachal.Pradesh, 8- Sohna, Haryana, 9- Takshing, Arunachal Pradesh, 10- Chitu Arunachal. Pradesh

10.1.7 Desertification, Cloud Bursts and Heavy Rainfalls

Nature has created cold desert type situation in a very large part of the eastern Ladakh, in the Tethyan Himalaya terrain and in the Tibet due to inability of monsoon winds to penetrate in deep interior. A fluctuation in global wind pattern may temporarily increase or decrease the areal extent of this desert area.

Cloudburst, a popular but not meteorological term is a common phenomenon of rainy season, produces extremely loud bang with huge discharge of atmospheric electricity and sudden very heavy rainfall causing flash flood. The vibrations caused by thunderous sound and impacts of sudden down pour are so powerful that steep, fragile slopes of mountains fail and slip instantly. Precise mechanism of cloud burst is not yet known but factors attributed to it are the combination of monsoon and convectional rain, topography, vegetation and rocks types. This type of natural calamity triggers landslides particularly debris flow on large scale and causing huge loss. It is observed that frequency of cloudbursts is increasing. Cloud burst in higher altitude of Alaknanda Valley in 1996 wiped off several villages. Cloudburst of August 10, 2002 in the Balganga Valley of Tehri district killed 28 people, 99 cattle, and damaged 38 houses (Sah et al., 2003). July 16, 2003 cloud burst in Pulia Nal; Himachal Pradesh killed 33 persons and damaged property of more than 1 million rupees. 11 persons were killed and more than 200 vehicles were damaged and 20 houses and shops were collapsed in June, 2005 cloud burst in the Tuppani Nala in the Govindghat area of Uttaranchal.

Floods are due to excessive rainfall in short duration. Areas situated along the river bank are affected by the flood water most, in India 49 million hectare land in the plain region is affected by the flood annually that affect nearly ten percent population with average annual estimated damage of 1000 crores of rupees sometimes damage exceeds to couple of ten thousands crores of rupees. The quantum of damage is increasing every year. Severely affected areas are falling in catchments of the Himalayan Rivers. The devastating floods occured in the river Indus ocurred in 1780, 1835, 1839, 1841, 1926 and in 1929; in the river Gilgit in 1844 and 1865 and in the Hunza valley in 1974. In plains of the Assam, one of the factors for flood is aggradation of riverbed as much as by 4m due to deposition of the silts. The Kulu valley is prone to seasonal flash floods, the important events occurred in 1902, 1945, 1988, 1993, 1995, 2000 and 2003. These events have caused massive loss to property and human life. The maximum flood discharge in 1971 in river Satluj was 17,230 cubic m/sec, 8,210 in river Chenab in 1966, 14200 in river Ravi in 1988. Failure of temporary dam in the Indus River wiped off the Sikh army in June 1841 in Attock, Pakistan. The Indus River was blocked by landslide debris in the January 1841 in the Hatu Pir spur of the Nanga Parbat. The vertical drop of debris for this massive landslide was 1900 m and the debris was transported 4 km across the landslide toe region. Massive volume of debris created a 150 feet deep and 30 km long reservoir in upstream direction that extended to the confluence of the rivers Gilgit and Hunza. Due to filling of water opposite slope failed and huge seiche were generated.

10.1.8 Global Warming

Long-range increase in atmospheric temperature is described as global warming. Green house gases like H_2O, CO_2, CH_4, N_20 and CFC are considered as a cause of global warming and climatic changes. The role of CO_2 among these gases is most important. Human actions of last 50 years are perceived as the cause of present global warming. The present warming is attributed to increase in CO_2 content in atmosphere and its quantity in the atmosphere has increased from 280 to 365 ppm since industrial revolution (started in the middle of seventeenth centaury) and quantity is likely to reach 550 ppm by middle of this centaury due to burning of fossil fuels and deforestation. In Himalayan regions in addition to above factors, summer months forest fire adds significant quantity of CO_2 and aerosols (particulate matters) in the atmosphere. Due to global warming earth temperature has increased by 5° C between the Last Glacial Maximum and present warm period. Intergovernmental Panel on Climate Change has reported 0.6°±0.2° C increases in temperature in the last centaury and by the end of present centaury temperature might rise between the ranges of 1.4° to 5.8° C. It is estimated that by 2080, the rise in temperature particularly in India would increase between 3.5°-5.5° C and its notable effects would be 5-25% decline in winter rainfall, 10-15% increase in summer monsoon precipitation with uncertain time of arrival (Lal et. al., 2001). Global warming is likely to increase precipitation in South Asian tropical countries like India, Nepal and Pakistan. More rainfall will solve the problem of water requirements for domestic, industrial and agriculture sectors. At the same time it will accelerate the process of erosion, landslide, floods, shifting of river courses and would adversely affect the crops due to excessive and unseasonal rains. In last ten thousand years 50 m rise in sea level of western coast of India has been observed and in one thousand years duration (10000-9000 BP) sea level rose by 23 m ! and during the Pleistocene-Holocene transition period rate of sedimentation was 6-7 times higher than the Holocene rate (Pandarinath et al., 2001).

Geologically global warming is not a new phenomenon; palaeoclimatic researches have revealed that the worldwide warming of atmosphere is a rather frequent phenomenon. The level of concentration of CO_2 similar to present day was also existing, 300 ppm, in the Antarctica around 150 ka ago. These data clearly suggests unfair and biased attitude to blame human actions for present warming plight, which is in fact a periodic process. The impact of this climate change has been noted in the Beas and Baspa basins where average stream run off in December has increased by 75% after 1966 (Kulkarni et al., 2002).

10.2 Problems Created by Man

10.2.1 Slope Modifications for Civil Constructions and Agriculture

Himalayan slopes are dressed with ever increasing pace for various types constructive activities like road, tunnels, dams, canals, settlement of rising population, expansion of lands for tourists related activities etc., as a protective measure to do away with unsafe and precarious slopes and to meet rising demand of agricultural land for food for growing human and animal population. Such activities are more intense in the easily approachable Outer and Lesser Himalaya and in selected areas of Higher and the Tethyan Himalaya. The modifications are carried out either manually or mechanically with the help of explosives and modern excavation techniques. 40000 to 80000 cubic meters of rock and soil materials are excavated for each km road construction. The 44000 km long roads network in the Himalaya would have resulted in generation of 2640 million cubic m debris. The annual turnover of overhauled strata is likely much more then the natural processes generate the sediment.

10.2.2 Pollution of Rivers and Lakes

Not long ago Himalayan rivers and lakes were known for their purity and crystal clear waters and now these water bodies in their lower reaches have been rendered unsafe for drinking purpose due to discharge of organic waste like animal and human wastes and dead bodies, decaying of vegetable matters and microorganisms, pathogens and discharge of industrial and domestic affluent. This sorry state is results of increase in local population, development new villages and hamlets on the riverbanks and also due to increase in tourists flow. In hilly regions entire waste is drained into rivers or lakes causing increase in contents of dissolved and suspended organic and inorganic matter and pollutants. Waters of the rivers like Yamuna, Ganga, Jhelum and Brahmaputra even in relatively pollution free in higher reaches and are mild to moderately hard due to natural causes. In Yamuna River, every day 300 tones dissolved matter and 150 tones suspended solid are added. Biological demand of oxygen (BOD) in Ganga water at Hardwar varies between 2.50 mg/l in summer to 13.20 mg/l in post-monsoon period and total coliform ranges between 43 MPN/100 ml to 2400 MPN/100 ml.

10.2.3 Minerals Exploitation

Presence of ores show is recorded from many locations of the Himalaya but workable metallic mineral deposits are few. These deposits are generally of small size and are having limited reserves making large-scale mining operations commercially uneconomical. At present limestone, dolomite soapstone, phosphorite, gypsum, pyrite, rock salt, gemstone coal and gravel and sand of non-metallic category and copper, lead and zinc of metallic category are mined with scientific and unscientific methods in Himalaya. Mining activities generate huge amount of waste and in most cases it is dumped into nearby sites or in the adjoining river valleys. The negative impacts of mining on environmental are deforestation, land dressing, increase in sediment load in river waters, discharge of all types polluting waste in river and atmosphere, further weakening of strata due to use of explosive and vibration caused by movement of heavy vehicles and operation of machinery, land modification for construction of processing plants, storage sites, dwellings units and roads and drying up of springs and small streams and loss of forest cover and scenic beauty.

Limestone mining activities spreading over a strike length of 30 km in Dehradun and Mussoorie area had to be banned by the Supreme Court of India due adverse environmental impact in 1982. Mining in this region has caused large scale scarification of the hill scape, deforestation, instability in the hill slopes, landsliding, siltation of riverbeds, damage to public roads, drying up of springs, deforestation etc. (GSI, 1996). With proper environmental safe guards mining can be started in the Himalayan region (Biyani and Dimri, 2002)

10.2.4 Reservoir Induced Seismicity

Koyna earthquake (M=6.3) of 1967 in Maharastra has attracted the global attention to seismicity caused by impounding of reservoirs. So far nearly hundred cases of earthquakes caused by the impounding of the reservoirs are recorded. Earthquakes in Koyna (>160 M4 since 1967) are still recorded in 30 by 20 km area with focal depth restricted 12 km. Such earthquakes are also recorded from the Bhatsa dam located 200 km in down stream direction of Koyna dam.

The weight of dam and reservoir exert stress on the underlying rocks. This stress slowly strained the rocks and when critical limit of strain is crossed the underling rocks fail by rupture. This failure of rocks produces earthquakes of moderate magnitude. This mode of generation of earthquakes is described as reservoir induced seismicity (RIS). RIS is generally reported from large dams (having storage capacity of more than one cubic km water or height of water column > 100m) and so far more than 70 example of such seismicity are reported from different parts of the world (Gupta, 2005). In Himalaya, fear of additional seismicity caused by RIS is expressed time to time because 11 large dams are in operations and many more are in construction stage or being planned. Environmentalists often highlight the danger of floods caused by dam failure either by natural seismicity or by RIS and have opposed the concept of large dams. Progress of construction e.g. Tehri Dam on the river Bhagirathi in Uttaranchal is frequently interrupted on this count as well as other issues. RIS factor, as a matter of fact, is recognized by planning and executing authorities and possibility of such eventuality is fully taken care of during the designing and construction stages by incorporating necessary safety measures. A dam is designed in such a way so that it can withstand firmly and safely during an event of maximum magnitude like to occur in dam area. The RIS also depends upon tectonic setup of the area and availability of critical stressed rocks with in shallow depth i.e. less than 10 km (Gupta and Rajendran, 1986). Normal fault condition enhances possibility at the RIS while in thrust fault set up RIS is not encountered but normal seismicity also decreases. In Himalaya possibility of RIS is minimal as terrain is dominated by the thrust faults. In fact around Tarbela Dam in Pakistan normal seismicity has decreased, so far from eleven large dams including the Tarbela dam no RIS is reported, however, possibility of natural seismicity can not ruled out completely.

Recently distinction has been made in the term triggered and induced earthquakes, former term is used for the causative factor is responsible for small amount of energy and in latter causative factor is main supplier of energy, therefore, Koyna falls in the latter category

10.2.5 Loss of Vegetation Cover, Faunal Actions and Elimination of Fauna

In comparison to all India figure of 12.1% forest cover nearly 35 percent land in the Himalaya is covered with forest. In India loss of forest cover is taking place with a rate 0.16 m ha/year. In lower part of the Himalaya forests are ruthlessly chopped off to meet the requirements fuel, housing, new agricultural land, civil constructions etc. Legal attempts to check deforestation process have been miserably failed. A vast portion of the Outer and the Lesser Himalaya is now barren and the net impact of deforestation is visible through increase in intensity of land denudation, landslides, aggressive, behavior of wildcats and in long run adverse effects on hydrological and climatic regimes. This problem can only be tackled by people's involvement in forest conservation and development.

Livestock population has substantially increased in the Himalaya, as they are source of dairy product, source of lively hood for large number of local inhabitants and principal medium of haulage in inner areas. In search of food, cattles in summer months migrate to higher altitudes; their movement on grasslands completely crushes down the baby plants. This action largely rules out the hope of revival of forest cover on grassland. Spread of animals dung due to their roaming not only make walk unpleasant but also pollute the river waters in rainy seasons.

The total global estimate of biota varies between five to ten millions, which is result of 3800 Ma of evolution. This figure is one percent of total biota evolved on the earth, in other words during the course of evolution at least half a billion biota have become extinct. At present 68375 faunal species including 60000 insects, 1693 are fish and 372 mammals are recorded in India. The plant wealth of India is about 45000 species (twelve percent of global wealth) out of which flowering species number around 15000 (Khoshoo, 1991). Except men and livestock, the numbers of other species are decreasing rapidly and some of them are on the verge of extinction. It is now accepted that plants, animals and microorganisms form an interdependent system. And if one component of this system shows adverse change then others are bound to be affected.

10.3 Problems Created by Nature and Man

10.3.1 Landslides

The term Landslide includes all varieties of mass movements on hill slopes and can be defined as the downward and outward movement of slope forming materials composed of rocks, soils, artificial fills or combination of these materials along surfaces of separation by falling, sliding and flowing, either slowly or quickly from one place to another. It is the third largest natural agent of destruction in the Himalaya and on an average loss of two hundred human lives and destruction of property of several tens of millions to hundreds of million rupees occurs every year due to such movements. In addition to these, traffic movement and supply lines are also cut off for varying periods in landslide prone areas. The cost of traffic and communication loss in Indian Himalaya is awesome, in 1981 it was estimated Rs. 350 millions. Natural factors like undercutting or toe erosion by rivers, in general fragile nature of rocks induced by tectonic deformation, over steepened slopes, increase in weight and creation of slippery surface in rocks due to percolation of rain waters, violent seismic shaking and cloud burst; and man made factors likes deforestation, slope modifications, change in hydrological condition due to construction of dams, excessive use of explosives and vibration caused by movement of heavy vehicles and machineries are responsible for landslides in Himalaya (Biyani 2002 d).

On the roads of Darjeeling town, in one season 20000 minor to major landslides occurred. This is an example of extreme situation in Himalaya. In Srinagar –Rudraprayag area, 139 landslides have been reported from 444 sq km area, giving landslide density 0.313/ sq km. After Chamoli earthquake (March 29, 1999) 377 slides were recorded along the Alaknanda, Mandakini and Madhyamaheshwar valleys of Uttaranchal with a frequency of 0.54 slides per sq km. Majority of the slides (72%) were 5 by 5 m size and only 2.5 % slides were of 25 by 25 m size. The Average annual rate of debris generation along the roads in the entire Himalayan region is about 24 million cubic m from the seasonal mass movement, 724 cubic m/ km debris is generated in Jammu- Srinagar Highway, 411 cubic m/km in Tanakpur-Tawaghat road in Uttaranchal and 591 cubic m/km in Kameng Highway in Arunachal Pradesh and more than 75% road receives debris from Kulu to Manali in Himachal Pradesh (Sikdar 2002). A 200 m wide and 500 m length section of the Niangal valley was filled by landslide near Tijam in Kumaun region.

Sirmuri Tal and Telangi are recurring landslides prone areas in Himachal Pradesh. Sirmuri Tal landslide zone is located in the Giri valley in Nahan district. On southern bank Lower Siwalik sandstone and claystone are present while on the northern slope Subathu limestone and Infra Krol shales and slates are present. Krol and Giri thrusts are aligned parallel to Giri River. The slope of the valley is 55-60°. The reported history of landslide is more than a centaury old, however, most damaging landside occurred on October 23, 1956 that completely ruined the village. This landslide is attributed to dip slope of the strata, steep slope, undercutting of the riverbank, heavy rainfall and cutting of forest. Despite several checking measures, the landslide problem remained uncontrolled. Telangi landslide occurs on Pangi-Kalpa link road in the Kinnaur district. The link road was constructed in 1965 and first time landslide occurred in 1973 and since then landslides are repeatedly taking place in this section. The slide zone is located in rocks of Vaikrita Group covered with thick blanket of morainic material of extremely heterogeneous composition and fluvial deposits on toe. The landslide is spreading in an area of 25000 sq m, it is 130 m wide in toe region and crown is 222m above it. The failure of steep slope is attributed to loose sloping surface of moraines, coincidence of slope and dip directions, shaking of ground by earthquakes, presence of springs and seepage and leakage from contour canals and disturbance caused by road construction. Like Sirmur Tal landslide, controlling measures for this landslide is proved ineffective (Srikantia and Bhargava, 1998). The Chandmari landslide zone on Gangtok-Nathula road is situated in deeply weathered high-grade metamorphic rocks of Darjeeling Formation of Sikkim Lesser Himalaya; first time noticed in 1966 since then it is continuously expanding. The present width of the zone is 480 m and length from crown to toe is 1180m. This landslide is caused by percolation of rain water, deep weathering, rejuvenation of first order streams due to road cutting, weakening of rocks due to shear and joints, agricultural practices, blocking of streams and ground shaking by earthquakes. To bring landslide in control, catch water drains above road level, afforestation of area with trees of fast growth, deep fibrous roots, use of geotextile, bituminous mulching, inserting sub-horizontal drains by inserting perforated pipes at a distance of 3-6 m, removal of moisture from slope material and structures for control of erosion in toe region are suggested (Rawat, 2005).

Subjective Rating Technique (STR) and Objective Rating Technique (ORT) have been used to study landslides in Himalaya. In the former numerical rating are assigned to various factors like lithology, structures, topography, soils, drainage, slope etc., of the landslides affected area. Numerical value is given to each factor and within the each factor different values are assigned to different conditions or situations. In the latter, ORT, subjective assessment is scrutinized by estimating the grade for each factor. After this data are subjected to frequency analysis according to hazard class (Sarkar and Gupta, 2005) Quantitative and qualitative approaches have been used in the study of hazard assessment in many Himalayan valleys like Satluj, Jhelum, Ganga, Yamuna, Brahmaputra, Trishuli, Kali, Karnali etc. Many zones of high landslides activities are identified and suitable remedial measures have been suggested. Through suggested measures elimination of landslides is not possible but impact of landslides can be certainly mitigated or minimized the to great extents, if remedial measures implemented properly. Thus safe environment may be created for people living in dangerous areas.

Detailed investigation of landslides in the Himalayan region is carried through landslide hazard zonation (LHZ) methods. The LHZ map of an area divides the land surface into facets or small homogeneous regions or zones of different potential landslide susceptibility. LHZ maps are generally prepared 1:50000 or 250000 scale for regional planning, 1:15000 or 1:10000 for district level planning and 1:2000 or 1000 scale for site specific microzonation. These maps in mountainous terrain help in the planning and development by identifying and delineation of high landslides probable regions, helps in selection of ecology based mitigation techniques and provide input data for preparation of risk maps. In this method, a landslide hazard zonation map based 10 points landslide hazard evaluation factors (LHEF) scale, of the affected region is prepared by subjective evaluation of different causative factors namely lithology, structure, slope morphometry, relative relief, land use and land cover and hydrological conditions, The final LHZ map comprises of delineation of the region into five zones of different landslide probability is based numerical sum of rating of six causative factors i.e. total estimated hazards (TEHD) for each facets. The landslide hazard evaluation scheme proposed by Anbalagan (1992) is shown in the Table 10.2.

Landslide zonation on the basis of total estimated hazard (TEHD) is as follows:

Zone	TEHD value	Description of zones
I	< 3.5	Very Low Hazard Zone
II	3.5- 5.0	Low Hazard Zone
III	5.1-6.0	Moderate Hazard Zone
IV	6.1- 7.5	High Hazard Zone
V	>7.5	Very High Hazard Zone

A facet falls in any one of these zones. A landslide prone region is divided into large numbers of facets and each facet would fall in any one of these five zones. Thus adjoining facets may have vastly different TEHD value. Zone IV and V require careful attention for any planning and developmental activities. LHZ mapping is successfully attempted in many areas of the Himalaya; among many few examples are Bhagirathi valley, Kathgodam area, Sukhidang area, Tehri –Pratapnagar area, Kaliasaur region etc.

In theme based landslide mapping, danger (only locations of past slides), hazard (probability of occurrence of a landslide danger) and risk (nature of damage likely to be caused in case of a failure) maps are prepared.

10.3.2 Forest Fire

Forest fires are reported from all parts of forested land in Kulu valley of the Himachal Pradesh, important forest fire events have taken place in 1910, 1915, 1921, 1929, 1031, 1946, 1952, 1965 and 1970. In Garhwal region 40195 hectares reserve forest (27218 ha) and civil soyam (12997 ha) were damaged by forest fire by anthropogenic and natural causes while gross forest fire in 1996-97 was 1747 ha. As a result of forest fires concentration of CO_2 has increased from 338-341 ppm in 1996 to 338-382 ppm in same duration in 1999, however with the onset of monsoon the level of the CO_2 returned to original level (Joshi, 2003).

Table 10.2: Landslide hazard evaluation scheme (after Anbalagan, 1992).

Description Factor	Category	Rating	Remarks
Lithology			
Rock Type	Type-I		Correction factor for weathering
	Quartzite and limestone	0.2	Highly weathered - rock discolored, joints open with weather products, rock fabric altered to a large extent; correction factor C1
	Granite and Gabbro	0.3	
	Gneiss	0.4	
	Type-II		
	Well cemented terrigenous sedimentary rocks, dominantly sandstone with minor beds of claystone	1.0	Moderately Weathered – rock discoloured with fresh rock patches, weathering more around joint planes, but rock intact in nature; correction factor C2. Slightly Weathered – rock slightly discoloured along joint planes, which may be moderately tight to open, intact rock; correction factor – C3.
	Poorly cemented terrigenous sedimentary rocks, dominantly sandstone with minor clay shale beds.	1.3	
	Type – III		
	Slate and phyllite	1.2	
	Schist	1.3	For rock type I $C_1 = 4, C_2 = 3, C_3 = 2$
	Shale with interbedded clayey and non-clayey rocks	1.8	For rock type II $C_1 = 1.5, C_2 = 1.25, C_3 = 1.0$
	Highly weathered shale, phyllite and Schist	2.0	
Soil Type	Older well compacted fluvial fill material (alluvial)	0.8	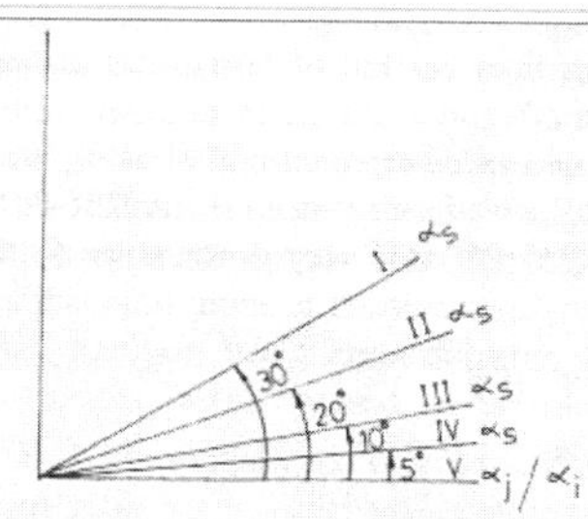 Parallelism between the slope and the discontinuity ($\alpha_j/\alpha_i - \alpha_s$).
	Clayey soil with naturally formed surface (eluvial)	1.0	
	Sandy soil with naturally formed surface (alluvial) Debris comprising mostly rock pieces mixed with clayey/sandy soil (colluvial)	1.4	
	Older well compacted	1.2	
	Younger loose material	2.0	
Structure			
Relationship of Structural Discontinuity with slope			
Relationship of parallelism between the slope and the discontinuity Planar Wedge	I >30°	0.20	α_j = dip direction of joint α_i = direction of line of of intersection of two discontinuities α_s = direction of slope inclination β_j = dip of joint β_i = plunge of line of intersecton of two discontinuities
	II 21°-30°	0.25	
	III 11°-20°	0.30	
	IV 6°-10°	0.40	

Parameter	Class	Rating	
Dip of discontinuity Planar Wedge	I <15°	0.20	
	II 16°-25°	0.25	
	III 26°-35°	0.30	
	IV 36°-45°	0.40	
	V >45°	0.50	
Depth of Soil Cover	< 5 m	0.65	
	6 – 10 m	0.85	
	11 - 15 m	1.30	
	16 – 20 m	2.0	
	> 20 m	1.20	
Slope Morphometry			
Escarpment/Cliff	>45°	2.0	
Steep slope	36° – 45°	1.7	
Moderately Steep Slope	26° – 35°	1.2	
Gentle Slope	16° – 25°	0.8	
Very Gentle Slope	<15°	0.5	
Relative Relief			
Low	<100 m	0.3	
Medium	101 – 300 m	0.6	
High	> 300 m	1.0	
Land Use And Land Cover			
Agriculture land/populated flat land		0.65	
Thickly vegetated forest Area		0.80	
Moderately vegeted Area		1.2	
Sparsely vegetated area		1.5	
Barren land		2.0	
Ground-water Conditions			Number of contour lines over one cm length (1:50,000) — Slope angle >25° — >45° 19-25 — 36°-45° • 26°-35° 8-12 — 16°-25° <7 — <15°
Flowing		1.0	
Dripping		0.8	
Wet		0.5	
Damp		0.2	
Dry		0.0	

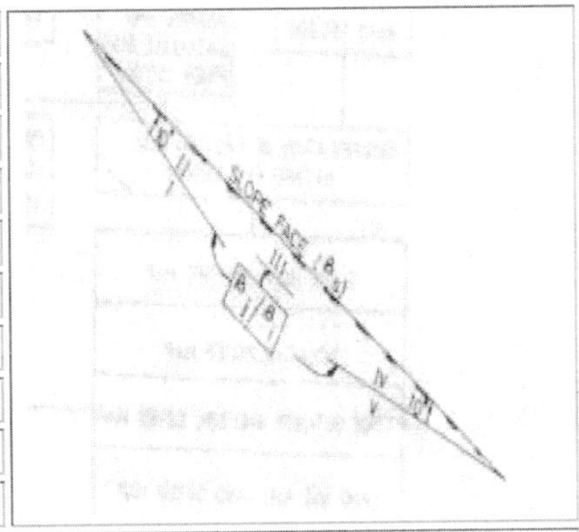

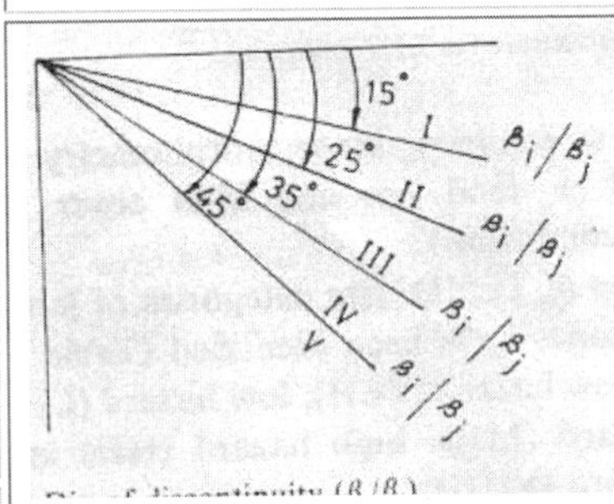

10.4 Palaeoclimatic Studies in Himalaya

The records of the global climatic changes of last some ten thousands of years are well recorded in the lacustrine sediments, deep-sea sediments, glacial cores, palaeosols (palaeosoils) and in organic matters. Sediments deposited in palaeo lakes preserves sudden environmental and climatic changes faithfully within their limited thickness. This record is decoded with oxygen, carbon isotopic analyses and this is correlated with suitable radiometrically determined age of the sediments.

Goting palaeolake sediments (42-28 ka old) of Garhwal Himalaya revealed six – 40.2, 38.2, 36.2, 34.2, 32.7 and 29.4 ka- short duration warm epochs of increased rainfall and three events- 41,5, 37 and 35 ka of poor monsoon on the basis of ^{18}O and ^{13}C of carbonate negative and positive shifts respectively (Ghosh and Bhattacharya, 2003). In Din Gad valley of NW Garhwal Higher Himalaya 1100 yr ago significant retreat of the Dokriani glacier was inferred, this phase was followed by progressive warm and wet phase with notable rise in SW monsoon during 10900-8400 (cal yr) before present (BP) again climate turned cold and dry with decrease in monsoon rain during 8400-7300 duration. Between 7300-5500 duration, climate turned into warm and wet with increase in precipitation this was succeeded by second phase of cold and dry climate during 5500 –4000 yr duration, during 4000-1600 cal yr BP step-wise warm and wet climate with dwindling monsoon phase occurred and in latest 1600 yr duration climate turned progressively warm and wet with increase in rainfall. In Pindar Valley of Kumaun Himalaya 1.25 m thick Dhakuri peat revealed palaeoclimatic history of last 3500 years. 3500-3300, 2300-2100,1700-740 and 640 to present were the wet phases and 3300-2300, 2100-1700 and 740-640 were relatively dry phases. The polynology based palaeoclimatic study of 284 drilled core samples contain quartz, feldspar, mica, Smectite, chlorite, kaolinite, gypsum and calcite of Kathmandu fluvial-lacustrine sediments indicate cool climate from 2.5-2.1 Ma, warm and relatively dry climate between 2.1-1.0 Ma and seven cycles of warm- and-wet and cold-and- dry climate from 1.0 Ma to present; the last cold maximum correspond to the last glacial age around 20 ka before present. The Last 450 ka climate of the basin closely related to global climate (Fuji and Sakai, 2001). Humid climate during 50-30 ka was recorded in the distal part of the western Ganga megafan. Palaeosols have also been used in palaeoclimatic study of Middle-Upper Siwalik section of Haripur-Kolar area of Himachal Pradesh; strongly red facies of palaeosols in the lower part of the Upper Siwalik inferred to be formed in warm, humid and strongly seasonal palaeoclimate. This climate was continuation of warm and humid climate of the middle Siwalik. This type of climate prevailed between 5.5-2.6 Ma and probably around 0.9 Ma. This was followed by development of yellow palaeosols that formed in cooler and dry climate of low rainfall (Thomas et al., 2005).

10.5 Health Problems in Himalaya

The annual fresh water consumption of the world and India is estimated around 4000 and 400 cubic km. However, this water in several regions has excessive concentrations of certain elements like fluorine, arsenic etc. and is causing health problems to consumers. The concentration of fluorine in parts per billion in river Alaknanda is 140, 287 in Bhagirathi, 131 in Dokriani, 1352 in river Bandal, 183-547 in river Yamuna, 116 in river Brahmaputra, 120 in river Satluj, 150 in river Indus, 100 in river Ravi, 3-6 in Dokriani snow and in glacier respectively, 285 in Himalayan rivers and 152 in average world's rivers (Madhavan and Subramanian, 2001). In Guwahati, lying in the alluvial plain of Assam, fluoride content in ground water varies from 0.18-6.88 ppm. The concentration of the fluoride increases with depth and shows positive correlation with sodium and potassium as well with total alkalinity. In Khiski area of Naushera, North West Frontier Province, Pakistan ground water contains up to 41.2 ppm fluoride content. The radon concentration in Garhwal and Kumaun varies between 1-196 Bq/l and in the soil it varies between 2.3-10.5 kBq/ cubic meter.

10.6 Geographic Information Systems (Gis) in Geology

Computers are used for wide ranging application in storing and interpretation of data in geology. GIS is one such very convenient software extensively used in daily life for storing and analyzing data in almost all walks of life. It is used in different geological studies as this technique offer great ease in interpreting surface geological information by permutations and combinations. GIS is a powerful set of tools for collecting, storing, retrieval and analysis, display of spatial data from the real world. GIS is a special purpose digital database in which common surface coordinate system is the primary means of reference.

GIS is extensively used in landslide, engineering geological and mining fields. The topographical and geological information of an area is first digitized with special softwares and stored in computer. Once the entire information is digitized, the individual information such as contour lines or drainage pattern or lithology or structure etc. can be retrieved as layered map depicting that selected particular information and this information can be superimposed on object of investigation. For example, in landslide study individual map layer showing contours, drainage, land use, slope, rain fall, road network, lithology, structures, hydrological setup etc. can be generated and subsequently super imposed over each other thus help in finding out precise identification of cause of landslide with possible area specific remedial measures.

10.7 Social Geology

In comity of science, geology had been dubiously treated at par with the established branches of science. This may be due to either its peculiar status resulting from its nil to limited exposure at school level science curriculum particularly in developing countries, late beginning of formal level education at university level, working of geology professionals in isolated or remote regions, therefore, having limited exposure to invisibility in the society, limited job opportunities in geology or large number of controversies arising out of the interpretations of geological data or facts and possibly apathy of professionals to popularize this subject in society. Basically geology is an applied science having twin faces, on one hand it is of full utility and on the other it has awesome destructive power, at times more than epidemics, hunger, limited war or major accidents. The lack of social mooring, still prevalent, resulted into widespread ignorance about the functioning of geological system as a whole and particularly of those which are affecting the society such as earthquakes, volcanic eruptions, landslides, fall of meteorites, availability of mineral resources etc. has and off the reality apprehensions about mining's, major constructions like dams and tunnels. Inadequate knowledge about safety, preparedness and mitigation measures related to geological hazards are further adding the confusion in society particularly where awareness level is low. Societal implications of natural disasters have been described in details by Bradford and Carmichael (2001).

The societal problems in geology are addressed by many established branches viz. mining geology, hydrology, mineral economics, resource geology etc. as well as by the many new branches. The branch specific problems are investigated and analyzed in utmost detail using available technology; reports are prepared and submitted to concerned authorities and to the government for consideration and implementations. Somehow or the other this aspect of societal approach is not percolating to societal level. Societal concern is an integral part of study matter in many subjects e.g. medicine, engineering etc.

Researches in geology have resulted into better understanding of operative geological system and as a result several new branches have emerged viz. societal geology, environmental geology, active tectonics, medical geology, urban- and rural geology etc. undoubtedly have established their utility and proved good in solving the problems faced by the society. For example, researches in palaeoseismology have successfully helped in decoding seismic history of any chosen area. The gulf between the society and geology can be bridged by inculcating habit of social interaction and educating the society through formal and informal means. The societal awareness programs can be launched through social geology which should exclusively aim at educating the society about utility and operation of geological processes concerning to it and societal response before, during and after adverse geological conditions and hazards

Social geological approach would benefit society in many ways such as it will remove prevalent misgiving or false or unfounded notions about operation geological and other natural processes. Further society can adopt right approach in calamitous situations to minimize the impact of calamity and can take necessary preventive and mitigation measures to face eventuality, to manage the natural resources in judicious ways, to check the trend of deteriotion and also extend help in improvement of existing geological system, society would not be carried away by irrational explanations, unverifiable predictions, warning and hypes suspicion about major constructions initiative like dam, reservoirs, tunnels could be tackled to great extent and above all a sense of partnership with nature can be developed by realizing the fact that we depend on nature while nature does not depend on us. The incremental strain generated by societal activities on geological system is unsustainable in long run; the unilateral activities of mankind would damage the system beyond repair. Had the society is involved in tackling the natural calamities loss resulting from them could be contained to great extent though not altogether eliminated. Education of masses play very important role in management of aftermath of the natural calamities. For example, Japan is often struck with devastating earthquakes (an earthquake of magnitude 8 struck in September, 2003), landslides or volcanic eruptions, we only read loss of properties and injury of the human beings in small prints as Japanese are successful in controlling the impact of the disasters. The credit for this feat goes to education of the masses and proper enforcement of prescribed norms. The societal awareness program can be launched through social geology, which should exclusively aim at educating the society. Keeping societal interest in focus, what is needed is to identify those geological processes that are directly or indirectly related to society and develop literature on them for education to professionals, students, and masses and also for media. Billions of years old geological system in last one century has come in severe strain due to progressive anthropogenic interference primarily to meet out the need of the society and at the same time in and outward controlled geological actions particularly in the form of natural hazards putting the expanding society to increasing amount of risks. These dual processes would have far reaching impact on future generations. Seemingly beyond control situation is out growth of ignorance of the masses about the operation of geologic system caused by deprivation geologic study in formal setup of education at every level. The net effects of these irreversible actions would be conversion of inherited and satisfactorily into unrepairable one with cascading impact on the users. To arrest the further deteriotion trend, society should be actively involved in the protection and management of the system by developing geoscientific awareness through societal education program. Societal awareness and public participation can be invoked by the sustained efforts by using every avenue available. The desired level of scientific understanding of masses can be obtained by propagating social geology. Social geology exclusively aims at educating the society about utility and operation of geological processes concerning to it, societal response before, during and after adverse geological conditions and hazards, judicious use of available geological and hydrological resources and proper maintenance or improvement of geological system (Biyani, 2001).

Social geology is basically a medium for minimizing adverse impact of geological processes through proper education, as geology is one of the subjects that are badly ignored from formal school education. Consequences of this ignorance are reflected in severe impact of natural disasters or improper use or misunderstanding about uses of geological resources. Severe disasters like earthquakes and landslides particularly in countries like Japan, Taiwan, and USA etc. and in Hong Kong were unable to affect the societies because of successful incorporation of geological inputs in civil constructions public awareness about use of proper safety measures in calamitous moments. In natural disasters prone Himalayan region social geological awareness can create safe living condition for the people.

Post disaster relief and rescue operation finds considerable difficulty in reaching to affected area owing to debris that blocking of the routes in badly damaged area or collapse of infrastructural facilities like roads, communication facilities, water supply, collapse of bridges, rail lines or air ports. Timing of most of the geological disasters can not be predicted in advance, therefore, in most of the instances they strike with out fore warning. A useful way to approach quickly in affected areas is to have in advance safety map of the habitated area. These maps could be prepared in advance for different disasters and in each case of disaster for different levels of furies or intensities. These safety maps will be based anticipated assessment of damage in each locality and would contain information about how to approach quickly and how to restore normalcy in shortest possible time duration and would also plan out alternative arrangement. These maps would be based on detailed survey of every locality and detailed assessment of every geological hazard (Biyani, 2005).

DIMENSIONS OF HIMALAYAN GEOLOGY
A. K. Biyani (2006) : pp.241-247
Satish Serial Publishing House, New Delhi
E-mail: ssp@ndf.vsnl.net.in, hkjain1975@yahoo.com; website: www.satishserial.com

Yogah karmasu kausalam (Geeta, 2.50)
Skill in action lies in yoga

CHAPTER - 11

ENGINEERING GEOLOGY

Safety of large and costly civil constructions like dams, tunnels, bridges, roads, mines, houses etc. primarily depend on various geological factors such as type of strata, soil, structural and tectonic setup, hydrological conditions, slope conditions, seismicity, behaviour of surficial processes and others. Engineering geology evaluates these factors thoroughly and is mainly concerns with stability and durability of civil constructions by identifying the safe and suitable sites or geological problems at the proposed sites and likely remedial measures to mitigate the impact of possible hazards.

Himalaya terrain is one of the most rugged mountain terrains in the world, despite this it is inhabitated by a large population and is rich in some biogenetic and abiotic natural resources. These resources deserve to be utilized judiciously for economic upliftment of the region. To utilize these resources for strengthening the economy without hindrance massive construction activities have been undertaken in post-independence period and will continue for many more decades. Water is one of the most important resources available in Himalaya. In this context the role of engineering geology is of tremendous importance. For long term benefits thorough assessments of ground as well as of subsurface geological conditions are very important as well as safety and stability of construction is of paramount importance.

A brief account of various constructional activities in Himalaya is as follows:

11.1 Dams

The hydropower potential of various rivers flowing in India and Nepal is estimated at 89200 and 83000 MW respectively and presently utilized capacity is less than 18000 and 300 MW respectively in both countries. The utilizable surface water resource of the Indus is 63 percent, 45-33 percent for the Ganga River and around 4 percent for the Brahmaputra River.

So far India has constructed 2900 dams of medium and large size, 700 dams are under construction and 450 dams are in proposal stage. In Himalaya more than 200 multi-purpose medium and major dams have been planned for power generation, irrigation, flood control, drinking water supply, industrial purpose etc.

The important Himalaya Dams in India are:

Bhakra Dam

Bhakra dam is constructed on the Satluj River near Bhakra village of Punjab and its 42 km long reservoir is known as Govindsagar Lake spreading into adjoining Una and Bilaspur districts of Himachal Pradesh. This concrete straight gravity dam is 226 m in height and 518 m in length with gross reservoir volume of 9867.8 million cubic meters (mcm) on the Siwalik rocks. 315 m long and 29 m high concrete Nangal Barrage is located 20 km in down stream direction.

Geologically Bhakra dam is located on the Lower Siwalik sandstone and clay stone, present both on upstream and downstream sides of dam, having strike parallel to dam axis and dips in downstream direction with 70° near the riverbed and 50° in the upper abutment. 82.8 m thick calcareous sandstone is well cemented while 31.39 m thick clay stones belongs to compact clay category. Approximately from 50 m depth clay stone was excavated off during construction stage. Structurally several shear zones were recognized in foundation area dipping in downstream direction and a thrust fault dipping in downstream direction is present about 700 m distance from the dam in upstream area. Below the dam, this thrust is present at 365 m depth. Shear zones were tackled by grouting or plug tunnels. As dam falls in seismic zone IV and epicenter of 1905 Kangra earthquake was located 80 km north-northeast of the dam, therefore, proper safety measures were incorporated in the design of the dam, 0.15 g for horizontal acceleration and 0.75 g for vertical acceleration. In the post construction period cracks were developed in the upheel claystone strut causing leakage of water in dam galleries. Initial effort to stop the leakage by grouting proved unsuccessful hence galleries were plugged by concrete to stop the leakage. Since commissioning of the dam in 1961, it is serving the nation very well.

Pong Dam

The Pong dam is constructed of the Beas River in Himachal Pradesh and has 123 m height 450 m length with reservoir capacity 8750 mcm. The construction of dam was started in 1962 and was completed in 1973. This rockfill dam is constructed on gently folded Upper Siwalik sandstone and clay-shales with a chute spillway on left abutment. Suitable remedial measures were taken for the presence weak plastic clay beds and for weak rock exposures.

Tehri Dam

The nearly constructed rock fill dam on the river Bhagirathi near Tehri town of Uttaranchal is known as the engineering marvel. The dam will have 260.5 m height and would generate more 2400 MW electricity and has irrigation capacity of 0.27 m ha land. The reservoir would be spreading over 5170-hectare land (submergence of 2687 hectare agricultural land) and would have storage capacity of 4345 million cubic meters. A powerhouse of 1000 MW capacity will be commissioned in May, 2006. Tehri pumped and storage plant will create another 1000 MW power capacity and a 400 MW powerhouse will be located at Koteshwar Dam (98.5 m), 23 km down stream of Tehri. Tehri dam is designed to withstand the earthquake shocks of 8.5 M.

Nathpa-Jhakhri Hydel Project

This engineering marvel is located on river satluj about 140 km from Shimla on the National Highway-22. A 60.5 m high concrete gravity dam is constructed at Nathpa and an underground powerhouse of 1500 mega watt installed capacity at Jhakhri. Water from Nathpa is brought by a 28 km long and of 10.5 m diameter headrace tunnel to Jhakhri. Entire project is located on Jeori-Wangtu Gneissic Complex

Thein Dam

This dam is situated on river Ravi. It is 157 m high and 878 m in length and has reservoir capacity 267 mcm.

Dhauliganga Hydel Scheme

A 56 m high rockfill dam using cut-off-wall technique on Dhauliganga River has recently been commissioned. This dam will produce 1134 million unit of electricity from four turbines of 70×4 mega watt capacity from an underground powerhouse located 5.29 km distance. Powerhouse is connected with dam by a headrace tunnel of 6.5 m diameter. The cost of entire project is Rs. 17510 millions.

The other important dams under construction or planned are at Panchshwar Dam on Kali (288 m height/17390 mcm reservoir capacity), Kishau Dam on Tons (238 m), Lakhwar Dam on Yamuna (Negi 1993), Srinagar Dam on Alaknanda (85 m) and medium dams in operation are at Ichari Dam on Tons (55 m), Maneri Dam on Bhagirathi (3 m), Kalagarh Dam on Ramganga river (126 m) in Uttaranchal. In Himachal Pradesh 51.5 high rockfill dam Baira-Siul project on river Baira near Chamba, Giri hydel project across the river Giri at Jateon, 300 MW Baspa hydroelectric project on river Baspa at village Kupa (10 m high barrage) and an underground powerhouse near Karcham in Kinnaur district, 120 MW Bhaba hydel project on river Satluj, three stage Parvati hydel project in the upper Beas basin, Larji hydroelectric project in Kulu district along with many minor projects are under constructions or in planning stages. In Pakistan, Mangla Dam on Jhelum is 138 m high, 3335 m in length and has 6358 mcm reservoir capacity while Tarbela Dam on the river Indus has 143 m height, 2743 m length and 13690 mcm reservoir capacity. Himachal Pradesh has 21000 MW hydel power potential, out of which only 6060 MW have been commissioned till 2005 while project to harness 4042 MW are being executed.

Hydroelectric potential of Nepal is nearly equal to India but feasible capacity for power generation is 42000 MW. At present power is generated from Marsyangdi (69 MW), Kulikhandi (92 MW), Trisuli (21 MW), Sunkoshi (10 MW), Gandaki dams (15 MW) and also through several minor dams.

The Himalayan dams are general gravity, masonry, earthfill, rockfill or buttress types. There are several problems encountered during construction of dams like fragility of strata (e.g. Tehri) alternate presence of weak and strong rocks, (e.g. Bhakra), high porosity of bed rocks (e.g. Tarbela) and dynamic tectonic set up (e.g. Tehri) and unstable steep slopes, high rate of sedimentation, presence of fault and shear zones etc. However, these problems can be tackled adequately by taking care of specific problem(s) during construction.

Large dams in the Himalaya are opposed by a section society on environmental, seismic, submergence of land and rehabilitation of displaced population's considerations. This section favors smaller dams working on run-off-the river principle as such small dams can not meet or substitute the large dams for the requirement of irrigation, drinking water demand of plain area and also can not generate the power consistently because large volume of monsoon water would flow off without being utilized. Small dams can be constructed in remote regions to meet the requirement of those areas where heavy investment is uneconomical.

11.2 Tunnels

A tunnel is an excavation in rocks, which is open at both ends. It is a common feature of mountainous terrain. It is constructed to cut short long road and rail routes and also as water conduit for power generation or drinking water purpose. Usually tunnels are circular, elliptical or semi-elliptical in shape and are of variable sizes. In Himalaya, many tunnels of several km lengths are excavated at several places for the purposes.

Nathpa Jhakhri tunnel in Himachal Pradesh is second longest tunnel in the world for transferring the water from the dam site to powerhouse site and is 27 km long. The water of the Beas River is diverted into the Satluj for augmenting the capacity of Govindsagar Lake through an 8 m wide and 24 km long tunnel. The tunnel is constructed in two parts; the first stretch between Pandoh to Baggi is 13 km long between Baggi to Sundernagar water is carried through 11 km long channel and from Sundernagar to Dehra by a tunnel of 11 km length. A 7.2 km long famous Jawahar Tunnel joins the Jammu region with Kashmir valley. The 6.2 km long tunnel between Ichari and Chhibro carries water from Ichari dam located on the river Tons to underground power house at Chhibro to utilize the drop of 124 m, again released water from power house is further transferred into an another nearby tunnel (5.6 km long and of 7.5 m in diameter) to carry it to Dakpathar for further power generation purpose. Another engineering marvel in Himalaya is Beas-Satluj link tunnel. This 25 km long tunnel joins these two rivers. The other important tunnels are Maneri Bhali (8.631 km long and 4.75 m diameter), Ramganga Tunnel, and Giri-Bata Tunnel in Himachal Pradesh etc. Several small tunnels for railways and roads are present in the every part of the Himalaya. Many Diversion tunnels have been also constructed to divert the river water for dams constructions–large excavations in the Himalaya are also made for hosting power plants like at Chhibro, Dakpathar and other places.

Stability of excavated structure is of prime importance and are largely dependent upon proper geological investigations and appropriate constructional measures as they have to cross through active faults, shear zones comprising of crushed rocks, joints of variable attitudes, regional lineaments, varieties of folds, weathered zones, clayey formations, presence of alternate hard and soft strata, highly porous rocks and small streams or rivers flowing on the surface etc. In many cases underground mobility was such that even one inch thick steel ribs twisted like ordinary rubber bends e.g. in Dakpathar power tunnel. In major tunnel construction activity, before the start of actual work detailed geological investigation, spanning over several years, is undertaken and the safe design is to be worked out according to situation of the area.

11.3 Roads, Bridges Canals and Houses

Roads

The total metalled and non-metalled road length in Himalaya is estimated at 44000 km in 1987. Himalayan roads are constantly subjected to degradation because of landslides, subsidence and collapse of roadbeds either due to tectonic condition or erosion along the side, heavy rainfall, seepage along the slopes, movement of heavy vehicles and digging for other activities. Road construction on steep slopes requires large-scale excavation for maintaining proper stability. It is estimated that for 44000 km long network of metalled roads needed excavation of approximately 2640 mcm. Debris is also generated due to regular landslides and other mass movement activities. The quantum of this secondary debris varies from place to place e.g. 724 cubic m/km/year on Jammu Srinagar highway and 411 cubic m/km/year on Tanakpur-Tawaghat road in Kumaun. The quantum of debris is also very high (5000 cubic m/km/yr) in certain pockets of Garhwal region because of recurring landslides. Excavation and disposal of debris on such a scale makes road construction, and maintenance extremely costly proposition. In many cases after some years of construction alternative road alignment has to be worked out due to recurring landslides e.g. between Barkot-Sayanachatti in Yamuna valley or due to increase in traffic pressure e.g. an alternate road is being constructed between Dehradun-Mussoorie. The road cuts can be stabilized by dewatering the slopes, constructing drainage along the inner side of the road; stabilizing the slopes by buttress, retaining walls, rock bolts, nets of steel wire, mortar etc. by slope modification or by afforestation.

Bridges

In Himalaya besides major rivers, larger number of small rivers, tributaries, streamlets and gullies are also present and to negotiate between banks of three millions of culverts and bridges of various sizes have been constructed. On an average for 1 km length road length 3 culverts of modest size are present in the Himalaya. Most of the bridges are of small size having length few meters. Bridges of simple beam, cantilever, arch and suspension types are common.

For stability of major bridges, site requires proper geological investigations as large load is transmitted on a small area. The foundation rocks and its load bearing capacity required to be properly investigated. Besides this slope stability nature of mass movement process, pressured water current, flood discharge, impact of drifting objects such as wooden logs, heavy vehicles, seismic condition of the region are also being taken into considerations. In many cases inadequate appraisal of geological conditions resulted into collapse of bridge e.g. near 100 m long Rispana Bridge in Dehradun collapsed in 1991 or a bridge across the river Ganga in Rishikesh swept off during flood in 1918.

Canals

The river and barrage water is diverted into canals for many purposes such as power generation in down stream location, irrigation and for requirement of drinking water in urban area. Canals are constructed along a predetermined route. The route for major canals should be geologically safe from landslides and mass wasting processes and the ground should be firm. In Himalaya all major canals are lined and minor Canals are unlined in nature. Canals drawn from the Satluj, Ganga, Yamuna, Ramganga rivers carry huge amount of sediments laden water flowing with swift velocity. Har-ki-Pauri in Hardwar is situated on the bank of canal.

Buildings

Proper study of mass movement and ground slope conditions is required for construction of multi story buildings in mountainous region. Due to inadequate appreciation of these facts in many multistory buildings in Mussoorie, Nainital, Shimla, Darjeeling, and Kathmandu and in other large towns cracks are developed in the buildings shortly after their constructions.

11.4 Engineering Study of The Himalayan Rocks

Various engineering properties of rocks and minerals have been investigated at different place. The rock mass rating and slope mass rating, strength characteristic i.e. uniaxial compressive strength, and point load strength, rapid quality designation have been studied in the Himalaya (Nainwal and Prasad, 2001).

11.5 Seismic Microzonation Study

The impact of earthquakes is particularly more pronounced in the urban areas because of high density of population. With increasing development and rise in population the seismic hazard is increasing. To mitigate the probable damage caused by the earthquakes in a particular urban area, proper understanding of that area is must. The damage potential of an area is function of impedance (product of shear wave velocity and density of medium) and thickness of the uppermost layer. Seismic hazard risk microzonation can be defined as criteria based discretizationof an urban agglomeration or geographical domain into units of likely uniform level of hazard and risk, guided by prognosis of scale of peak ground motion (peak ground acceleration, PGV, PGD, duration and number of cycles of earthquake loading) and nature of induced processes (liquefaction, land deformation/landslide etc.) and their manifestation in built environment (Mishra, 2003). Seismic microzonation study involves comprehensive multidisciplinary geological, geomorphological, geophysical, hydrological, engineering geological, pedological and anthropological/ administrative response of an area to an earthquake of maximum possible intensity likely to strike in the chosen city or region for providing relatively safe living condition to people living in the area. In this multidisciplinary study the end geoscientific map characterized as hazard map with ground motion defined in terms of frequency dependent amplifications and response spectra. This hazard map is finally transformed into risk microzonation map. One of the very important components of this microzonation is to estimate site-specific dynamic response of soil or soil amplification study and underlying rocks. The dynamic response of the soil is dependent on shear modulus (G) and damping ratio (D). These properties are controlled by soil type, soil thickness, variation in the soil, number of cycles of dynamic load application, magnitude of shear strain and confining pressure, horizontal extent and depth of the soil deposits overlying the bed rocks, slopes of the bedding planes of the soil overlying the bed rock, changes of soil types horizontally, topography of bedrock and deposited soils and faults crossing the soil are some important factor for ground response analysis. The ground response also depends on strain-dependent nonlinearity of soil properties at soft soil sites during destructive earthquakes. Soil behavior during an earthquake changes from linear material to non linear as induced shear strain grows from 10^{-6} to 10^{-3} and finally reach failures at strain larger than 10^{-2} as a result of this shear modulus G and damping ratio do changes gradually during earthquake. Dynamic properties of soil such as shear modulus are evaluated in low and high strain conditions in the laboratories as well as in field. In low strain lab tests resonant column, ultrasonic pulse, shaker table, sample shear and piezoelectric bender elements test are usually conducted and in the field seismic reflection, steady-state vibration or Raleigh wave, spectral analysis of surface wave, seismic cross hole, seismic down hole and seismic cone tests are conducted. In high strain lab tests shaker table, cyclic triaxial, cyclic torsional shear tests are conducted and standard penetration test and dilatometer, pressure meter and cone penetration tests are common for high strain field tests.

Liquefaction is another very important factor responsible for damage during the earthquake. It is governed by intensity of earthquake and its duration, location of ground water table, soil type, soil relative density, particles size gradation, particles shape, depositional environment of soil, soil drainage conditions, confining pressures, aging and cementation of the soil, historical development of the soil deposit and additional load on soil in the form of civil structures like building etc. Liquefaction potential is evaluated by Standard Penetration Test or by shear wave velocity method. The factor of safety against liquefaction is worked out to find the liquefaction susceptibility. The ratio of cyclic resistance to cyclic stress ratio provides factor of safety against liquefaction (Booinathan, 2004, Dinesh et. al., 2004, Sitharam et al., 2004, Kokusho, 2004).

Seismic microzonation study is in progress for Dehradun city and it is also being planned for other Himalayan and off Himalayan towns and cities. Delhi falls in region of high seismic activities and an earthquake of IX intensity on MMI scale struck on July 17,1720. In Delhi and surrounding region of 250-300 km radius 20 faults capable of producing earthquakes of 4.5-8 magnitude are present with peak ground acceleration 0.24 g. Microtremor measurements in Delhi for seismic hazard zonation revealed that resonance frequencies vary from place to place and variations in thickness and types of the soils. Amplification of ground motion is high in soft or loose soil locations and less in quartzitic region (Iyengar and Ghosh, 2004).

DIMENSIONS OF HIMALAYAN GEOLOGY
A. K. Biyani (2006) : pp.249-280
Satish Serial Publishing House, New Delhi
E-mail: ssp@ndf.vsnl.net.in, hkjain1975@yahoo.com; website: www.satishserial.com

Bhunkswa rajyam samrddham **(Geeta, 11.33)**
Enjoy the prosperous state

CHAPTER - 12

RESOURCES OF HIMALAYA

All geological investigations directly or indirectly are concerned with search of mineral deposits for the benefit of mankind. Geological investigations in Himalaya are no exception to it. As a result of this we have fairly accurate catalogue of metallic, non-metallic or industrial minerals or of fuel resources bearing locations. However, Himalaya is richly endowed with industrial minerals. This potential is largely untapped either due to remoteness of deposits or environmental factors or for paucity of suitable market conditions. Mineral resources of Himalayan region are divided into metallic, non-metallic and fuel mineral categories. Locations of the deposits or surface shows and a short geological description of important deposits are given in the following text.

12.1 Metallic Mineral Deposits

Metallogeny events are related with the global tectonic events and their time and space aspects are correlated with tectonostratigraphic and magmatic processes. Economic geologists are of view that Himalaya is an orogen of collision tectonics; it is poor in metallic mineral deposits in comparison to Andean type orogen or island arc type convergent boundary terrain. Latter hosts voluminous deposits of metallic ores. Major ore bodies such as Cu-Mo and Cu-Au deposits; manto and breccia-pipe Cu deposits, and contact metasomatic and Kurako-type poly massive sulphide deposits are found in the Andean-type and Island-arc convergent plate boundaries (Sillitoe, 1969). For this purpose Himalaya is divided into four divisions namely Karakoram, Indus and Shyok Sutures, Main Himalaya and Foot Hill Belt. The Karakoram Belt has convergent type setting and contains I-type granites that are supposed to host promising tin-tungsten and copper-molybdenum deposits. Its highly inaccessible topography and glacier cover over large area prohibited the search of ore bodies. Indus-Shyok Suture belts are promising regions for mineral deposits as it has also I-type granite and ophiolites or ophiolitic melange with ultramafic rocks. Despite several shows of chromite, nickel, platinoids, gold and fluorite promising deposits are still elusive because area is highly tectonized and dismembered. Tethyan, Higher and Lesser Himalayan zones constitute the Main Himalayan Belt; have large number of sulphide (Pb, Zn, Cu, Sb, Ag, Au) occurrences either of syngenetic or epigenetic origins. A few of these deposits are mined regularly. The Foot Hill Belt is very poor in mineral deposits as this terrain evolved at comparatively late age, however, presence of uranium and gold placers are reported from several places in the Siwalik terrain. Mukerjee (1983) has classified mineralizations in Himalaya in the following categories:

1. Mineralization associated with ophiolite belt and Trans-Himalayan granitoids e.g. Dras and Nidar areas, Karakoram fluorite
2. Mineralization associated with Neogene and older granites of the Central Crystallines and Lesser Himalayan belts e.g., Bi and polymetallic ores in Manaslu Granite of Nepal, Paddar sapphire mine of Kashmir
3. Pre-Tertiary metalliferous deposits: these deposits are further sub-divided into-

   Type-1 stratabound sulphides usually associated with low-medium grade metaargilllites of the Lesser Himalaya and Lower Palaeozoics of the Tethyan Himalaya e.g. Rangpo Sikkim, Gorubathan, and Darjeeling.

   Type-2 polymetallic (Pb-Zn-Cu) stratabound to highly lenticular and pod like bodies of sulphide mineralization associated with carbonate horizons of the Central Crystallines e.g. Ganekha, Bhutan

   Type-3 polymetallic (Pb-Zn-Cu) disseminations or thin stringer type sulphide mineralization in the Lesser Himalaya e.g. Riasi-Kotli Jammu, Shilai Himachal Pradesh
4. Other mineral deposits of the Lesser and Sub-Himalaya e.g. Mussoorie Phosphorite of Uttaranchal.

Joshi (1985) has identified six tectono-minerogenetic provinces namely the Trans Himalayan ophiolite melange mineralizations with chromite, metallic sulphide and pegmatite sub provinces, Central Crystalline-Lesser Himalaya granite mineralization with polymetallic and pegmatite sub provinces, Tethyan Himalayan sulphide mineralization, Lesser Himalayan mineralization with carbonate, sulphide, iron ore, phosphorite, evaporite and coal sub provinces and Sub-Himalayan mineralization in the Himalaya

Mahadevan (1971); Chaudhri (1972,1973); Gokul (1981); Geological Survey of India (1983,1984); Krishnaswami and Sinha (1988); Ghose et al., 1989; Sen (1981); Nayak (1993) have compiled the list of locations of mineral deposits in the Himalaya.

12.1.1 Sulphide Mineralizations

Sulphides of copper, lead, zinc, iron, molybdenum, arsenic, antimony, nickel, cobalt and iron are reported from large number of locations in the Himalaya. The deposits are present in all the geological divisions of Himalaya except the outermost one. Theoretically, subduction related I- type granitoids are associated with calc-alkaline volcanism are favorable target for porphyry type Cu-Mo and other related mineral deposits; these deposits show either structural, stratigraphical and lithological controls. Majority of these deposits are of syngenetic origin either formed along with sediments deposition or along with the emplacement of magmatic mostly granitic masses. Many of these deposits have remobilized during the Himalayan metamorphism and deformation. Some deposits are epigenetic in nature. Majority of sulphide deposits show polymetallic characters with fairly equal or unequal proportion of different metallic ores and some are dominated by the ore(s) of one metal. Many of these deposits were mined locally in the past and even now on small scale some ore bodies are mined even in very remote areas. The quality and quantity of these deposits are highly variable. Quite a few of them are extremely rich in metal content e.g. Ganesh Himal zinc deposit of Nepal. However, overwhelming majority is of subeconomical to uneconomical nature and many of them are occurrences only. Their remote location, inaccessible topography, forest and snow cover, and very high cost of prospecting and exploration prohibit the detailed investigations of these deposits. Following description gives a brief account of these sulphide deposits.

12.1.1.1 Copper-Lead-Zinc Deposits

Jammu and Kashmir

In Jammu and Kashmir sulphides are reported from many locations. The important locations are Riasi and Katli (Pb-Zn), Buniyar (Fe-Zn-Pb-Cu-As), Chiche (Pb), Shumahal (Cu), Ramsu (nickelliferous pyrrhotite) and from the area of Shyok suture. Among these most important is Riasi deposit, which contains galena, marcasite, pyrite and sphalerite present in the Late Proterozoic Great Limestone Formation. The mineralizations have taken place along the bedding plane and fractures in synsedimentary condition. Porphyry copper deposits are reported from the Basti (Manugah) and Nomal in Pakistan.

Himachal Pradesh

Uchich (argentiferous-pyrite-galena-arsenopyrite), Kulu (Cu), Naraul-Dhanla (Cu-Ni-Co), Purbani (Fe), Taradevi (Fe-As), Aiyur (Pb-Zn-Fe), Ambakhala in Sirmur (Pb-Zn), Bara Shigri (Sb-Pb-Zn), Garsha and Rangbar in Kinnaur (Cu), Charadra Tal in Lahaul (Cu), Panuh, Shangri, and Mangna in Solan (Pb-Zn) and along the river Tons in Chamri-Ambota (Pb-Zn-Cu-Fe) are notable sulphide occurrences in the Himachal Pradesh. Local people on small scale mined the Rangbar copper deposit. The Garsha valley deposit extends for over 4.5 km and is mined by local inhabitants. A sulphide zone of 55 m length and 6 to 8 m thickness at Lalgi present in a shear zone, confined in Early Proterozoic conglomeratic quartzite of the Naraul Formation is a promising occurrence in the Garsha Valley.

Stratabound deposits spreading over 10 km length along the Tons river between Chamri and Ambota in Himachal Pradesh and across Uttaranchal containing galena, sphalerite-chalcopyrite, pyrite are attractive deposits of considerable potential. The surface exposures of deposits occur in detached zone at Chamri (200 m strike length), Amtiyar (150 m), Aniyar (700m) and Ambota (125 m) in Himachal Pradesh and Kwanoo in Uttaranchal in Proterozoic Deoban carbonates.

Uttaranchal

Sulphides are associated with carbonates (limestone, dolomite and magnesite), talc schist, quartz porphyry and porphyritic granite, chlorite schist and metatuffites of Berinag Formation.

Sulphides are reported from large number of places such as Pokhri (Cu-Pb-Zn, Sb-As), Askot (Cu-Pb-Zn), Kanalichhina (Fe-Cu), Dunda (Fe-Cu), Dewalthal (Cu), Sisakhani near Hanumanchatti (Pb-Cu-Zn), Rishiganga Valley near Banas (Cu-Pb), Janakichatti (Fe), Yamunotri (Fe), Rain Agar (Cu-Fe), Tusrani and Dal in Bora Agar (Cu-Fe), Bhaldeo-Dewaldhar (Cu), Dhanpur-Dobri in southern Chamoli district (Cu-Fe), Maniknath near Devprayag (Cu), Ramgarh, Nagnath areas of Pokhri district (Cu). Dudhkhamba and Harishankar (Cu-Sb) and at Byansghat south of Pauri (Cu). Lead in association with other ores are reported from Sisakhani-Chhanapani-Bhaldeo and Gaul south of Bageshwar, Rain (Dasoli), Chamdale near Pithoragarh, Tachhira (Dhanpur), Bhainskhet (Gori Valley), Gwar, Matha and Dhanpur (Chamoli). Among these promising deposits are of Pokhri, Ascot Sisakhanis of Hanumanchatti and Bageshwar areas and extension of Himachal deposits in Uttaranchal along the Tons river between Kwanoo and Tiuni and the Pokhri deposits occurs between fault planes and contain mainly chalcopyrite-pyrite, -pyrrhotite with minor amount galena, sphalerite, stibnite and arsenopyrite as veins and disseminated grains in chlorite schist.

The Askot deposit occurs between Barigaon and Gadeli and is mined on small scale. The deposit is confined in sheared schistose zone comprising of chlorite-actinolite-epidote-sericite schist derived from alteration of volcanics and also quartzite present in the synformal structure. The ore body is 600 m long and 4 m thick containing veinlets, stringers and disseminated grains of chalcopyrite, sphalerite, galena, arsenopyrite, pyrrhotite, stibnite, cubanite, chalcocite, marcasite, molybdenite and bornite of post-thrusting mineralizations. Granitic rock is believed to be the source of mineralizing solution. The estimated reserve of Askot deposit is 0.77 Mt and has 8.9 percent total metal content (Cu-2.32%, Pb-2.64% and Zn-3.95%).

Sisakhani deposit of Bageshwar area is traceable over 4 km length between Sisakhani and Bhaldeo; here galena occurs as veins, streaks and patches in dolomite, magnesite and talc schist. Another Sisakhani location near Hanumanchatti in Yamuna Valley has notable galena concentration in quartz vein of 7-8 m length and 1.0-1.5 m thickness. Besides galena, chalcopyrite and sphalerite also occur in small quantity. This low temperature epigenetic deposit occurs at the contact of chlorite schist and biotite schist in the proximity of Yamuna Mylonite Zone. Possibly same deposit extends into Rishiganga Valley near Banas. Detailed exploration of this deposit is necessary (Biyani, 1997 b).

Nepal

Nepal has exceptionally large number of sulphide locations and has been thoroughly explored by a combined team Department of Mines and Geology (DMG) of Nepal government, United Nations Development Programme (1981, a and b) and United Nation Economic and Social Commission (UNESCO) experts.

Sharma (1990) has mentioned 98 polymetallic occurrences (Cu-Pb-Zn) in Nepal, out of which 66 are dominated by the copper and rest by, lead and zinc. Out of these, DMG and UNESCO experts have investigated 27 copper and 11 lead-zinc occurrences (UNESCO, 1993). Investigated sites for copper are Wapsa and Kalitar deposits and Dhusa, Devrali, Kurule, Bhutkhola, Pandav Khani, Golkhalta, Chhirlingkhola, Sanotar, Gyazi, Baisekhani

(Okharbot), Ningre, Bhorle, Nangre, Ipa-Baraghare, Banspani, Minamkat, Jantrekhani-Tamakhani, khola Khani, Agrakhola, Siddhikhani, Rukum Khani Chandauli, Kottham, Mulkhani-Jokumkhani and Anbu prospects. While important lead-zinc investigated sites are Ganesh Himal and Lalbang - Khairang deposits and Pangun, Damar, Phakuwa, Khirauli - Bhanjyang, Kolchebang Kandebas, Solendanda, Phulchowki and Shalimar Valley prospects.

Most of the copper deposits are epigenetic in origin and show stratigraphical or structural (fault) controls. Epigenetic deposits pre-date the Himalayan orogeny and some e.g. Phakuwa are remobilized during Himalayan metamorphism (Tuladhar, 1991). Devrali and Chandauli deposits are stratiform in nature and are of syngenetic origin. Mineralogically, three trends have been observed in copper deposits namely the simple chalcopyrite mineralizations with minor presence of pyrite and magnetite and by and large absence of solidification e.g. Kalitar, Ipa etc. while in other categories chalcopyrite occurs in association with other polymetallic sulphides e.g. Agarkhola, Kurule, Bamangaon etc. and chalcopyrite with strong silicification e.g. Wapsa, Bhutkhola deposits.

The lead and zinc occur mainly in the Higher Himalayan Crystallines, Bhimphedi Group and Palaeozoics of the Lesser Himalaya. Mostly they are syngenetic in nature and had been remobilized during metamorphism and deformation.

Ganesh Himal Zinc Deposit: It is situated at or above 4450 m elevation in the Higher Himalayan crystallines of Ganesh Himal region. On the surface deposits are exposed at Lari, Suple, Serkaping and Poktanjoo areas and under ground deposit is exposed through one 1.6 km long edit at Pengutand by the Nepal Metal Company Ltd. Granular sphalerite beds of 1.0 m to few millimeter thickness having 22% Zn, 4% lead, 300 ppm cadmium and 27 ppm silver occur in white, granular, highly metamorphosed dolomite. This area depicts synclinal structure. This exceptionally rich sphalerite zone also contains in minor amount galena, rarely chalcopyrite and pyrite. A magnetite vein of nearly 1 m thickness is also present at Lari. The area is metamorphosed in amphibolite facies condition. The present estimate of reserve is calculated at more than 1 Mt. With average 13% Zn and 6% lead contents. Besides silver and cadmium, other elements are also present in the traces in the ore minerals. The syngenetic mineralisation shows evidences of later remobilization and due to severe deformation ore beds also migrated into small fault planes and minor fold axes suggesting secondary structural control (Chakrabarti, 1982, Ghose, 1996, Ghimire et al., 1996). 875-785 Ma age is obtained by lead isotope analysis for sediment deposition and sulphur isotope data suggests 650 Ma age for lead-zinc deposit and this deposit has considered equivalent to Vindhyan Supergroup of peninsular India, the rocks were metamorphosed in 500°-769°C temperature range (Ghosh, et al., 2005)

Labang-Khairang lead-zinc deposit is situated 50 km west of Kathmandu, on the either side of a marble hill. Disseminated, stringers and lenses of galena and sphalerite are present along the foliation planes in dolomitic or light gray marble. 21 Mt Pb-Zn Ores of 1% metal are estimated till the 400 m depth while tonnage decreases to 0.20 Mt only for 6% metal content.

The Wapsa copper deposit is situated on spur between the Rue and Guekhola on the left bank of the Dudh Kosi River. Chalcopyrite-pyrite along with minor presence of bornite, covellite, pyrrhotite magnetite, sphalerite and galena occur as pods, lenses, stringers and dissemination in a domal structure of chlorite schist or phyllite of Kuncha Group. Preliminary exploration data reveals the availability of 1.74 Mt. copper ore of 0.88% grade. Presently a private producer yielding around 20 tons copper every year above 5% ore.

Kalitar copper deposit is situated on Tribhuvan Highway south of Kathmandu. The copper mineralisation is present in Kulikhani micaceous quartzite and schist of Bhimphedi Group. Chalcopyrite, chalcocite and magnetite pods and dissemination occur in shear zones coinciding with the axis of regional overturned syncline. The strike length of mineralized zone is 1300 m and width is 20 m. The reserve is estimated at 1.39 Mt. and 0.09 Mt of 0.5% and 1.5% copper respectively.

Dhura copper deposit of Dhading district has reserve of 1.1 Mt of 0.65% Cu while in the Curdle (Udaipur) and Bhutkhola (Tanachu) copper deposits have 10.5 Mt (0.2% Cu) and 0.21 Mt (1.0% Cu) respectively. The Beringkhola pyrite deposit of Jhala district has reserve of nearly 3.0 Mt. The pyrite contains 0.3 to 0.7% copper, lead and zinc and 12 percent sulphur and gold in minor amount.

West Bengal

In West Bengal sulphide deposit occurs at Gombathan locality in Darjeeling district. Magnetite sphalerite and galena occur as uneconomical stratabound deposits in Late Proterozoic Daling Formation.

Sikkim

Sulphides are reported from several locations namely Rangpo, Dikechm, Dejong, Ben, Barmaik, Monghru, Namphab, Chongbong, Jugdum, Roathale, Richingpong, Khanikhola and Subbule. Temi Rangpo is only base metal producing mine in Indian Himalaya. Stratabound zinc-lead-copper mineralisation occurs in Daling Formation. Genetic link has been observed between mineralisation and andesitic magma, which was metamorphosed in green schist condition in the island arc setting. The mineralizing fluid seems to be derived from deep-seated magmatic source (Rai and Rao, 1988).

Bhutan

Sulphides in Bhutan are reported from the rocks of Paro Group at Ganekha in Thimpu district and copper in Chekha Formation near Nobji in Tongsa district. In Shumar Group mineralisation is observed at Gomchu in Tashigang district and in Daling Group between Jaldhaka in the West to Chamurchikhola in the east, in Samchi district. Uneconomic mineralizations occur parallel to bedding as dissemination, fracture fillings and irregular lensoidal bodies. The Gangkhola copper deposit in Tonga district contain 2.24 Mt ore of 1% copper while Ganekha deposit contains 3.63 Mt lead-zinc ore of 1.4% Pb lead and 6.06% zinc. Other occurrences in Bhutan are at Samstse, Sarpong and Ratepani.

Arunachal Pradesh

Khetabari Formation of Bomdila Group hosts sulphide deposits in Arunachal Himalaya and contains polymetallic sulphide mineralizations at Potin and Yazali in Ranga Valley, Pakro,

Bara Rupak and Sododoke. At Pofin, mineralisation occurs in the veins, bands, stringers and disseminated grains in a zone of 500 m wide and 3.0 km long. Test drilling reveals the presence of copper (0.31-0.61%) at 113 to 132 m depth. In Chillepam Formation sulphides are reported from Shergaon in West Kameng district and Menga in the Upper Subansiri District. The lead-zinc mineralisation at Shergaon occurs in dolomite interbedded in phyllite and schistose quartzite. 1 km long and 50-150 m wide zone is enriched in lead (4.72-84.02%) and zinc (0.1-0.3%). The strata bound ores occur as dissemination, veins, blotches and stringers in carbonate rocks.

12.1.1.2 Arsenic

Uneconomical realgar (AsS) and orpiment (As_2S_3) ores of arsenic are reported from the Chitral area in Jammu and Kashmir and glacial moraines of Kumaun. Arsenopyrite (FeAsS) is reported from Harakoti in Chamoli, Askot in Pithoragarh and from many polymetallic sulphides described earlier.

12.1.1.3 Molybdenum

Molybdenite is described from quartzofeldspathic metasediments of Tidding Formation of Arunachal Pradesh. Here it occurs with sphene, pyrite and chalcopyrite and pyrrhotite. Presence of molybdenite in insignificant quantity is also recorded in many sulphide deposits.

12.1.1.4 Nickel and Cobalt.

Presence of economical nickel ores like heazlewoodite, pentlandite and awasuite are reported from Alpurai and Kishora area of Swat in Pakistan Himalaya. In general Himalaya is not known for mineable nickel and cobalt deposits. However, economical deposits may exist in ophiolite sequences of Ladakh region, as these are present in the ophiolite rocks of Nagaland and Manipur states of India. Occurrences of nickel and cobalt are reported from Tamghas, Samarbhamar, Neta and Nangre area of Nepal. In Tamghas area cobalt - copper mineralisation occur along contact zone of dolomite and phyllite in the form mm to cm scale seams. 0.14% cobalt is assayed in the rocks while in Samarbhamar area cobalt is 0.4%. In these areas old workings are also present.

12.1.1.4 Antimony

Antimony is reported from Rohtang Gneissic Complex of Vaikrita Group near Bara Shigri glacier of Himachal Pradesh and is present between 4250 and 4650 m heights. Stibnite bearing fracture filling veins are 2-40 cm wide. The antimony content is assayed at 2.88 percent and besides antimony; pyrite, galena and sphalerite are also present. Antimony is also associated with many other sulphide deposits in negligible quantity. Uneconomical sheared lenses of stibnite (Sb_2S_3) in biotite schist are reported from Pokhri- Mohankhal area of Chamoli district.

12.1.1.5 Silver

Native silver is reported from Chargaon area in Kinnaur District. Argentiferous galena is present in Manikaran Quartzite at Uchich in Parvati Valley and Saughtan (Jhasi) locality in Himachal Pradesh. As trace element, silver is associated with most of the sulphide deposits.

12.1.1.6 Mercury

Presences of natural mercury drops are reported from quartz veins of shear zone present in phyllite of Suparitar Formation at Tirechepani-Takura area of Nepal. Shear zone may contain considerable deposits of mercury.

12.1.1.1.7 Pyrite

Pyrite is invariably associated, in small quantity, with all sulphide deposits. As such significant presence of pyrite is also reported from many areas and many of these have been mined on small scale in the past. Volcanogenic massive pyrite is reported from Jutial area in Pakistan. In Himachal Pradesh, pyrite is reported as lenses from Jutogh Formation at Tara Devi in Shimla district. Haripur Dhar and between Chamri and Ambota in Tons Valley of Sirmur district also contain in notable quantity of pyrite, Mandhali Slates, Blaini, Nagthat and Krol formations of the Lesser Himalaya and Yamuna Formation of Higher Himalaya contain disseminated grains of pyrite at several locations in Uttaranchal, but none of these are workable. Hydrothermal Beringkhola pyrite deposit of Illam district of eastern Nepal is most promising one in the entire Himalaya. This deposit has probable ore reserve of 2.2 Mt and occurs in two veins in fine-grained greenish black quartzite. Veins having 1.8 and 1.2 m maximum thickness extends over 2 km along the strike of bedrock. Pyrrhotite and chalcopyrite are additional phases in the pyrite veins. The average sulphur and copper content in pyrite are 11.65 and 0.21 percents. Disseminated grains of pyrite are also reported from the Daling and Baxa rocks of West Bengal and at the head of Ganje rivulet in Arunachal Pradesh.

12.1.2 Non-suphide Ores

12.1.2.1 Sulphur

Native sulphur is associated with gypsum in Puga valley of Ladakh and sulphur encrustations are reported around thermal springs of Vihi and Lashitial districts of Ladakh. Sulphurous water emanates from many springs of Himalaya e.g. Sahastradhara (Dehradun) and Syabrubensi (Dhunche) in Nepal.

12.1.2.2 Tin-Tungsten

Tin (Sn) and tungsten (W) usually found together but they may also occur separately. Their non-commercial occurrences are reported from many areas of Himalaya and in many cases they are associated with sulphide mineralisation.

Considerable presence of tungsten is described from Dadeldhura granitic massif of farwestern Nepal. At Bamangaon, 15 km east of Dadeldhura, tungsten mineralisation occurs at the exo contact of granite with siliceous phyllite and quartzite. The epigenetic mineralisation is present as replacement veins and dissemination of hypothermal to mesothermal origin. Mineralisation is confined in 2.5 to 71 m thick and 800 m long area. The ore body contains wolframite, chalcopyrite, chalcocite and pyrite. The reserve of 0.2% W and 1.72% Cu is assessed in 1.54 Mt out of which 0.16 Mt. belongs to 0.43% W and high-grade copper ores. Cassiterite is also reported from Meddi situated 22 km east of Dadeldhura. Over 0.13% metal occurs in quartz-cassiterite or quartz-cassiterite-pyrite association in phyllitic quartzite.

Chalcopyrite is also noted in this area. 360 ppm tin and 200 ppm tungsten are also reported from Palling and Ipa granites.

In Uttaranchal Himalaya, scheelite ore is reported from Almora and Chamoli districts and cassiterite from Jalali, 15 km northwest of Ranikhet in Almora district. Cassiterite is associated with pegmatites emplaced in mica schist and granite.

In Sikkim, scheelite-cassiterite mineralization is reported from pegmatites and calc-silicates of the Paro Group at Chitre. Quartz veins in Daling phyllite of Monsong area of Darjeeling-Sikkim region also contain hypothermal Sn-W mineralisation. The uneconomical wolframite contains 66 to 77.7% WO_3 and 0.09 to 0.53% SiO_2. A 4.6 Mt deposit of tungsten ore of 0.22 to 0.26 WO_3 content is located in Dhalpani in Sarpane district of Bhutan. The mineralisation occurs in calc-silicate rocks near Thimpu Thrust.

12.1.2.3 Platinoids

Elements of platinum group such as palladium (Pd) and iridium (Ir) are reported from the Tidding serpentinites in Arunachal Pradesh. Good deposits of these elements can be expected in the ultrabasic and basic rocks of suture zones in Ladakh, Arunachal Pradesh, Nagaland and Manipur and also in Pakistan.

12.1.2.4 Uranium

Himalayan uranium occurs in granite-pegmatite and syenite, sandstones and placer sands. In Pakistan, uranium mineralisation is reported from Bagrian and Thakal area of Hazara Complex. Uraninite and pitchblende are present near Chalt and also in sands of Basha Braldy and Indus River near Darband, Chilas and Olding. Lahor Granite contains uranium in association with lead-zinc-molybdenum (Shams, 1983). Ladakh Batholith, Karakoram Granite and Salkhala Formation also reported to have shown the presence of uranium in Pakistan. In adjoining Jammu and Kashmir these rocks have wide presence; therefore, presence of uranium cannot be ruled out here. Insignificant presence of uranium is reported from the Jammu lignite and pink granite.

In Manikaran Quartzite, uranium occurs as yellow brown encrustation at Chhirijra and Shakirandhar in Kulu district of Himachal Pradesh. Here percentage of U_3O_8 varies between 0.1 to 0.78% (Narayan Das, 1981). Shear zone mineralisation of uranium occurs at Kasha area in Rampur window. Pitchblende, uraninite, atunite and their secondary products and sulphide minerals occur in this region. Parvati Pat, Pune, Sharan Banjar and Tirhar Valley also contain patchy occurrences of uranium. Uraniferous Lower Siwalik sandstone of Hamirpur contains epithermal (<90°C) koutekite (Cu_5As_2) and domeykite (Cu_3As), rare copper arsenide minerals. Uranium is reported from Krol Belt exposed in Giri and Tons rivers. Lower and Middle Siwaliks of Himachal Pradesh contain 0.001 to 0.6% U_3O_8 with low quantity of thorium in strata bound tabular outcrops of few sq m to 6000 sq m area. Upper Siwalik is also reported to contain uranium in association with phosphatic minerals in Himachal Pradesh.

In Uttaranchal, presence of uranium is reported from Chamoli and Pokhri districts. In Mussoorie syncline, phosphorite occurs at the contact of Krol and Tal formations in Dehradun and Tehri districts contain U_3O_8 in traces. Siwalik rocks have also been investigated for uranium.

A 500 m long and 1-5 m thick uraniferous pebbly arkosic sandstone bed of the Upper Siwalik Formation located at Tinbhangali, 45 km from Hetaura in Makwanpur district of Central Nepal is a promising location of uranium in Nepal. Another presence of uranium in the Siwalik is reported from Chandikhola. Gneissic rocks of Thumbi, Jagat, Kakani, Panchmane and Malung and Higher Himalayan granites have also shown the presence of uranium in Nepal.

In Arunachal Pradesh, epigenetic U mineralization in association with iron occurs in greenschist to lower amphibolite volcano-sedimentary sequence in Badak- Gamkak-Kau and Maro- Baririjo sections of Middle- Late Proterozoic sequence of the Siang Group. In these locations uranium occurs in strike length that varying from 50 to 600 m and width of mineralized zones varying between 1-10 m. The uranium content varies between 0.008-0.20%. This epigenetic mineralization is 471–585 Ma old (Bisht et al., 2005).

12.1.2.5 Chromite

Chromite deposits occur in basic and ultrabasic rocks such as gabbro, dunite, peridotite etc. These rocks occur in suture zones of Himalaya, podiform uneconomic chromite occurrences are described from Dras, Shyok and Nidar areas. Concordant to semi concordant bodies of chromite occur as pods, lenses, pencil like streaks, schlieren in peridotite and dunite. Economical deposits Mayu come into light on detailed investigation of Ladakh area as reported from the ophiolite belts of Nagaland, Manipur and Magalia.

12.1.2.6 Iron

Magnetite and hematite are principal ores of iron used for metal extraction.

Iron ores are reported from Matah near Riasi and Khandli Devi in Jammu and Kashmir. At Matah 2 m thick-banded ferruginous chert occurs in the Jammu Limestone Formation.

In Himachal Pradesh iron ore mining on small scale were carried out at several places like Charayandhar, Thang, Sangalwara and Jhugie in Mandi district. Magnetite and hematite are present in low grade Salkhala and once huge deposit was falsely reported from here. Several old workings are reported from Dharmsala in Kangra district, Gargi and Garsha in Kulu district, Pujarli in Shimla and from Mangsala area in Kinnaur district. In Spiti region, lenticular, hematite quartzite beds are reported from Thango (130m x 1.5-2.5m) and Shitikar areas. Ores occur in the Ordovician Thango Formation.

1-3 m thick clinton type iron ore bed is reported from the base of Spiti Shales and magnetite-hematite concentration in Rautgarha, Deoban and Nagthat formations were mined at small scale at Bhukunda, Ali, Pokhri-Agar and Kala Agar (Nainital).

Small-scale iron mining and smelting in Nepal is as old as her history. This practice is now stopped, as it was increasingly proving uneconomical. Several modest size iron ore deposits of sedimentary, metamorphic, hydrothermal, skarn and magmatic types are present in different places of Nepal. Despite good potential they could not be exploited commercially on account of mining, geological locational and technical factors. The important iron bearing occurrences are:

Phulchowki Iron Deposits

Phulchowki iron ore deposit is located in Phulchowki Hill, 26 km southeast of Kathmandu Valley. The massive hematite bed of 4.5 m to 22.0 m thickness is overlain by the Chandragiri Limestone of Cambro-Ordovician age and underlain by the fossiliferous Godawari Limestone of Devonian age. The steeply dipping ore body has potential reserve of more than 10 Mt out of which nearly 3.9 Mt is proved. Due to steep slope of ore body (65° N) and limited width of deposit it could not be exploited commercially. Those Iron deposit is situated 2 km east of Those Bazar in the Janakpur Zone. The iron mineralisation occurs in the Lesser Himalayan carbonaceous phyllite schist and quartzite. Tens of centimeters to 2.5 m thick bands of hematite occur in three horizons 10 Mt ores of 40% iron is estimated here. Jelbang and Labdikhola-Bhulkhola are other sedimentary type iron bearing area in Nepal. Uneconomical hydrothermal type iron ore body is found at Iron Kachali, Khani Bagad and Patiuri. Skarn type deposits occurring at exo contacts of granitic bodies containing magnetite-hematite along with sulphides are reported from Beringkhola and Baraghare area. Totally uneconomical magmatic type deposits are associated with gabbro and diabases are reported from Khanidanda and Kalche.

A deposit of 2.7 Mt having 25-46% iron is located at Mauri along Sankosh River in Sarpang district of Bhutan. Iron ores are reported from the Tenga Formation at Bana, Pichang and Laggi Gamkak areas in Arunachal Pradesh. Iron oxide mineralization occurs in the form of bedded ironstone having strike length over 30 km and width varies between 20-40 m the ironstone is massive, gray, hard and medium to coarse grained showing metallic luster. Ironstones comprise of granular mortitized magnetite, hematite and quartz. The iron ore contains 44.03 to 63.26% iron. The origin of ironstone involved transportation and deposition of iron rich solution produced by erosion and weathering of older rocks followed by sedimentation in the Proterozoic Siang basin, the oxides were converted into magnetite and hematite by the diagenetic and metamorphic processes and in later phase they were deeply weathered (Bisht et al., 2005)

12.1.2.7 Gold

Presence of bedrock and placer gold is recorded from several locations of the Himalaya. Manikaran quartzite of Rampur Group contains primary gold and uranium in Kasha pat area of Shimla district. In this commercially non-viable area, highest gold concentration 19 gm per ton has been recorded. In Himachal Pradesh, Uttaranchal and Uttar Pradesh gold is also present as palaeoplacer and neoplacer in Siwalik rocks. The highest concentration gold flakes 0.2 gm per ton is assessed in palaeoplacer present in Middle and Upper Siwalik Subgroup of Himachal Pradesh Neoplacer gold is described from point bar and meander belt deposits of rivers draining Siwalik terrain and also draining the Lower and Higher Himalaya. In Satluj Valley noteworthy places are Nogle, Powari and Morang. In Uttaranchal, uneconomical concentration gold in reported from Soma Nadi, Ramganga, Ganga near Lacchmanjhula, Kali and many other small rivers and rivulet draining the Siwalik Group.

In Nepal, gold occurs both in primary and placer forms. As a constituent of rock mass, it occurs in a ridge running between Swayambhu Nath and Halchoke, NW of Kathmandu. Here Phulchowki limestone contains as high as 342 gm per ton gold on surface but metal vanishes at a few metre depths. Lungrikhola Gold Prospect, another promising area, is situated in the Rolpa district of western Nepal. The 30 km long discontinuous zone of mineralisation occurring in Lesser Himalayan crystallines of Jaljala Nappe, chlorite schist and quartzite of Proterozoic Jelbang and Telkhola formations. Disseminated flakes of gold are present in the magnetite-hematite, hydrothermally altered carbonates, quartz veins, mylonite, augen gneiss and chlorite and quartz-chlorite schist. The low grade auriferous zone contains on average 0.19 gm per ton gold but values as high as 6.7 gm per ton is also reported and gold particles varies from submicroscopic to 1.5 mm in size. Occasionally gold nuggets up to 6x3 mm size are also recovered. Gold placers are also present in the area. Gold is believed to be of hydrothermal origin. Another primary gold bearing uneconomical location is reported from Nehal in Sunmay - Beringkhola. Here gold is associated with host of sulphide ore minerals like arsenopyrite, antimonite, bismuthinite, chalcopyrite etc. Placer gold is reported from the Kali Gandaki River and its tributaries like Murigurja, Khibang, Kulikhani, Rabhar, Rahughat, Ching, Pandar, and Mistri etc. Local people regularly panned Maradikhola and Renkhola placers in the past. Buri Gandaki, Trisuli, Marsyangdi and Modi basins also contain uneconomical gold in Nepal.

Gold is reported from Gangkhola, Dholpani, Bhurkhola and Kuru chu area of Bhutan; the content varies 0.2 to 3.909 ppm. In Arunachal Pradesh, gold placers are reported from several tributaries of the Brahmaputra River. The gold bearing sands are present in Subansiri, Rohit, Dihang, Bury-Dihang and Bhareli rivers. Palaeoplacer in Siwalik are located in Pikrang River. Gold mining in Arunachal was flourishing activity in last and early years of present centuries.

12.1.2.8 Alum

Alum is a hydrous salt of aluminium and potassium sulphate. Alum efflorescence of white, yellow and orange color is reported from graphitic shales at Kapur and also from north of Kathmandu in Nepal. Slates of Kosi Valley near Almora and Jakh and Khairana area of Nainital in Uttaranchal also contain patches of alum of secondary origin.

12.1.2.9 Bauxite

Insignificant presence of bauxite is reported from the Jammu and Himachal Pradesh. In Jammu area bauxite is reported from Salal (3-4 m thick) and Mutal (1.5 m thick) areas in the Udhampur district. The elongated to circular pisolites of bauxite is composed of gibbsite, goethite, diaspore, boehmite, kaolinite, hematite and quartz. Earlier these bauxites were considered as karst bauxite now they are believed to be of reworked origin (Singh et al., 2005). Occasional beds or lenses of bauxite are reported from Deharu and Chabla in Shimla district; Bench and Nahan in Sirmur district and Deothal in Solan district of Himachal Pradesh.

The pisolitic bauxite beds are high in silica and have length and thickness less than 10 m and 1 to 3 m respectively. Geological and climatic condition of Himalaya is not suitable for bauxite formation.

12.2 Non-metallic Mineral Deposits

Non-metallic mineral deposits are a wide term including minerals and rocks, which are used as whole or after their beneficiation. These minerals and rocks can be divided into industry wise use such as abrasive minerals, refractory minerals, construction or building minerals, fertilizer minerals, cement minerals, paint and pigment minerals, gemstone etc. Himalaya is exceptionally rich in some of these minerals. Many of these minerals or rocks are mined at several places all over the Himalaya. However, many minerals have just shown their presence here and are totally uneconomical. A short compiled accounts of these minerals and rocks are as follows:

12.2.1 Asbestos

Asbestos is a group of fibrolite minerals having thermal insulation applications. Chrysotile is the best fibrous variety and other fibrous minerals varieties are tremolite, anthophyllite, crocidolite and amosite of amphibole group and are also belonging to the asbestos family. Asbestos is formed by low temperature metamorphic alteration of amphibole minerals or dolomitic limestone. In Himalaya presence of asbestos is observed at many places but large-scale deposits are not found so far. In Jammu and Kashmir, asbestos is recorded from Manaskot-Bandipura area in Baramula district and from the Panjal Traps in Shams and Abari mountains. In Himachal Pradesh, presence of tremolite asbestos is observed at Mahasu near Dharampur and Kami in Kangra. Chrysotile asbestos occurs in Shali ridge in Shimla district. Presence of asbestos in Uttaranchal is observed in metamorphosed basic igneous rocks of Bhatwari thrust sheet in Chamoli district at Jagat, Rarwa, Goniala, Parkandi, Satynana, Jalai Bamoori, Kund, Kandhara and Nagalsu. It is also reported from Mussoorie and around Badhangarh near Pithoragarh in Kumaun region.

12.2.2 Barite

Synsedimentary barite ($BaSO_4$) is mineralized in Riphean- Vendian (Proterozoic), Cambro- Ordovician and Carboniferous- Permian periods in the western Himalaya. Barite is used in paint, paper, textile and leather industries, as mud in drilling, photography, and barium meal for medical purpose and for the formation of alloy. Its uneconomic presence is recorded from Jangalgali and Kharikot area of Udhampur district in Great Limestone and in Ladda Coal field of Riasi district in Jammu and Kashmir. It is regularly mined in Himachal Pradesh in Tons Valley in the neighborhood of Kanti and Tatjana in Sirmur district. Kanti barite is pure white in color and coarse grained in nature while Tatyana contains impurities. Kula-Andhra, Batoware, Kheel and Banaur are other barite locations in the Tons Valley and in Shimla district barites are recorded at Subathu, Haripur and Sair. Strontium isotopic study of this barite indicated highly radiogenic old crustal source (Sharma et al., 2003) Chamba (Shimla), Haripur, Sair, Mangla Thah, Korga, Sakando, Ghurwa, Juwan ki Nala and Rajpur are other locations of barites in Himachal Pradesh. Across Tons River these deposits pass into Uttaranchal and occurs as fracture filling veins near in Tons Valley, Rajpur (Dehradun), Maldeota, Khairsain, Satpuli and Barmatiya- Milam (Sharma et al., 2003). Barite is present in Krol-Tal contact at Maldeota and in Rishikesh in Dehradun district and 1.0 to 1.5 m thick bed at Khairsain in Nayar valley.

12.2.3 Beryl

Beryl ($Be_3Al_2Si_6O_{18}$) is used for extraction of beryllium, which is used as moderator in atomic reactors, in making lightweight alloys, and its gemstone varieties are known as aquamarine and emerald. It is mostly associated with pegmatite and granites. Minor pale blue crystals of beryl are reported from Wangtu Bridge and Shapka area of Kinnaur and from pegmatoids of Chor Mountain area in Himachal Pradesh. In Nepal, beryl of 3-4 cm size is extracted from pegmatite veins at Kulphukhola, Jyamine and rarely up to 7 cm size from boulders of quartzite and sandstone in Ikhukhola. In Bhutan, well-developed crystals of 0.5 to 4 cm size are present in pegmatite and quartz vein at or near Kosela, Chirang, Gopini, Lapse and Bhole.

12.2.4 Borax

Borax is reported from Puga valley of Ladakh. Here it occurs as 5-8 cm thick layer traceable over 3 km distance between Rupsa to Hundes. Several lakes containing borax-bearing water are situated in the neighborhood of Rudak, Thak, Jalung, Raksum and Challa in Ladakh region of Jammu and Kashmir.

12.2.5 Clay

The term clay is used for very fine grained (0.002 mm diameter) material having layered silicate structure in which SiO_4 tetrahedra combines with sheet containing Al $(O, OH)_6$ and/ or Mg $(O.OH)_6$ octahedra. Kaolinite, montmorillonite (smectite) and attapulgite are common mineralogical varieties of clay family. Clays have large industrial applications so these varieties are known by different names according to their usage. Clays are formed mainly by the prolonged action of weathering or by hydrothermal alterations. In Himalaya good clay deposits occur at few places because steep slope of surface did not retain the weathered product of long time.

Kaolinite clay also known as kaolin, china clay, ball clay, plastic clay, refractory or fire clay and flint clay. In Jammu and Kashmir impure variety of kaolinite occurs at Sikhwalgali, Jangalgali, Chirayeh, Mutal, Painthal, Sangarmarg and in Salal areas. The estimated reserves in Jangalgali are 0.1 Mt, 1.04 Mt in Panhasa and in Salal area 1.4 Mt.

In Himachal Pradesh, clay deposits are classified in lacustrine and fluviatile clays. Clay is also associated with granite, and with the Siwaliks. Lacustrine clay occurs in Kinnaur, Lahaul and Spiti districts at Shiasu, Ganga, Chango, Shakar, Sumdo, Hurling Attargo and Kioto; in Sirmur district at Kalidhang and in Sundernagar area. Fluvial clays are reported from Sirmuri Tal, Nimba-Ka-Khala and Rati Khark. Clay beds on granite occur in Kinnaur and Mandi districts. Few pockets of pure clay occur in Kathar and Hatli (montmorillonite type). Combined reserves of these deposits are of few million tons only.

In Nepal, high-grade clay occurs over Palung Granite in Makwanpur district. Naliban and Kharka prospects situated over Palung Granite contain 0.56 and 0.92 Mt, weathered granitic mass having 10.16 and 10.79% clay content respectively. Panchmane and Dalcap deposits occur at 16 km distance from Kathmandu on Kathmandu-Trisuli road. Clay is also reported from the Daman area also.

In West Bengal, china clay are present in Darjeeling and Jalpaiguri districts. These deposits are formed by decomposition of Darjeeling Gneisses and Daling rocks. In Arunachal Pradesh, plastic clay occurs in Upper Brahmaputra plain in Lohit district. Nearly 5 Mt clay occurs in Bekuliana-Paya area. Pozzolanic clay deposits are reported from Doimukh (Lower Subansiri district), Seijosa in East Kameng district and in Bahlukpong-Tippi area in West Kameng district.

Bentonite or Fullers' earth is commercial name of montmorillonite, is reported from Uttar-Baini to Parmandal in Jammu and Bhimkar to Thandi Choi in Mirpur District.

Ochres

Ochre is reddish yellow clay having substantial iron oxides content and is used as cheap pigment. It is reported from Phulchowki, Gorati and Kharidhunga area of Nepal.

12.2.6 Construction Materials

Construction or building materials are the most significant geological resource of the Himalaya. These widespread, inexhaustible and renewable resources have vast potential of direct and indirect employment and revenue generation. Slabs of roofing slates, dimensional stones, boulder, pebbles, gravels, sand, silt, and clay fall in the category of construction materials. All these materials either occur in loose form or can be mined with little efforts. Himalayan rivers bring loose and weathered materials in rainy season from higher altitude to lower altitude and deposit this carried mass in foothill or terai area as well in the valley itself. These materials are used for varieties of purposes with in or outsides of mountain for construction of foundation to roofing of houses and in final plastering of walls, for bricks manufacture, in making roads, in construction of retaining wall, for filling depression, canal, dams etc. It is the nature of requirements, controls the size and shape of construction material for example for roofing thin slabs of slate or schist are preferred, for wall construction medium size rectangular boulders or blocks are preferred, for aesthetic work dimensional stones are preferred for flooring smooth thin slabs are utilized, for road bed angular aggregate of 2-4 inch size are preferred and in case of dam construction everything of every size can be utilized depending upon the stage of construction.

Everywhere in Himalaya till recent past locally available materials are utilized for all types of constructions. People used nearby available materials, after selection and processing, for constructions. Their dependence on local materials was and still now in many areas controlled by absence of roads, high cost of transportation of outside material and cost of outside material. Mechanized processing activities are slowly making forays in by and large this mainly manual industry/activities. This segment provides job to large number of workers and has many times more job potential if granite mining and polishing get boost in the Himalaya.

12.2.7 Fluorite

Sporadic occurrences of light green fluorite are recorded from Wangtu granite in Mahasu district of Himachal Pradesh.

12.2.8 Gem Stones

Gemstones are transparent or opaque minerals possessing properties of 'water' (capacity to sparkle) and 'fire' (production of spectrum colors from natural light). Besides these essential characters, they should also be hard, durable, rare, and capable of cutting and polishing. Cut and polished gemstones are called gem. Gemstones are divided into precious viz, diamond, ruby, sapphire, emerald, pearl and precious opal and semiprecious viz. topaz, garnet, aquamarine, emerald, chrysoberyl, zircon, olivine, jade, agate, rock crystal or quartz, amethyst, opal, turquoise, lapis lazuli, moonstone, epidote, kyanite, tourmaline etc. categories. Micro diamonds have been reported from the Ladakh-Karakoram Himalayan region.

Ruby and Sapphire

Ruby and sapphire occur in various shades of red and blue colors respectively and are varieties of corundum (Al_2O_3). Ruby is reported from Paddar (4500 m) in Jammu and sapphire from Samsam in Kishtwar tehsil of Doda district and its minor occurrences are also reported from Dangal, Shasho and Hangshula area situated 15 km northeast of Sumsam; in pegmatites and as placer deposits. In Uttaranchal, minute size of ruby and sapphire are present in the north of Yamunotri and near Badrinath. In Nepal, slightly inferior variety of ruby and sapphire occur as crystals of few mm sizes in metamorphosed dolomite of Ganesh Himal area. Local people mine these gems.

Emerald and Aquamarine

Emerald and aquamarine are varieties of beryl. Emerald is brilliant or light sea green in color and aquamarine is blue to sea green in color.

In Himalaya, emerald occurs in talc-carbonates of Amarkot and Barang in Kohistan, in Mingora area in Swat in Pakistan, near Dompal and Manikaran area of Himachal Pradesh and aquamarine is reported from Daso, Rondu, Broldu and Basha Valleys of Ladakh. In Doda district, aquamarine occurs around Sumsam, Dangle, Kahan, Hamur and Chishote areas. In Nepal, aquamarine is regularly mined in Hyakule, Phakum, Naje and also in Langtang valley in Central Nepal. Aquamarine occurs in pegmatitic rocks.

Tourmaline

Common tourmaline occurs in jet-black color, however, its red and blue color varieties known by rubellite and indicolite are used as gemstones. Gem grade tourmaline mostly occurs in pegmatitic rocks. Red and blue tourmaline is reported Sumsam area and Zanskar Range in Kashmir. In Nepal, considerable deposits of tourmaline are present in the Hyakule, Phakuwa, Naje, Langtang, Jajarkot, Taplejung and Sankhuwasabha areas. Good quality tourmaline is mined in these areas. Hyakule area of eastern Nepal is most promising one and more than 4500 kg tourmaline has been mined out since 1934, when mineralisation was exposed due to landslide. Tourmaline occurs in pegmatites in the crystallines of Higher Himalaya and in the Main Central Thrust Zone. The pegmatite is schorl-biotite type. Here 226.1 kg tourmaline with an assay value 2.38 kg per 100 tones occurs in the ore material. Phakum, 5.5 km southeast of Hyakule, is again another promising area for aquamarine and polychrome tourmalines. Here again tourmaline occurs in pegmatitic bodies. Local prospectors mine it for unknown time in the past. Actual reserve position is still not known yet. In Naje village of Manang district of western Nepal tourmaline occurs at 2900 m elevation in pegmatites lying to the north of MCT. Two partly exposed promising pegmatite bodies have 24x10 m and 12x6 m dimensions. 0.28 kg gem tourmaline in per ton of ore material is estimated in Naje area. Gem tourmaline is also mined from the Ikhabu pegmatite north of Taplejung, Chokle pegmatites near Mayum village, Thurbu pegmatite near Ranidunga, Danda and Mangsima pegmatite east of Num village in the Sankhuwasabha district and in Langtang Valley in Dhunche district of central Nepal

Almandine

Almandine garnet of red colors occurs widely in the metamorphic rocks and pegmatite of the Higher Himalaya of Sabaya Khola of Nepal (Andrews, 1985). It is used as semi-precious stone and as abrasive mineral.

Among semi-precious gemstones, yellowish green to bright green epidote occur in Puga Valley of Jammu and Kashmir, kyanite of blue color in Kinnaur and Bhabeh areas of Busher district of Himachal Pradesh. Rock crystals (quartz), agate, carnelian and amethyst occur in many parts of the Himalaya.

12.2.9 Glass Sand

Ordinary glass is made by melting high purity silica, calcium carbonate (lime or limestone) and sodium carbonate. Other substances are also added for making glass of special quality for example, heat resistant glass is made by adding boron trioxide or flint glass by adding potassium carbonate, Silica, main constituent of glass, is obtained from sandstone, quartzite, quartz vein or from the river sand.

Glass grade silica of 97-99% SiO_2 content is available in many Himalayan states. In Jammu and Kashmir, glass sands occur in the Gulgaldar-Haptnar area in Anantnag district. Two bands of white quartzite of thickness 25 to 100 m and 0.78 Mt reserve to a depth of 5 m occur here. In Himachal Pradesh, glass sand is reported from many locations namely Jaijonkhad (97.4% to 95.7% SiO_2) and Kunkhad (98.4 to 98.9% SiO_2) in Una district. White quartzite suitable for glass manufacture is reported from Aruali, Banda, Sieha and Solag in Bilaspur district, Barhwaini and Bhadroa in Kangra district and Badhuri - Palkwa area of Una district. Manikaran, Nagthat, Muth and Subathu quartzites, at some places are also suitable for glass making. In adjoining Uttaranchal, glass sand or quartzite of Lachhmansidh, Rajpur and Kalanga Hill of Dehradun district are found to be suitable for glass making; Berinag quartzite a very wide, thick and extensively developed formation is also suitable for glass making. In Yamuna Valley, between Wazri and Kuthnaur, a 10 km wide belt, this milky white quartzite is fine-grained, massive in nature and contains 95-98% silica. Here pure limestone is also available for glass manufacture.

In Nepal, nearly 12 Mt sand suitable for glass manufacturing is reported from Karakhola near Hetauda in the Makwanpur district. In Bhutan, 15.71 Mt glass grade quartzite occurs in mineable form.

12.2.10 Graphite

Graphite is theoretically a pure carbon. Very soft, lead gray colored mineral finds extensive use in varieties of industries like foundry, paint, pigment, and atomic energy and in pencil making.

Graphite occurs at Samjam and between Moshand and Atholi villages in the Bhutna valley in Doda district. Khadyanar to Islamabad graphite deposits in Baramula district are amorphous in nature. Here 2.45 Mt graphite reserve is present till 30 m in down dip direction and 15.3 Mt in 152 m down dip depth. Because of poor quality of deposits they are not of any major economic activity. Graphite is also reported from Salkhala Formation at Reshian and in the vicinity of Uri, in Jammu and Kashmir. In Uttaranchal, minor deposits of graphite are reported from Kalmatiya, Sisar, Petsal and Banim Devi areas situated in 15 km long belt containing 9% graphite, in Almora district. Kalmatiya-Sirar deposits contain 10.7 Mt reserve. Minor presence of graphite in Gumalikhet Formation is reported from Dudatoli to Kali valley near Gumalikhet, Balt, Diuri, Ukkakot and Chhira area and in Munsiari Formation near Munsiari, Garjoli, Palsima and Palon.

In Nepal, presence of minor pockets of graphite is reported from Kharentar, Marthum Pass, Patibhanjyang, Yaijo, Soki, Pandum, Parmejung and Sweng areas. Lumpy and flaky graphite associated with graphite schist and marble at Chitre and Dareli occur in 30-80 cm wide band. At Chitre 6000 tons graphite is estimated. Due to high altitude (3000-4000 m) and inaccessibility of area, these deposits cannot be exploited. Suitable for economic exploitation, graphite deposits occur in Khepchishi and Chilaila hills of Paro and Ma districts and Donga, Depchara and Duna areas in Bhutan. A reserve of 53 Mt graphite is estimated in Khepchishi Hill. These deposits are present in Paro Group and require beneficiation. In Arunachal Pradesh, graphite deposits occur in Subansiri, Siang and Lohit districts. These deposits occur in Khetabari Formation of the Bomdila Group, Se La Group and Yang Sangchu Formation of Lohit complex. Graphite of Khetabari Formation occurs at Dodo saran, Khetabari, Lamdak, Bopi, Pomluknala, Tai and Lalpani villages. The amorphous graphite occurs in pockets or lenses. More than 23000 tons reserve is estimated in two graphite pockets present on Daporizo-Ziro road while reserve positions of other deposits are yet to be published.

12.2.11 Gypsum

Gypsum ($CaSO_4 . 2H_2O$), an industrial mineral, rarely occurs in pure form. It contains argillaceous and siliceous impurities. It is formed by the evaporation of seawater or evaporation inland lake water. It is mainly used in cement industry as it retards the setting time of cement, in making plaster of paris; its pure crystalline variety is used in making optical accessories of microscope, in ayurvedic medicines, ceramic industry and in fertilizer manufacture. Himalaya has rich deposits of gypsum in almost all states and it is mined at several places.

The state of Jammu & Kashmir is rich in gypsum deposits. In Doda district of Jammu division gypsum deposits found at Khothujara, Marsu, Barias, Kundgram, Dhansal and Gagla in Assar area, at Parlanka, Thalwa and Ganga in the Ramban area. In these areas snow-white gypsum is massive to crystalline type and its purity ranges between 83-96%. In Baramula district of Kashmir division, gypsum occurs at Bijhama-Dhanni-Saiyidan and Bogna-Nalusa areas. 97% purity is recorded here in some places. Gypsum also occurs at nearly 4 km upstream of Kurgiakh village (5190 m elevation) in Ladakh division. In Jammu and Kashmir more than 100 Mt reserve is estimated.

In Himachal Pradesh, gypsum is reported from the Tethyan and Lesser Himalayan divisions. In Tethyan Himalaya, huge deposits of gypsum are reported from Early Carboniferous Lipak Formation exposed in Kinnaur, Lahaul and Spiti districts. The snow-white color gypsum of plaster of paris grade occurs in 20 km long and 10 m thick bed in Liwa Thatch - Murling-Losar-Dhuma region. At Shalker, 1.25 Mt reserve is estimated in a small area. Impure gypsum from Korga (0.55 Mt) and Bhareli (0.1 Mt) in Krol Formation of Sirmur district and minor pockets in Mahasu district are other occurrences in Himachal Pradesh.

In Uttaranchal, pockets of gypsum are reported from Pauri, Tehri, Dehradun and Nainital districts. Gypsum deposits of Kharichatti, Khamsar and Kirnolakhal area near Lacchmanjhula, Rishikesh contains nearly 0.18 Mt gypsum occurs in 250 m long and 30 m thick bed and gypsum is also reported from Garurchatti-Rangangaon area.

Another 0.14 Mt deposit occurs at Sera-Dhaulsot-Amsot area in Tehri district. In Sahastradhara area of Dehradun, gypsum occurs in patches at Silkoti, Jharipani, Manjhera, Kalighat, Kiarkuli, and Khalgaon and Bhatta area situated on the Krol Formation. Manjhera-Kalighat deposits are estimated to contain 0.35 Mts gypsum.

No deposits of gypsum are reported from Nepal and Sikkim. Gypsum may be discovered there in future. In Bhutan, high-grade gypsum, in large deposits, occur in Khothakpa (14.8 Mt), Urichu and Shongar districts.

12.2.12 Kyanite and Sillimanite

Kyanite and sillimanite are important constituents of metamorphosed pelitic rocks of medium to high grade. They are used as refractory minerals. Blue color kyanite crystals are reported from Kanaur and Bhabeh areas of Kinnaur district and also from eastern part of Kulu district. These deposits are not mined at present Sillimanite is mined in Meghalaya state.

12.2.13 Limestone and Dolomite

Himalaya is fortunately placed in limestone and dolomite. In nature every gradation exists from pure limestone to pure dolomite. What is designated as limestone generally contains some amount of magnesia. These two rocks are used in bulk or consumed entirely in industry for cements to other chemicals making. Limestone and dolomite are reported from all divisions of Himalaya. The quality and quantities of deposits are variable in wide limit. Almost pure to highly impure rocks are reported from different parts of the Himalaya.

The government agencies and private producers mainly for cement, chips, lime, polished slabs, metallurgical and chemical and other industrial and constructional purposes extensively mine these rocks. The important Himalayan places of limestone and dolomite occurrences, in brief, are as follows:

Himachal Pradesh

Vast reserve of limestone and dolomite occurs in Shali, Deoban, Larji, Mandhali, Basantpur-Kunihar and Krol Group of the Lesser Himalaya, Kalhet; Lipak Formation and Lilang Formation in Tethyan sequence and Kakara and Subathu of Early Tertiary age. At present Tethyan limestone is not exploited due to remoteness of the area and high cost of production and transport. In 1996 nearly 4.2 Mt limestone is produced from 76 quarries working in Bilaspur, Solan and Sirmur districts.

Chemical grade Jutogh marble has been investigated at Chunvi, Mathana-Olana, Nohra and Bulain Dhar areas. Presently known 44 Mt reserves of these blocks is likely to be multiplied by many times if entire belt is explored. Shali limestone and dolomite partially explored in Bilaspur, Kangra, Mandi, Shimla and Solan districts. Aut Formation of Larji Group contains cement grade limestone between the village Thalot and Aut. Deoban Group limestone is reported from Shimla and Sirmur districts. Late Proterozoic limestone of uncertain geological affinity that occurring between Sataun and Roj in Sirmur district is high grade to dolomitic in nature.

Basantpur and Kunihar formations of Shimla Group contain fairly thick beds of limestone in Solan, Shimla and Chamba districts. It is mined at Darlaghat and Arki and Kakarhathi in Solan district and Naldehra in Shimla district. Cement, chemical and flux grade limestone of the Mandhali Formation of Jaunsar Group is exposed in a 15 km long belt of 40 to 195 m thickness in the eastern part of Shimla district. Good quality bluish gray limestone of Krol Group occurs in huge quantity in Solan and Sirmur districts. Tethyan limestone of Chamba and Hahant Spiti-Kinnaur region occurs in many Palaeozoic-Mesozoic formations. Huge reserve could not be assessed due to remoteness of area and high altitude. Fossiliferous Kakara and Subathu limestone of Palaeocene-Eocene age occurs in Sirmur and Solan districts.

The limestone reserve in Himachal Pradesh may be in several thousand million tons.

Uttaranchal

Extensive limestone deposits occur in all districts of Uttaranchal belonging to either Tethyan or Higher Himalayan and Lesser Himalayan sequences. Details are known about Lesser Himalayan limestone, as it is readily accessible area. Gangolihat, Mandhali and Krol Group are main belts of limestone. In Pithoragarh district Mandhali limestone occurs Bashti to Naularg in Ramganga Valley, Adhichaura, Garjia-Balmara in Gori Valley Jauljibi-Dhung in Kali Valley. In Almora district limestone occurs from Parkat to Darimhel and beyond and in western Ramganga valley north of Chaukhutia. Gangolihat limestone is reported from Postala-Kulior Gad in Pithoragarh, Nargol Gad through Bagaur and Chaunala Shuintola in Ramganga Valley and across Ramganga Valley Jakh Puran and Tharkot and in further east in Kali valley in Major Kanabhater-Jhulaghat area. Mandhali limestone also occurs in the Deoban peak, north of Chakrata, Kalsi area. Krol limestone occurs in Nainital hills, Lansdowne, Mussoorie, Dehradun area. Limestone of high purity occurs in Hathipaon and Lambidhar area. Limestone of inferior quality also occurs at several places. At many places limestone is metamorphosed into marble of various pleasant color with pleasing flow banding. These marbles can be utilized for making floor files of high value. At few places pure white marble is also reported. At many places limestone is dolomitic in nature. Deoban and Upper Krol carbonates are particularly dolomitic in nature and occur in Nainital, Tehri, Uttarkashi and Dehradun districts.

Nepal

Limestone dolomite and marble of Proterozoic to Tertiary age occur in many areas of Lesser Himalaya. Cement grade limestone occurs in Trijuga Formation of the Bhimphedi Group at Sindali, 22 km north of Gaighat in eastern Nepal, Bhainsedevan and Okhare 12 km north Hetauda and Chobar limestone of Cambro-Ordovician age of Chandragiri Formation, north of Kathmandu. These limestone deposits are used for cement making. The other notable occurrences of limestone are Nigar limestone of Dhankuta area in eastern Nepal, Katari-Galtar, Udaipur in Russi and Sakando Limestone, Kabre, Jagimara and Dhading, in central Nepal, Arghakhanchi western Nepal, Gandari and Surkhet in western Nepal, Syaldanda, Lakharpata, Sewarkhola, Bandrekhola, Kanchikot, Sabdu, Tarsikot, Kakrighat, Nadchede, Hathisundh Phariyatar and at several other places. Marble occurs at Godawari in Phulchowki hill, near Kathmandu. This marble is used for flooring purpose after polishing. Pandal marble is of poor quality. Calcareous tuffs are reported from north of Kailash ridge and Bakekhola. Ganesh Himal dolomite exposed at Lari, Suple, Taplejung and Pengutang is pure white, granular type. Limestone reserves of Nepal are expected to be in several billion tons.

Sikkim

Limestone and dolomite occur extensively in Sikkim. Gray limestone occurs in Rishikhola south of Namgaon and Pink limestone around Nayabazar in western district. High grade, massive and low-grade flaggy dolomite occurs around Rishi in Rishikhola, between Namehi and Phong and between Namchi and Jorethong and also in the Rangit valley. Deposits of limestone and dolomite are believed to be of several million tones. In adjoining West Bengal, limestone occurs in Darjeeling and Jalpaiguri districts.

Bhutan

Limestone, marble, dolomite and calcareous tufa are present in Baxa and Shumar groups. Important deposits of cement grade limestone occur at Pagli and Korung Ri in western and eastern Bhutan respectively. Paro marble of white to gray color occurs at Khankhu in Paro district. Calc-tufa occurs at several places but most important cement grade deposit is found at Kalapani in Samchu district. Dolomite useful for steel industry is found between Khanabharti and Narphung in large volume. The limestone reserve is estimated at 181.5 Mt while proven reserve of dolomite is 51.33 Mt and inferred reserve is 15759 Mt in Bhutan.

Arunachal Pradesh

Limestone and dolomite of industrial purposes are available at many places in Arunachal Pradesh. Nearly 4 Mt limestone reserve of high-grade occurs in the Manga area of Subansiri district, in the Chillepam Formation. Tidding Limestone occurs between Tidding and Huinli and reserve is estimated to be around 160 Mt. A marble band of 20 m thickness also occurs between Lohit and Dalai valleys. Dolomite belonging to Chillepam Formation is exposed in Rupa and Dedza areas in the Tenga Valley in West Kameng district. This crystalline white to gray color flux grade dolomite is estimated around 300 Mt. Dolomite of cherty type and light to dark gray color is exposed between Sipi and Mara villages of Menga area in Subansiri district.

12.2.14 Magnesite

Magnesite ($MgCO_3$) is an industrial mineral used for making refractory bricks, white cement and magnesium chemicals. It is available in small pocket size to huge deposits at many places.

In Jammu and Kashmir, magnesite occurs as eleven small irregular pockets near Shergol on Kargil-Leh road and as small pockets with in dismembered rocks of suture zones. Large size boulders of crystalline dolomite occur in the Balganganala in Udhampur district. A magnesite deposit of 240 m length, 80 m wide and 30-70 m thick with estimated reserve of 2.5 Mt is reported from Katra area near Jammu. Workable magnesite deposits occur in Batal Formation in the Chamba district of Himachal Pradesh. 4 to 9 m thick bed having length of 350 m present at the confluence of Muchhetarnala and Ravi River is fine grained massive to coarse crystalline variety. Several minor lenses also occur in down stream areas. The reserve is assessed at 0.05 Mt with average 39% MgO content. Magnesite is also reported from Pasankot, Dunei, Bindrabani, Bhujund, Patengot, Kav, Tunda Mani Mahesh, Pangladnal and Suri areas of Himachal Pradesh.

Magnesite development in Uttaranchal has taken place in a 150 km wide zone situated between the Mandakini and Kali valleys. Coarsely crystalline lentiform pockets of magnesite, carrying reserve in several thousand million tons, occur in three zones namely northern, central and southern. Important locations of the northern zone are Panyan-Pauri and Imkhola in Gori Valley; Gol, Kholi and Thal; Kirauli, Khuna, Binsar, Tuper in the Punger Valley; Kanda Masauli, Jakh, Harbar, Daurgiri, Bajarbhanga, Harap west of Sanju Valley in Kumaun region and at Kilbara, Dolam, Wan, Durmi, Birahi, Gulabkot and Bamru in Garhwal. In the central zone deposits occur it Panthsera, Gaindali, Dundu, Dewalthal between the Kali and Ramganga rivers; Bora Agar, Rain Agar and Kanda between Ramganga and Saryu rivers; Pana, Bhurgaon, Bhaldeo, Chahana and Someshwar west of the Saryu river, Bagoli in the Pindar Valley and Bangathat - Kimotha, Latudhar - Laldhara in the Alaknanda and Mandakini valleys. In southern region deposits occur at Bisabajer, Chandak, Dharigaon and Deopala in the district Pithoragarh; Dol-Tusrani, Ganai, Tachhini, Bauri, Asaun and Salia. In the Almora district and Jhirauli area, west of the Saryu river. The estimated quantity of magnesite in Patri-Jhiroli-Naini area is nearly 12 Mt; in Agar-Chahana area the estimated reserve is 350

Mt. Other locations of magnesite in Uttaranchal are Palla-Jakhola-Kimona, Tapowan-Dwing-Langi, Huyana-Pokhri, south of Helang, Mandara, Taraktal and Ramani in Garhwal region.

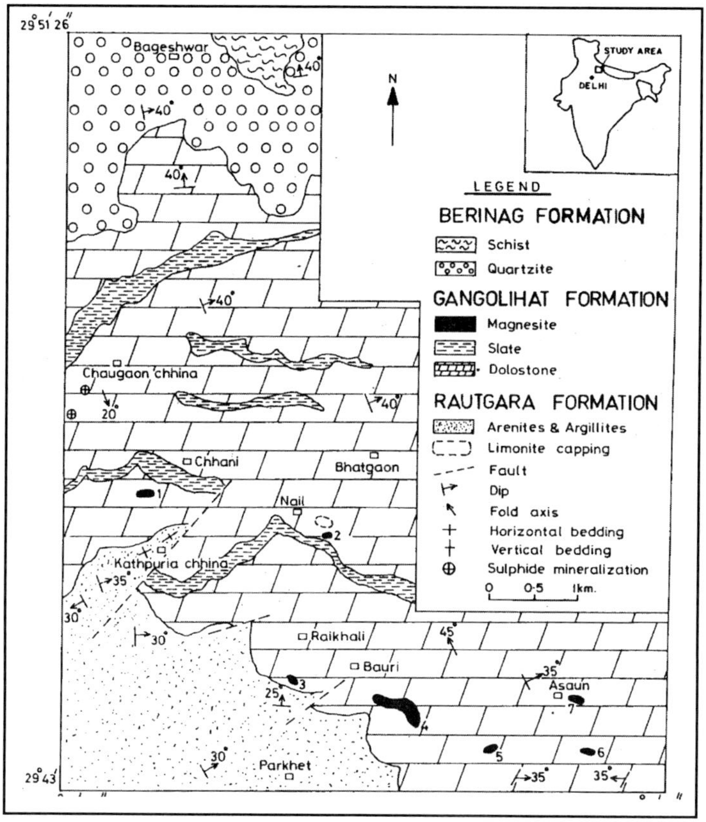

Figure 12.1. Bauri-Almora Magnesite, Uttaranchal (after Joshi, Bhattacharya, Anantharaman et al., 1992)

Origins of these magnesite are debated between syngenetic and epigenetic models. Joshi (1995), Joshi and Bhattacharya (1979), Joshi et al. (1993) after detailed scrutiny proposed digenetic replacement origin of Bauri magnesite (Fig No. 12.1) in particular.

In Uttaranchal, the private companies regularly mine some of these magnesite deposits. The state is second largest magnesite producer in India. Academicians and professionals have investigated the quality and quantity extensively.

Nepal

Magnesite occurs in Lesser Himalaya in Nepal at several places as small, isolated irregular shaped lensoidal bodies. They occur in dolomite and in calc phyllite. Kharidhunga, and Kampughat are major deposits of magnesite and are being exploited; other bodies are located at Kubuide, Dharapani, Duguragari and Basamachi in Sindhupalchoke and Dolakha districts and at Phaplu in Jhiku Carbonates of Benighat Slates of Nawakot Group. In western Nepal, magnesite occurs in Baitadi area. The Kharidhunga deposit of Dolakha district occurs at 2700 m altitude. The crystalline breunnerite type magnesite is metamorphosed sedimentary type and deposited along with dolomite in sedimentary basin. Out of 180 Mt reserve, 66 Mt is of refractory grade and is mined by Nepal Orind Magnesite (P) Limited. The Kampughat magnesite deposit of 20 Mt reserve is located in Sunkasi Valley in Udaipur district. Like Kharidhunga, Kampughat magnesite also occurs in Jhiku Carbonate of Benighat Slate of Nawakot Group. This metamorphosed magnesite deposit is formed by alteration of dolomite.

Small pockets of magnesite are also located in the Rangit Valley in Sikkim. Magnesite is also reported from Hindubagh area near Quetta in Pakistan. This magnesite occurs in serpentinized dunite and also in Kumhar and Kulhar area of Sherwan.

12.2.15 Mineral Water

Cold water springs are abundantly present at most of the places in the Himalaya while hot springs are restricted in number. Underground water emits out from these springs. During underground circulation of water, a minute portion of rock(s) is dissolved in the water and becomes its dissolved constituent. Some of these constituents are beneficial for health and at times have therapeutic value. Excessive concentration of some elements like sulphur, boron or iron either gives disagreeable odor or color and make the water unfit for use. At many places, people use waters of known medicinal properties and its bottling and selling is relatively new phenomenon in the Himalaya. Mineral spring water at Kalath near Manali in Kulu district is considered as good mineral water and is sold in bottle. In Garhwal area of Uttaranchal, the Ganga water is bottled and sold. Sulphur water of Sahastradhara in Dehradun is also one well-known area. In Nepal, spring water near Dhunche is also bottled and exported to USA. Ganga water is well known from time immemorial for its hygienic value, imperishable qualities and unstinking nature. Himalaya has tens of thousands cold water springs. Their detailed chemical investigations are necessary to judge the water quality and it is certain that many of these would prove to be commercially successful.

12.2.16 Muscovite

Muscovite [$K_2 Si_3 O_{10} (OH)_2$] is transparent colorless to light colored variety of mica. Though mica denotes a mineral family but in commercial parlance the term is used for muscovite only. Muscovite is an excellent electrical insulator and is unaffected by temperature, therefore, it is widely used in electrical industry. Muscovite sheets or books occur in siliceous pegmatite and granitic rocks.

In Himachal Pradesh, muscovite is reported from upper Chandra and Parvati valley of Kangra district. Green color and stained muscovite is reported from Badnikhal, Kunet, Majholi in Garhwal and Mukteswar and Paharapani in Nainital district of Uttaranchal. Muscovite occurrences of Uttaranchal and Himachal Pradesh are uneconomical. In Nepal, mica is reported from Higher Himalayan pegmatites of Godikang and Chindringkhola of Rasuwa district. The size of mica sheets varies between 1 to 10 sq inches but due to small recovery percentage (0.2 to 0.3%) in pegmatite and high altitude (3330 m). These deposits are uneconomical.

12.2.17 Phosphorite and Phosphate

Phosphorite is a sedimentary rock containing more than 19.5% P_2O_5 i.e. having nearly 50% apatite (main mineral phosphorous) while the term phosphatic rock is used for more than 7.8% P_2O_5 or 20% apatite (Pettijohn, 1974). Phosphate also occurs as bone beds, guano, phosphatised limestone, residual phosphate, pebble phosphate and nodular phosphate. Phosphorite contains several species of apatite; collophane, group of submicroscopic minerals and various rare, trace and common elements occur as replacement or substitute in phosphorite besides these common rock forming minerals like quartz, calcite, dolomite, feldspar, chert, clay etc. are also present.

In Himalaya, low-grade phosphorite and phosphate are located at many places and are professionally mined for fertilizer purpose. In Jammu and Kashmir, uneconomical phosphatic material (collophane) occurs as pellets, stringers and clots in the Upper Permian limestone around Aru village. In Himachal Pradesh, uneconomical phosphatic nodules occur in several formations like Batal - Katargali, Gungri, Salooni, Spiti, Kakara and Subathu. Low grade, phosphatic shales occur in Khatpul and Tatapani formations, Krol and Tal formations also have phosphate between Rajgarh and Deosia (8-10% P_2O_5) and in Nigali Shyan La area.

In Uttaranchal, economical deposits of phosphorite occur in the Mussoorie Syncline, at the contact of Krol and Tal formations. Here phosphorite is associated with black shale, chert and limestone. The important phosphorite deposits occurred at Maldeota and Durmala and uneconomical at Khamsar, Pari-Tibia, Chaunba-Kumali, Mathet, Dubra, Chipaldi, Bhusti, Masrana, Jalikhal, Kimori and Kaudiala. The Maldeota and Durmala deposits have strike length between 300-2000 m and thickness varies between 1-9m. They contain P_2O_5 10-35%. The cumulative reserve of these deposits is 19 Mt. The working mines of Maldeota and Durmala have been recently closed due to high production cost. In Krol-Tal setup phosphorite also occurs in Lansdowne Hills, Kiurki-Mohanchatti in Ganga Valley and Kasani Sot, Kakarwargad and Jabarsot (5-10% P_2O_5) in the Nainital area.

The algal limestone of Gangolihat Dolomite of Pithoragarh district is poorly phosphatic in nature. The 15 km long phosphate-bearing horizon is traced between Bisabajer to Deopala and in the Gangoli area. Dewalthal and Dungri-Pali limestone area is also mildly phosphatic.

Several low-grade phosphorite-bearing horizons are identified in the Nepal. Most important Baitadi phosphorite deposit is present in Baitadi carbonate formation and is exposed at Dhikgad, Junkuna, Morgaon, Sanagaon and Dhaubisaune areas in Patan district of farwestern Nepal (Bashyal, 1985). Phosphorite occurs above black slates in massive cherty dolomite. Baitadi phosphorite mineralisation occurs in two types namely stromatolitic (32% P_2O_5) and sandy and pelletal type. The strike length of phosphatic horizons varies between 200 m to one km and thickness between 0.7 to 4.7 m and P_2O_5 content ranges between 11-32%. Bajhang phosphorite prospect in stromatolitic phosphorite bearing carbonate horizon, occurs at Tarugad, Juilgad and Goichan-Kandachaur areas and is exposed on the ridge. Here P_2O_5 content varies between 5-20%; thickness between 1 and 23 m and maximum strike length is 6 km. The richer part containing 20 % P_2O_5 has thickness only 0.65 m. Among the other occurrences are 2 m thick with 4 % P_2O_5 Bajaura in farwestern Nepal,

Dang-Pyuthar (5% P_2O_5) in Sewarkhola in Dang and Marikhola in Piuthan districts and in the vicinity of MBT in Barahakshetra - Dharan - Takure area of eastern Nepal.

Rock phosphate of only academic records occurs in Nudaling Groups at Tatapani and Suburk areas of western district of Sikkim.

12.2.18 Pyrophyllite

Pyrophyllite, $Al_2Si_4O_{10}$ $(OH)_{2,}$ is a soft mineral like talc and occurs in white, gray or light blue or green color and has pearly luster. It is used in refractories, ceramics, steel making and as filler. It is formed by the hydrothermal alteration or metamorphism of kaolin. A 5 km long and 20 m thick bed of pyrophyllitic quartzite, containing 20-40% pyrophyllite, is recognized in the Higher Himalayan Yamuna Valley in Uttarkashi district of Uttaranchal. This bed is situated near MCT and is overlying the Sayanachatti Mylonitic Granite. The concentration of mineral can be improved by beneficiation techniques.

12.2.19 Talc and Steatite

Talc, Mg $Si_4O_{10}OH_2$, is the softest industrial mineral used widely from cosmetic to filling purposes. It is formed during low-grade metamorphism of magnesium bearing rocks like dolomite, dunite, peridotite etc. Steatite is slightly massive and compact variety of talc. Soapstone is an impure variety of talc or steatite. Talc and steatite are present at many places as pockets and veins and are exploited commercially at several locations.

Talc and steatite occur in Great Limestone in Riasi and Udhampur districts near the Riasi, Kunian, Nangal and Purana Darubur in Jammu and Kashmir. In Himachal Pradesh, uneconomical talc is reported from Asrang and Lipa in Mahasu district and steatite from Nahan. Talc is widely distributed in Uttaranchal and commercially exploited in few areas. Important soapstone deposits occur at Gate Chhinna, Julykhel, Badolisera, Shishakhani, Chaunra, Khakoli, Jakh, Nandigaon and Asaun in Almora district. Steatite is mind at Gari. Largest talc deposit occurs at Surag and of modest size at Kharidhar, while presence is noted at Malgad, Harap, Garjia and Birkhal. In Pithoragarh district talc is located around Dalyali, Gairar, Masauli, Rain Agar and Imkhola and as discontinuous bands in Pungar Valley. In Alaknanda Valley of Chamoli district, talc occurs at Torosi, Belakuchi and Huina 2 Mt reserve is estimated at Girichhina, Badolisera, Chinite and Kanda areas and 4 Mt in Punger valley. Unassessed huge reserve may present at Surag (500 x 1.25 m), Birkhal (200 x 6 m), Dyali (2000 x 5 m), Maljhula (3000 x 60m) and in Barmori and Reya Talla areas. These deposits are formed either by hydrothermal replacement of dolomite or by mild metamorphism of carbonate rocks.

In Nepal, mineable talc occurs at Kharidhunga (0.3 Mt) in Dolakha district, Gaighat (0.013 Mt) close to Mugling-Narayanghat road, very high-grade talc at Phaplu (0.007 Mt) in Dolakha district and Selan Katti in Sindhupalchok. Talc is also reported from Rattikhola, Bandre-Dhamja, Aale, Maldung, Banderkhola, Kalikhola, Sew Gaon and Kammate areas.

In West Bengal, talc deposits are located in Karmi estate, Ramsukhola and Tirukkhola in Darjeeling district and steatite at Lepachako in the Jalpaiguri district. In West district of

Sikkim, talc is reported from metamorphosed Daling Group in Ranikhola, Rishikhola and Roathakkhola. At Budheni, Tindoban and Lepchekina area of Samchi district and at Pachu, Setikhola, Khembu, Suktikhola and other areas in Bhutan talc occur in small pockets. The reserve is estimated at 1.4 Mt.

12.2.20 Salt

The common rock salt or halite (NaCl) is formed by evaporation of saline water and occurs as sedimentary beds in the Salt Range of Pakistan, Mandi Salt Beds in Himachal Pradesh and as encrustation around many mineralized springs in Himalaya.

Rock salt occurs in dolomite-shale beds of the Ropari Formation of Shali Group in Mandi and Kangra districts of Himachal Pradesh. At Guma Darang a salt bed, containing 70% NaCl, 3% KCl and rest as impurities of CaO, $CaSO_4$, MgO and insoluble are mined regularly by underground method. More than 7.5 Mt salt reserves are estimated here and annual production is less than 4000 tons. Megal, Harkakan and Dewal Khas are other salt bearing locations in this belt. This belt further extends to Mandbari in Kangra district (Srikantia and Bhargava, 1998).

Salt in the form encrustation or brine water occurs in the Tethyan rocks of western Nepal. In Nursingkhola, salt occurs as encrustation or in brine water (3% NaCl and 1% KCl). Brine seeps out from cracks and crevices of limestone and annual production of dry salt varies from 80 to 500 tons. In Chhidingkhola brines containing 1.14% NaCl and 0.5% KCl oozes out vigorously due to hydrostatic pressure. Brine water of Chorang area at Chharkab Hotgaon is also used for production of salt. In Ladakh, salt encrustations occur around brackish water lakes.

12.2.21 Sheelajeet

Sheelajeet is black color, soft tar like bituminous substance oozes out from quartzite of Higher Himalayan Yamuna Valley and dolomite and limestone of Gangolihat Formation in Pithoragarh district in Uttaranchal and from several localities of Nepal and Ladakh. It is used as ayurvedic medicine.

12.3 Fuel Resources

Coal and Petroleum, conventional fuel resources, have been explored in great detail in many parts of the Himalaya. Presence of these are recorded at many places in the Outer and Lesser Himalaya Current status of these resources are as follows:

12.3.1 Coal

Pakistan and Jammu and Kashmir

Two types of coal deposit namely the Gondwana and the Tertiary coals are present in the Himalaya. Peat to anthracitic varieties of coal is reported from the foothills and adjoining Lesser Himalayan rocks. Palaeocene coal occurs in Hangu and Kachai area in Kohat in Pakistan. Few coal seams of 2 m thickness are present here. Coal is also reported from

Khilla and Kotli area near Muzaffarabad in Salt Range. In Jammu and Kashmir coal seams are reported from Khuiretta (1.25 m) in Mirpur district, Ber Goa (5 m) in Punch district, between Chakar and Chenab (1.5 m) in Riasi and Udhampur districts, Kalakot (maximum thickness 5.5 m), Metka (1.5 m), Mahagala (4 m), Sangarmarg (1m), Jangalgali (3 m), Pancharigali, to Mutal, Siro Valley and Ladda areas. These deposits are semibituminous to bituminous but in most places they have been crushed due to deformation. Lignite occurs in an 80 km long stretch between Shaliganga River and Nichahom. Lignite occurs as streaks to 5 m thick seams and 36 Mt reserve is estimated. The lignite contains 27% fixed carbon, 28% volatiles, 15% moisture and 30% ash contents. 0.4 m thick subsurface peat bed is reported from Pingaling in Anantnag district.

Himahal Pradesh and Uttaranchal

In Himachal Pradesh, carbonaceous matter is reported from Mandi, Kangra, Bilaspur and Shimla districts. However, worth mentioning locations are Mansai where 2 and 1 m thick seams are traceable for less than 100 m and other important location is on right bank of Dehar where two seams 180 and 90 cm thickness are present. Small uneconomical seams are also reported from the Siwaliks. Lignite present in Subathu Formation is reported from Delag and Bandla areas in Bilaspur district.

Streaks of lignite are described from Siwaliks rocks near Mohand and Rajpur village in Dehradun district; Kotdwar-Lansdowne road in Pauri Garhwal and in the valley of Nihal Nadi in Nainital district.

Nepal

Gondwana and Tertiary coals are present in Nepal. Tertiary coal is present in Quaternary lacustrine deposits of Kathmandu Valley. Seven lenticular subhorizontal peat/lignite beds of 0.1 to 1 m thickness are present in the Lukundal area of the Kathmandu. Sporadic occurrences of coaly lenses, stringers and seams are reported from Koilabas-Chorkhola, Sakti, Kanoharkhola and many other places situated in the Middle and Lower Siwalik terrain. Eocene coal occurs at 4000 m altitude in Tosh area, 18 km north of Ghorai Bazar. Three lignitic to subbituminous horizons of 1.1 m to 2.1 m average thickness have cumulative reserve of 2.9 Mt. Anthracitic Gondwana coal occurs in highly sheared form at Kokohakhola, Kosi river area, Kampughat and Barahakshetra area of eastern Nepal.

West Bengal

Gondwana coal is present in the village Jainti in Jalpaiguri district and in a 48 km long belt between Pankhabari-Daling stretches of Darjeeling district. In Darjeeling district important locations of coal deposits are Tindharia, Kalikhola (3.5 m, thick seam with 30 m length), Nalikhola, Naksalkhola, Bagrakot, Manakhola and Dhobikhola. Lignite from Lower Siwalik is reported Baxa Fort and Jainti area of Jalpaiguri and in Sivok River in Darjeeling district.

Sikkim

Thin coal seams of Gondwana coal have been located around Putkhala, Roathakkhola, Richingpong area and Jorethang - Gyalshing highway along Rangit River in west district. In

South district 0.14 Mt coal of bituminous to semi-anthracitic variety is located in and around Namchi.

Bhutan

In foothill area of Sarbhang district, a 75 km long coal belt is present. Poor quality coal due to admixture of sandy matter is present in numbers of isolated outcrops.

Arunachal Pradesh

Gondwana coal is reported from Bhareli Formation resting over Bomdila Group to the north of Main Boundary Fault. Impersistant coal seams of 0.6 to 1.5 m thickness occasionally up to 7.5 m thickness are present in the Amatulla and Garu-Gensi area of the western part of the State. Tertiary coal occurs as small pockets and lenses of 1.0-0.1 m thickness in the Upper Tertiary Siwalik sandstones Pasighat- Rengging, Likabali- Garu, Kimin- Zero, Doimukh-Kheel and Bhalukpong- Sessa road sections and outside of Himalayan territory, in the Naga-Patkoi Ranges and Namuchik-Namphuk area. In Namuchik-Namphuk coal field total reserve (proved, probable and possible) is estimated at 90.23 Mt.

12.3.2 Petroleum and Natural Gas

Sub Himalayan Tertiary succession of huge thickness along the southern boundary provide ideal situation for vast reserves of oil and gas and hopes are further increased by the presence of several gas seepages or shows. Commercial geological, geophysical and drilling prospecting and exploration efforts on large scale in many promising segments have yet to yield positive results in this 10-50 km wide and 2400 km long foot hill belt, in some favorable interior locations, in the Lesser Himalaya and in the vast expanse of southern Quaternary plain of the Ganga-Sindhu-Brahmaputra rivers.

More than 20 wells have been drilled in the foothills and plains of India. First well was drilled in 1957 at Jawalamukhi in Himachal Pradesh. The well was targeted to 3200 m depth in Lower Dharamsala Formation but due to technical difficulties bottomed at 3067 m. Two gas indications were obtained between 870-890 m and 919-923 m intervals. Another deep well (5047 m) one km away from first location and three structural wells altogether failed to show oil and gas indications. Another attempt between Beas and Satluj rivers in the Janauri anticlinal structure (4531 m) was also unsuccessful. So was the case with Bahl anticline (4475 m) situated 6 km south of Jawalamukhi, except minor show of gas between 3610-3616 m interval. Mohand deep well near Dehradun, did not encounter any oil or gas evidence in the course of its 5264 m depth. Next attempt in Surin-Mastgarh anticline in Jammu region was bottomed out 3665 m depth without any evidences of hydrocarbon. In the Indo-Gangetic Plain wells are drilled in Hoshiyarpur area in Punjab (3439 m, 2513 m and 718 m), Kasganj and Tilhar in Uttar Pradesh and Raxaul (4100 m) and Purnea (2845 m) in Bihar but without any success.

Gas seepages near Sopore and Naugam in Kashmir Valley have also attracted the prospectors. The shallow wells drilling in these area encountered methane bearings water. Following lead provided by the seepages, the Karewa and adjoining area was mapped with expectations and favorable areas were recognized at Raitham, Watergam, Chattargam and Pattan. Subsequently, wells were drilled at Chattargam and Narabal and feeble oil shows were observed during drilling.

Story is more or less same in the eastern sector also. Gas seepages are noted at Nampong in Arunachal Himalaya and subsequent investigations of this area have not encouraged the prospectors.

In Nepal, continuous efforts are being made by the Department of Mines and Geology in collaboration with international players to unearth the oil and gas deposits in the Siwalik, Gondwana and Lakharpata groups (Bashyal, 1997). In Dailekh area of western Nepal 45 seeps have been observed in a zone 14 km length and seeps are also observed in Kathmandu valley. An exploratory well drilled in Biratnagar (3500 m) was proved to be unsuccessful. In Kathmandu valley, 26 sq km area has been delineated as gas bearing. In this region a reserve 300 million cubic meter gas (methane 75-80% and CO_2 14-23%) of 7200 kcal/m^3 calorific value is estimated. Gas seep is also recorded in the Jurassic rocks of Tethyan succession in Muktinath area of northern Nepal. Recently government has divided entire southern area into 10 promising blocks namely Dhangari, Karnali, Nepalganj, Lumbini, Chitwan, Birganj Malangwa, Janakpur, Rajbiraj and Biratnagar for private prospectors for further exploration.

12.4 Water Resources of Himalaya

The global water reserve is estimated at 135.5 M km^3 out of which 132.2 M km^3 is confined in the ocean, 0.292 M km^3 in ice caps and glaciers and the remaining quantity is present in rivers, dams, lakes, atmosphere, soil and sub surface condition. 4000 km^3 rains is annually received by the entire land surface of India. India receives highest rainfall among the countries of comparable size in the world. Annually, between 1645 to 1880 km^3 water flows through her rivers. This is nearly 3 percent of world's total surface water resource while 16 percent global population resides here. 12 major and 46 medium size river basins have been demarcated in India. Out of the flowing waters only 690 km^3 could be utilized for various purposes when all basins are fully tapped. The present available and under construction storage capacity are of 180 and 70 km^3 respectively and 20-25 km^3 is stored in minor dams and tanks. Thus more than 400 km^3 water is simply remained unutilized and flows into the seas. Nearly 418.5 km^3 water annually seeps into ground.

The Himalayan rivers contribute nearly two-third of annual flow of water. Brahmaputra is most powerful river through which 510.5 km^3 water flows annually and second and third in the ranks are Ganga system (493.4 km^3 at Farrakha) and Indus System (207.8 km^3). In Brahmaputra System, the Brahmaputra contribution is nearly 90% i.e. 455 km^3 and remaining contribution comes from its many minor tributaries. Same important contributors in the Ganga system are Ganga (24 km^3 at Hardwar), Yamuna (12.3 km^3 at Saharanpur) Gandak (53.2 km^3 at Patna) and Tista (19.4 km^3 at Anderson Bridge). In Indus system, the contribution of Indus is (73.3 km^3) while Jhelum, Chenab, Ravi, Satluj and Beas contribute 27.9, and 29.0, 8.0, 16.6 and 15.8 km^3 respectively. Rainfall is the main supplier (70 percent) of waters to various Himalayan rivers, next in rank is glacial melt (20-30 percent) and rest comes from lakes and underground sources.

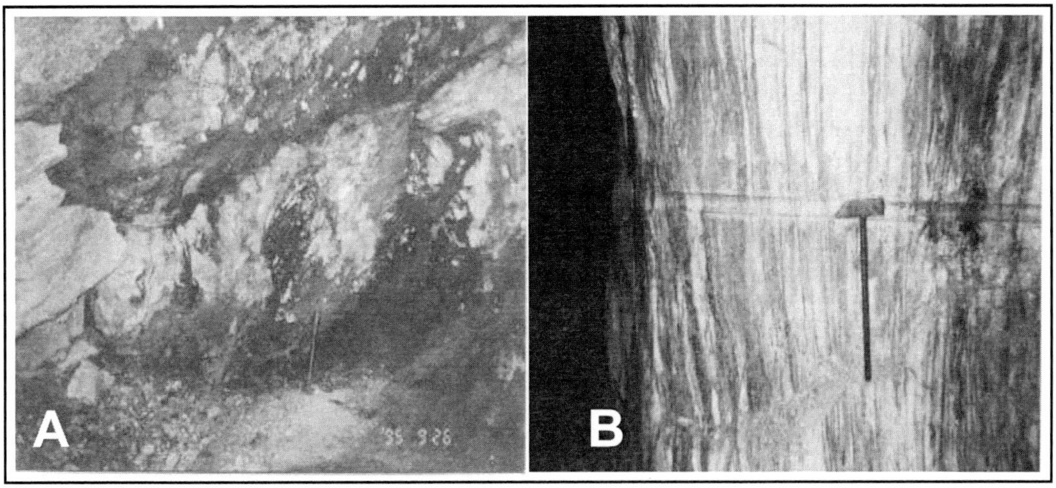

Plate 12.1 A Mineralization of poly-metallic sulphides in quartz vein at Shishakhani, near Hanumanchatti, Yamuna valley, Uttaranchal. B- Remobilized sphalerite deposit in dolomite at Pengutang, Ganesh Himal, Nepal.

Technoeconomic study indicated that Bhutan has hydro resources potential equivalent to 20000 mega watt capacity at 30% efficiency. Recently commissioned 60 m high Chukha Power Project is generating 370 mega watt electricity. The under construction Tala Hydropower Project of 1020 mega watt capacity was expected to be commissioned in the year 2004. It is expected that 3000 mega watt electricity will be produced in the Bhutan by 2007. A major part of electricity will be exported to India.

The Himalayan water is used for various purposes like drinking, irrigation, and power generation, in industry, for transportation in plain areas. Several multipurpose dams have been constructed at various places in the Himalaya. In true sense Himalayan water is a backbone of Indian economy (Biyani 2002 b).

River Linking- A grandiose plan of Government of India to link all Indian major rivers is in progress. The plan envisage of transferring surplus water of perennial Himalayan rivers and two peninsular rivers to water deficient southern regions of the country through a network of interlinked rivers and canals. Presently most of the waters of rainy season of Himalayan rivers are going into the sea. In rainy season, this water in down stream regions brings floods and every year loss of billions of rupees occurs with substantial human toll. Water of trained rivers would be diverted to the network first by impounding in dams and ponds at several places. Beas-Satluj link is an example of successful transfer of water transfer in the Himalayan region and in Gujarat many smaller rivers have been linked in a network. Transferred water will be utilized for power generation, irrigation, transportation, pisciculture and for industrial and domestic requirements. An investment of Rs. 10000 billion is expected by the time scheme completes (Biyani and Gupta, 2004).

Thermal Springs - Himalayan thermal springs have large energy potential but that remains practically untapped so far. The extractable heat from 113 Indian major thermal system including of peninsular India and Indo Gangetic Plain is estimated at $40x10^{18}$ cal. which is equivalent to 10600 MW power potential, From this nearly 2000 MW electricity can be generated annually and rest energy could be utilized for other purposes at 0.025 percent efficiency. In terms of oil and coal, it is equivalent to 27.6 billion barrel petroleum or 5730 million tons of coal. Rupee wise this amount to a sheer loss of 60 billion every year. At present annual utilization of energy is equivalent to less than 1 MW electricity (Gupta, 1984). Power can be conveniently generated with water of reservoir temperature >200°C but water above 120°C can also be used. Besides this, unassessed potential, in volume, likely to exist of hot dry rocks (HDR) in Himalaya from which heat could be extracted with suitable technological advancement in future. These relatively hot and impermeable rocks are expected to be situated 4 to 5 km depth.

DIMENSIONS OF HIMALAYAN GEOLOGY
A. K. Biyani (2006) : pp.281-298
Satish Serial Publishing House, New Delhi
E-mail: ssp@ndf.vsnl.net.in, hkjain1975@yahoo.com; website: www.satishserial.com

Yadyadvibhutamatsattvam srimadurjitameva va
Tattadevavagaccha tvam mma tejomsasambhavam (Geeta, 10.41)
Every glorious, brilliant or power creation know that to be a
manifestation of a spark of my efflgence.

CHAPTER - 13

ORIGIN AND EVOLUTION OF HIMALAYA

An orogenic mountain is visualized as narrow topped, elevated land having steeper slopes in comparison to its surrounding area. Mountains are highly variable in their shape, size and heights. Mountain refers to relatively elevated area while hill refers to the less elevated part of the earth surface. The use of these terms is more based on convenience rather than precision. The Aravalli Range of Rajasthan is described as mountain but high regions of the Lesser Himalaya e.g. Shimla, Mussoorie, Nainital etc. are described as hills, despite the fact that their elevations are more than the Aravalli. Mountains are classified on the basis of their shape, height, length, locations and origin. Geologists prefer genetic classification – erosional, depositional and tectonic - of mountains. Mountains of first two categories are simple, small and occupy less areas on the other hand mountains belonging to third category show gradation in every respect be it their shape, size, length, origin, time factor in their evolution etc. Distinction is often made between the origin and evolution, former refers to creation of a new thing that was not in existence prior to its creation and evolution refers to development, continuous or discontinuous, of a thing with respect to time. In case of geological objects like mountains, basins etc. the process of origin is very long and time consuming, involving in some cases many millions of years. The geochronological data are as a rule continuously being revised, thus making it difficult to pin point the precise time of origin and already existed materials have been involved in the origin, therefore, term evolution is preferred.

The diastrophic movements of endogenous origin produce the tectonic mountains. These movements in simplest term cause either upliftment or subsidence of the land. Diastrophic movements are further divided into two broad categories namely orogenic and epiorogenic movements. Formers are comparative swift, operate in a linear region and are responsible for severe deformation of the rocks; in contrast to this latter movements also referred to as continental making movements are slow, less severe causing simple upliftment or subsidence of land and are operate over a region of considerable length and breadth. Modern studies have shown that these movements may occur independently as well as rhythmically in an area.

Many large mountain systems or cordilleras (a broad assemblage of ranges) or a large massif (now isolated broken block of once large mountain) are formed by complex process of orogenesis or tectongenesis. Orogeny (Greek-Oros means 'mountain' and genes means 'stemming from') in general sense describes the period of mountain building but is specifically defined as penetrative deformation (folding, faulting, overthrusting etc.) of the earth's crust accompanied with metamorphism and igneous activity taking place in restricted linear zones and time interval. The portion of the earth's crust subjected to such type of activity might have undergone to mountain building process but not necessarily subjected area is uplifted; sometimes area may subside too. Hence, orogenesis or tectogenesis and morphogenesis are two different and independent aspects of a mountain. Morphogenesis imply upliftment and acquiring physical shape of a mountain, Orogenesis is followed by the morphogenesis and the time gap between these two events may be as much as of several hundreds of millions years. The N-S Trending Transantarctic mountain, 2000 km long and 3000-4000 m high, situated along the western coast of Ross Bay in the Eastern Antarctica continent was subjected to Early Palaeozoic Ross orogeny but is uplifted after 400 million years along the normal faults in the Tertiary and Quaternary epochs. Similar examples can be cited from Mesozoic Sierra Nevada orogenic mountain of North America or Hida Range of Japan. However, case of the Himalaya is different, here tectogenic phase is immediately followed by punctuated morphogenic phase means periodic, very short, phase of rapid upliftment followed by the long period of stability. The present average height of the Tibetan plateau is above 5000 m but it was only 1000 m high around 40 Ma ago, 1400 m high 20 Mao, 1800 m high 10 Ma ago. After this very slowly rising plateau started gaining height very rapidly, it was 2500 m high about 5 Ma ago, 3300 m high 2 Ma ago, 3700 m high 1 Ma ago and 4300 m high around 0.5 Ma ago. This estimation was done on the basis ecology of fossils (Xu Ren, 1981). Tectogenic processes are still operating with same vigor in the Himalayan region.

Orogenesis is extremely complex tectonic process helping ultimately in the growth and evolution of continent through welding of its creation, i.e. mountain to continent. This process is a matter of intense scrutiny by geologists and geophysicists and highly diverse and contradictory opinions have been expressed about how it occurs in the earth. Beloussov (1962) and Miyashiro, Aki and Sengor (1982 have given the historical account of various hypotheses, based on the horizontal and vertical tectonics, and complexities in orogenic evolution are highlighted by Burg and Ford (1997). Among the notable hypotheses or theories of mountain building are geosynclinal, continental drift, mental convection current, sea floor spreading, plate tectonics, sheet tectonics, spiral or zipper tectonics and wedge tectonics belong to horizontal or compressional tectonics in mobilist category and contraction and vertical or gravity tectonics -undation, oscillation, blister, oceanization and surge tectonics (Meyerhoff et al., 1996) of fixit category are important. Each hypothesis when applied to a particular orogenic belt begins with a specific geotectonic and paleogeographic setting; which proponents of other hypotheses normally do not accept.

13.1 Important Characteristics of the Himalaya

The salient characters of the Himalaya are that it is a part of Alpine-Himalayan mountain chain; 2400 km length of *sensu stricto* mountain is convex towards the south and takes sharp turns on northwest and northeast ends. The mountain has long and complicated geological history; the oldest dated volcanic rock yielded Late Archaean or Early Proterozoic age and the youngest sedimentary rocks as well as sediments are of Quaternary age. Sedimentary rocks in the Himalaya prominently occur in three divisions i.e. in the Outer, Lesser and Tethyan Himalaya. A 30-50 km wide belt of Higher Himalayan crystalline rock separates the Lesser and Tethyan divisions. Thirteen acidic and four basic episodes of magmatism have been identified in the Himalaya. Though Tertiary deformation is most dominating tectonic feature of this mountain range, but the signatures of earlier deformations like Hercynian, Caledonian and other orogenies including Proterozoic are also identified in the rocks at many places. Evidences point out its connection with peninsular shield of India. A series of thrust faults like Indus, Trans-Himadri or Tethyan, Karakoram Fault, Main Central Thrust, Main Boundary Thrust and Himalayan Frontal Fault played very important role in the evolution of this mountain range. Indus-Tsangpo Suture Zone containing ophiolitic rocks and mélanges, characteristic of ocean floor, deep-water sedimentary rock and high- pressure metamorphic rocks joins the Indian and Tibetan landmasses. The morphogenic upliftment of the Himalaya has mainly taken place in Quaternary period (Kizaki, 1994). Geophysically, Himalaya is seismically very active, has double crustal thickness and is still gaining height. The models dealing with origin and evolution of Himalaya should explain these salient characters.

13.2 Origin of Himalaya

Comprehensive account of Indian stratigraphy is available in highly popular textbooks written by D. N. Wadia (First written in 1919 and was revised in 1939, 1953 and 1973) and by M.S. Krishnan (first published in 1943 and its sixth edition came out in 1982 with several reprints). These authors were having thorough field experience and have described the geology and evolution of the Himalaya. Wadia, a pioneer Himalayan geologist, proposed that intermittent upheaval of the Himalaya started from the Permian period. Upheaval gradually obliterated the Tethyan Sea. A geanticline barrier started surfacing in the Eocene period with in the Tethyan basin. This geanticline divided the geosyncline into sub-basins. The Himalayan upliftment occurred in several phases. Among these phases of upliftment he identified three most important phases of upliftment in this mountain system. The first phase was of post-Nummulitic (Cretaceous) which culminated in the Oligocene, in this phase ancient Central Crystallines or Higher Himalayan rocks rose up. This was followed most violent or catastrophic phase of the Middle Miocene period; this phase elevated the central part of the Himalayan range and also formed a foredeep in front of towering mountain in which Siwalik sediments were deposited in the ensuing period. The last major phase of upliftment occurred between post-Pliocene and Middle Pleistocene, causing upliftment of the Siwalik and formation valleys in southern side of the mountain range.

On the other hand, Krishnan described the rise of Himalaya from the Tethys by tangential compression, which began first time in the Late Cretaceous and continued intermittently throughout in the Tertiary. He described the growth of mountain through a series of five or more phases, which were highly violent or stupendous. These violent phases were separated by periods of comparative quiescence or geological calm. The first upheaval occurred in the Upper Cretaceous, this phase furrowed the Tethys in several basins and ridges and also in formation of island arcs, now reported from in the Ladakh-Karakoram region. The Tethys also became deep in some regions, in which radiolarian chert was deposited. The second upheaval recorded in the Late Eocene; the main effect of this phase was observed in Sind and Baluchistan regions situated in Pakistan. In the Middle Miocene upheaval, the Himalaya acquired its major features, the Tethys Sea was completely vanished and a foredeep was formed in the south in which Himalayan sediments (Siwalik) were deposited. The penultimate upheaval occurred towards the end of the Pliocene, bringing an end of the Siwalik sedimentation. The last phase of upheaval occurred in the Early Pleistocene, the present shape of the Himalaya is acquired in this upheaval.

Both authors were of view that Himalaya is still tectonically active and is growing continuously.

In post-sixties period, Himalaya attracted incisive attention of the Indian as well as global geological community as it is the site of the youngest orogenic creation. Himalaya offers well-preserved ready documents of the processes involved in tectonic evolution. The burst in investigations- field, geological, geophysical, geochronological, photogeological and remote sensing; experimental-, thermal-, mathematical- and computer modeling- have brought out many new facts which were hardly known to early pioneers. As a result of spurt in research activities many new ideas based on horizontal and vertical tectonics and their combinations were formulated and tested and new models were advocated for origin and evolution of this great range.

13.3 Hypotheses Based on Horizontal Tectonics

13.3.1 Geosynclinal Origin

The idea of geosyncline for origin and evolution of mountain is given by Hall and Dana. This concept received wide support with tremendous modifications till the advent of plate tectonic theory. Plate tectonics categorically opposes the concept of geosyncline, however, some attempts were made to fit various types of geosynclines in the plate setup(Mitchell and Reading,1969). The term is still frequently used in the literature because of 'practical conveniences' (Miyashiro, et al., 1982).

Geosynclinal origin of Himalaya was favored by many early workers including Wadia and Krishnan who have considered the Tethys sea as a single basin which existed between the Late Proterozoic to end of the Mesozoic. Contrasting characters of sedimentary rocks across the north and south of the Higher Himalaya were recognized as early as 1878 by Waagan and by Burrard and Hayden in 1907, this observation later crystallized in modification of single Tethyan geosyncline into two separate basins of deposition. Pande (1966, 1979, 1991) named the northern basin as Tethyan geosyncline and southern basin as Himalayan geosyncline. The Central Crystalline Axial Zone separated these two geosynclines, the northern (the Tethyan) and southern (the Himalayan) geosynclines and is described as eugeosyncline and miogeosyncline respectively. Caledonian orogeny (390-410 Ma) of Early Palaeozoic age has uplifted the crystalline ridge from the sea floor. Besides horizontal compression of the Cenozoic age, reactivation deep-seated faults and basement also played valuable role in the evolution of the Himalaya and later he modified his view of polygeosyncline in favor of shifting geosynclines. According to this view geosynclines have originated in the Central Asia. Reactivation of deep-seated mantle reaching parallel or subparallel faults have simultaneously uplifted and subsided the large portion of this region. The uplifted parts formed the mountain and the subsided blocks formed the geosynclines. In favor of this model he cited the evidences of progressively southwards younger ages of fold belts and also of associated magmatism in Central Asian region.

Gansser (1964, 66,77) expressed opinion that only a small part of the Himalaya i.e. Tethyan rocks evolved from the geosyncline while major part i.e. the Higher and the Lesser Himalaya are extension of peninsular India. The northward drift of peninsular India pushed the geosynclinal as well as peninsular part into a series of folds that gradually rose higher and higher as the landmass moved onward. Later, he favored plate tectonic model of origin.

Saxena (1978) considers evolution of the Himalaya from a polygeosynclinal set up. The Central Crystalline Axis separated two geosynclines the Tethyan in the north and the Himalayan in the south (Saxena, 1971). These geosynclines underwent many synchronous orogenic and epiorogenic events occurred in Late Proterozoic, Early Palaeozoic, pre-Carboniferous (Caledonian) Permo-Carboniferous (Hercynian) and main phase in Cenozoic; thus the Himalaya has a polyorogenic history. The evolution of the mountain range is attributed to mental diapirism, geochemical control in tectonic activities and gravity tectonics.

He suggested melting in mantle is caused by rhythmic contraction and expansion of the earth. The time of contractions and expansions would correspond the periods when geochemical differentiation and melting reached the peak and these events also coincide with major diastrophic events of the earth history. Large scale melting produces mantle diapirism. Large-scale movement of diapers in turn initiated the gravity tectonic deformations. Movement of upper mantle material, *sensu lato* gravity tectonics, uplifted the continents in general while localized mantle diapirism affecting the lithosphere, which he called *sensu stricto* gravity tectonics, are responsible the oscillatory movements in the blocks. These oscillatory movements were of compression and expansion types. The compressional movements have produced crustal shortening and upliftment in the intracontinental island arc sedimentary prism i.e. in polygeosynclinal region. Himalaya is evolved from such compressional and extensional movements of the Cenozoic age.

13.3.2 Sheet Tectonics

The concept of sheet tectonics, somewhat similar to continental drift, is given by Rode (1986). The sheet model envisages deposition of volatile constituents of atmosphere, submarine volcanic pyroclasts and chemical precipitates in the 2.4 km deep primitive Panthalessa or Pan-Pacific Ocean in the form of sheets or layers. Entire earth was covered with water till the first emergence of land in the Silurian period. The present day continents were borne under water as a part of sheet packs. These continents were laying one over the other in the water of the Panthalessa and were situated in the present day Central Asian region with nucleus in the Tibet. The slow emergence of the sheets comprising of the Angaraland and the Gondwanaland began in the Silurian and continued till the late Tertiary. The magma generated in the mantle pushed upward along fractures and faults and spread laterally. Thus overlying sheet got detached from the underlying sheets of the Gondwanaland and the Angaraland and started migrating laterally as a float on magma under the influence of gravity because of sloping conditions existing on the Pangaea. Thus large-scale sheet movement started in the Late Tertiary. According to this hypothesis, Tethyan Sea or geosyncline first developed as a narrow rift due to wedge action of Permocarboniferous magmatic intrusions in the Central Tibet. This rift gradually widened into a medium size gulf and finally developed into a sea or geosyncline by pushing the continents aside. Tethyan geosyncline received sediments brought by many rivers, thus a thick pile of sediments were deposited in it. This widespread volcanism of the Quaternary period caused differential upheaval of a portion of continental crust including Himalaya and extensive migration of other continents in different directions with attendant rotations.

13.3.3 Experimental Modeling

Dubey and Bhat (1986) and Dubey (2004) have successfully generated tectonic model of the Himalaya in laboratory. They considered Himalayan terrain is a northern extension of peninsular shield and strongly refuted the theory of plate tectonics. This model envisages evolution of Himalaya in three stages namely the rift phase, the curvature phase of basement faults and the orogenic phase. They proposed rifting process of the Late Archaean and in Early Proterozoic age first formed Himalayan basin. Rifting has produced a set of parallel running listric normal faults and these faults were reactivated many times till the onset of Cenozoic orogenesis (Fig.No.13.1). As a proof reactivation they offered evidences of spasmodic basic magmatism. Each reactivation of the faults progressively deepened the basin towards northern side. These basement faults are converted into folds and thrusts – MCT, MBT, and MFT – due to post-Cretaceous compressions and inversions. In orogenic phase, they recognized three episodes of folding. Bhat (1987) stated that Himalayan ophiolites are upwelled, ocean floor. This upwelling occurs due to mafic – ultramafic diapirism in an extremely attenuated lithosphere.

Continental Drift and Plate Tectonics

Continental drift hypothesis considered as precursor of plate tectonics. Early advocates of continental drift hypothesis like Argand, du Toit, Holmes, Ahmad and others have supported the closure of Tethys Sea by impact of northerly drifting India shield. The Angaraland jammed northerly drift of Indian shield in the Cretaceous period and continuous exertion of compressive pressure by Indian landmass folded, faulted and uplifted the sediments of Tethyan Sea and also a northern part of the shield into the Himalayan Mountain.

13.3.4 Plate Tectonics

The theory of plate tectonics says that rigid lithosphere of 70-150 km thickness is divisible into 6 major and 12 or more minor spherical caps or plates which are separated from each other either by mid-oceanic ridges or by trenches or by transform faults – and are moving tangentially with respect to one another over the underlying upper mantle or asthenosphere. The supporter of this concept hails it as a revolutionary theory which has unified many hypotheses e.g. geosynclines, continental drifts etc. and is capable of giving rational explanations of majority of geotectonic activities like earthquakes, volcanism and mountain building, island arcs and other similar large scale geotectonic activities. Even though plate tectonics is widely acclaimed as most rational geotectonic theory but since its formulation in 1968 it was constantly undergoing in process of revisions and modifications and in many cases explanation based on plate theory are as controversial as explained by any other hypotheses.

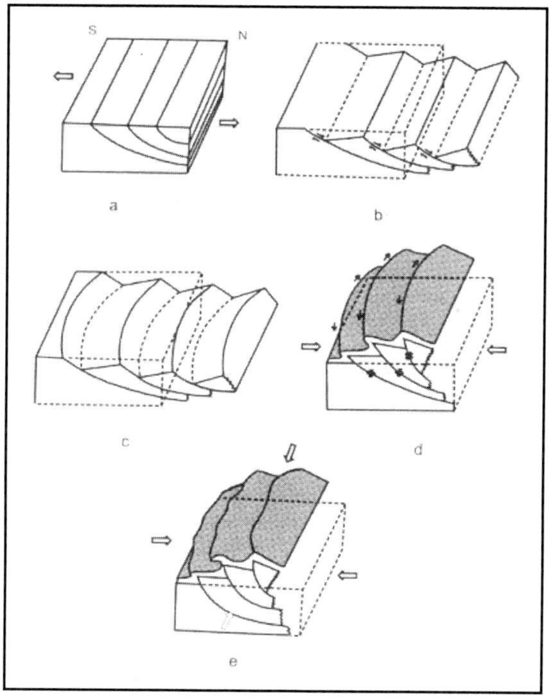

Figure 13.1. Tectonic evolution of Himalaya (after Dubey and Bhat, 1986)

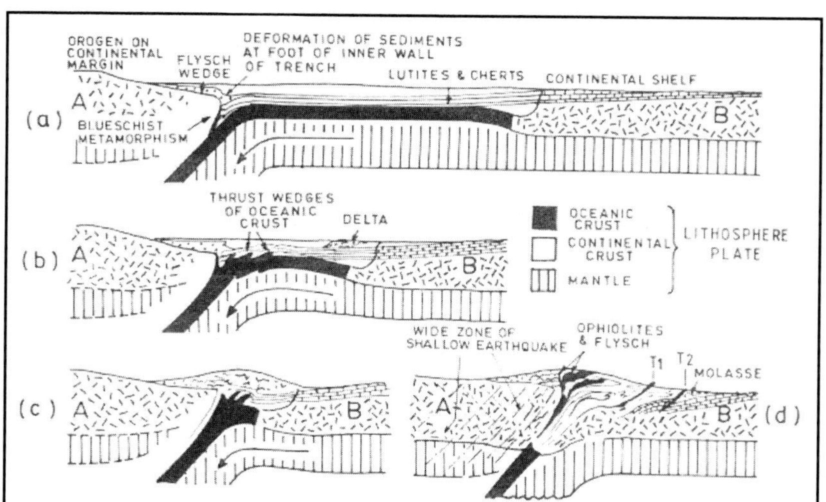

Figure 13.2. Dewey and Bird model of evolution of Himalaya. Note differences in interpretations

The concept of plate tectonics was initially applicable to the Cenozoic geotectonic processes later it was extended to entire Phanerozoic time and now it is even applied to entire Proterozoic or even in Archaean time. Extrapolation of the concept into geological older ages gave conflicting results particularly in case of direction and rate of drift of continents and difficulties were also experienced with respect to physical and mechanical properties of plates existing in the distant past (Burg and Ford, 1997). Different workers have given different interpretations about plate tectonic evolution of Himalaya, Fig No. 13.2 based on Dewey and Bird, 1969; 13.3 after Valdiya, 1998; 13.4 after Sinha and Mishra, 1997 .

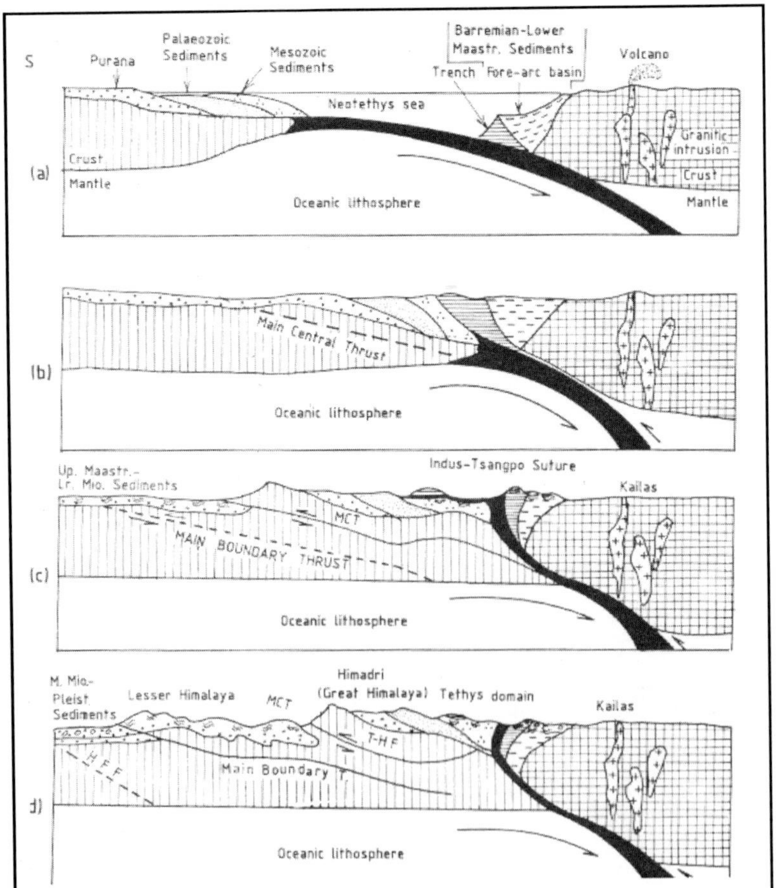

Figure 13.3. Plate-tectonic evolution of Himalaya (Valdiya, 1998)

13.3.5 Superassembilies Configuration

Supercontinents are assemblies of all or nearly all of the earth's continental blocks. Three times in the global geological history all continents were united into single landmass. These supercontinents were named Columbia (~1800-1500 Ma), Rodinia (~1100-800 Ma) and Pangaea (~250 Ma). A rough time gap of 750 Ma is observed between Columbia and Rodinia and also same time gap is present between Rodinia and Pangaea, suggesting some type of periodicity in the mantle processes that is responsible for assembly and dispersal of continents. In addition to super assemblies, there were several minor assemblies of continents (Fig. No. 13.5) formed in different periods, namely the Arctica (~2500 Ma), Kanorland (2500), Atlantica (~2000 Ma), Nena ((~1800 Ma), Mawson continent (~1700 Ma), Palaeopangaea (~1100-1000 Ma), Gondwana (~600-500 Ma) and Laurasia (~250 Ma)(Rogers and Santosh, 2003).

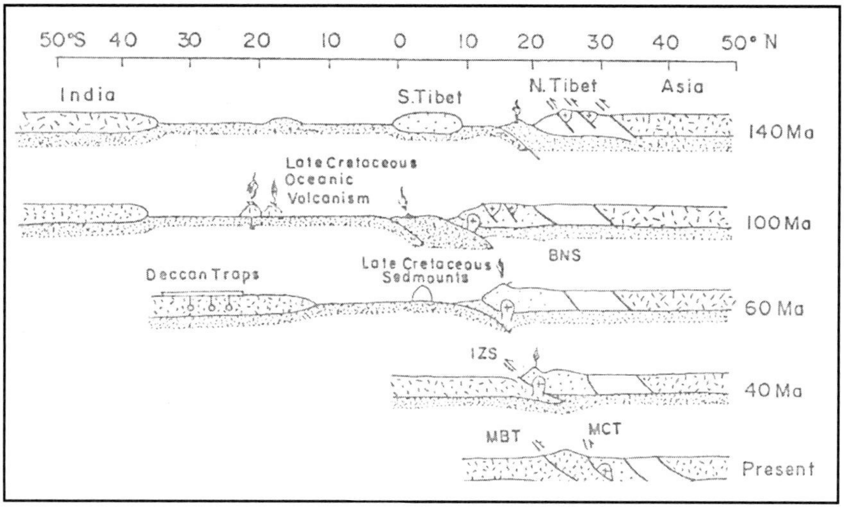

Figure 13.4. Geological evolution of Himalaya (after Sinha and Mishra 1997), abbreviations; BNS: Bangong Nujiang Suture, ISZ: Indus Zangbo Suture, MCT: Main Central Thrust, MBT: Main Boundary Thrust,

Ur (German prefix describes the original and also the oldest city of the Chaldees) name is proposed to the Archaean (~3000 Ma) stabilized continents assembly of India, Africa and Australia located in the region of the Pangaea (Roger, 1993). Unrug (1995) described Rodinia to Middle Proterozoic assembly of continents. Unrug stated that between 1600-1000 Ma, now ~1100-800 Ma, duration many mobile belts were welded into a single landmass. During Late Proterozoic and very Early Palaeozoic (1000-540 Ma), Rodinia was fragmented in several continents and towards the end of the Proterozoic or in the Early Cambrian these fragmented continents partially assembled once again, thereby formed the Gondwanaland. By Middle Palaeozoic, Gondwana united with Laurasia and formed the Pangaea. Geological configuration of the Gondwanaland suggests its division into two units, which are named as Eastern and Western Gondwanaland. The Eastern Gondwana is a surviving portion of Rodinia and comprises of present day Madagascar, Antarctica, India and Australia. The Western Gondwanaland comprises of South America, Africa and Arabia and shows history of accretion and was docked into the Eastern Gondwanaland. Eastern and Western Gondwanaland were sutured along Mozambique Belt of East Africa, Arabia and Neubian shield.

Another framework proposes that in the Early Cambrian, continents belonging to the southern Gondwanaland were loosely clustered but continental units of Laurasia – North

America, Europe and Asia – were separated from each others, particularly Europe and North America were separated by the then Iapelus Ocean. The Caledonian, Appalachian and the Hercynian fold belts performed suturing in the Laurasia. In mid-Permian (250 Ma) all continents were knitted in one land mass called Pangaea. The Pangaea remained intact for next 50-100 Ma. The sign of fragmentation in the Pangaea appeared in the mid Late Jurassic (200-180 Ma), first between the Northern America and the West Africa-Europe. South America and Africa started fragmenting in Early Cretaceous (130-120 Ma). Antarctica and India parted company around 105-100 Ma ago while Australia broke away from the Antarctica around 45 Ma ago and South America from the Antarctica around 30 Ma BP. Thus the Pangaea was completely fragmented into present continental masses by the Middle of Tertiary (Powell and Conaghan, 1973). The Indian Ocean came into existence around 45 Ma BP through separation of Australia from Antarctica. The Burmese microplate was created in the Middle Miocene through opening of the Andaman Sea.

Drift of Indian Landmass

Drift itinerary of India, located in the southern part of the Gondwanaland is deciphered out with the help of apparent polar wandering curves based on geomagnetic study, geochronological data and geological evidences. The curves suggest that around 510 Ma BP, the southern tip of India was situated near equator and remained there for considerable period of time. Between Devonian (380 Ma BP) and Early Carboniferous (340 Ma BP) the southern tip drifted further southward and docked near 50° south of equator. In the Permian, Indian landmass was rotated in northeast-southwest direction and India wandered further in south to the 60° south of the equator. In the Triassic and Jurassic due to polar wandering the southern tip started migrating in the north and occupied position near the 40° south of the equator. In the Cretaceous (100 Ma BP), India separated itself from Antarctica. Rajmahal volcanism (105-100 Ma) initiated this rift and an independent identity of the Indian subcontinent was carved out. This time the southern tip of India was located near 20° south of the equator. Movement of southern tip of India between 40° south of the equator to present 8° north of equator i.e. southern tip of India, implies a travel of 7000 Km.

Between 75 and 40 Ma BP, Indian plate moved comparatively fast with a velocity of 18 cm/year. Around 70 Ma ago, the Central India was situated around 30° south of the equator where as this area is now located around 20° north of the equator. At the time of collision, northern tip of India was close to the equator. In the Late Cretaceous Lhasa block was located near 12°5-30°N of the equator and it was around 13° 5-N in the Palaeocene-Eocene. As a consequence of collision of India with Eurasia, the plate velocity drastically dropped to 5 cm/year. The northward drift of India initially facilitated by the Chagos-Laccadive transform fault in the west and the Ninety Degree East Ridge in the east.

13.3.6 Collision Tectonics

Dewey and Bird (1976) identified three types of orogenic settings namely the island–arc, Andean and the continents collision or the Himalayan type. The Himalayan range is evolved through continent–continent collision and the participating continents in this process were the subducting Atlantic type continental margin of India and the Asia (Coward, 1986).

The collisional history is marred with several controversies. The date of collision is suggested from the Palaeocene to the Eocene i.e. between 55 or even 63 Ma to 45 Ma. The collision is debated between single stage and double stage collision. In case of former one Tethyan Sea was present thus recognizing one suture zone i.e. Indus-Tsangpo Suture Zone, a 2500 km long Indus or Himalayan Suture. In the single stage subduction, Indus-Tsangpo Suture Zone (Valdiya, 1984 b) comprising of ophiolitic mélanges, blue schist, magmatic rocks, flysch, deep-water sediments evolved in the Early Tertiary. In case of latter, presence of two Tethyan seas are recognized namely the Palaeo-Tethys and the Neo-Tethys, hence two suture zones were identified namely the northern or the older one i.e. the Shyok Suture and is also described as Pre-Himalayan Suture and the southern or the younger one i.e. Indus Suture. Younger part of Late Proterozoic to entire Palaeozoic epoch was the life span of the Palaeo-Tethys. The Palaeo-Tethys Sea was vanished in the Late Triassic by collision between the Eurasian and the Tibetan plates and as a result of this closure the Kunlun range evolved in the southern Europe. The northerly drifting of the Tibetan landmass on one hand was responsible for the obliteration of the Palaeo-Tethys and as well as for opening up of the Neo-Tethys Sea. The remnant of the Palaeo-Tethys is the 300 km long Shyok Suture. The Neo-Tethys Sea vanished in the Early Tertiary. Closure of Neo-Tethys formed the Indus Suture. The Ladakh magmatic arc, a remnant of ancient island arc, separates these two sutures. Himalayan orogenesis is discussed by the various workers viz. Powell and Conaghan (1973), Molnar and Tapponnier (1975) Dewey and Burke (1973), Le Fort (1975), Gansser (1977), Bird (1978), Bordet et al. (1975) Fuchs (1976), Valdiya (1984 a, b), Thakur (1992) and others. Molnar and Tapponnier (1975) proposed that India has subducted or advanced nearly 2000 km northwards into the Eurasia landmass. Other workers revised this estimate between as less as 250 km to 3000 km towards higher side. In NW part of Pakistan Himalaya, convergence between 550-1100 km has been estimated. Of 2000 km convergence, one-third convergence occurred below the Himalaya and of the remaining equally distributed between the southern Tibet and the Tarim basin and along the east-west trending strike-slip faults in western China and Mangolia.

Thakur (1992) has discussed evolution of the Himalaya in the detail. He considered Lesser Himalaya is an extension of peninsular shield. The Higher Himalaya is an uplifted part of mid-crustal segment whose upper surface or overlying rocks are eroded and thus produced sediments that were deposited in foreland basin and the Tethyan rocks were deposited on the northern portion of the Indian shield. Trans-Himalayan region including the Indus and Shyok sutures and the Karakoram region is part of the southern Asian plate. India was a part of the sprawling Gondwana paleogeographic province, which also included the Tibet, Karakoram, Afghanistan, Iran and Turkey. Rifting in this province began in Mid-Permian and drift started soon afterward. As a result rift and drift, Neo-Tethys Sea opened up in the Mid-Jurassic that formed a Cimmerian continent comprising of the southern Tibet, Afghanistan, Iran and Turkey. The short-lived Neo-Tethys obliterated in the Early Cretaceous due to northward movement and collision of Indian landmass. As a result of the collision first oceanic lithosphere subducted under the margin of the Karakoram block of Asian plate. An island arc geotectonic setting in the Kohistan, Dras-Chalt areas came into existence in the Cretaceous and Early Eocene. Trans-Himalayan magmatic arc of Ladakh-Deosai and

Karakoram were formed in Andean type of setting over staggering time duration spanning between 100 and 40 Ma BP, on the southern margin of Asian plate. Towards the south of this magmatic arc in deep-sea Indus sediments, 5000 m thick, were deposited in a 500 km long basin or in the arc-trench gap setting or in the fore arc basin. Shallowing of this basin began in the Early Eocene and finally the gap was obliterated in the Miocene. To the south of this gap, Himalaya started uplifting through a series of intracrustal faults like the Indus Thrust, Trans-Himadri Thrust, Main Central Thrust, Main Boundary Thrust and the Himalayan Frontal Fault. These faults not only imbricate the rocks but also thickened the crust and particularly movement in the MCT produced metamorphism and also formed the Miocene leucogranites. A characteristic feature of these faults is that they are progressively becoming shallower in the south direction and also of younger in age. MBT is now consuming a part of subducting Indian plate. According to Thakur, collision began around 55 Ma ago and completed around 40 Ma ago. In the plate tectonics based evolution of the Himalaya thrusting, metamorphism with inversion and leucogranites formation played important roles.

The orographic evolution of the Himalaya is described by two models namely the evolutionary or step and steady state models, and also through modified versions of these models. The evolutionary model of Lyon-Caen and Molnar (1983), Ni and Barazangi (1984) and of others believe in two phases of orogeny. In first phase, the oceanic lithosphere of Indian plate subducted beneath the Eurasia in the Cretaceous to Early Tertiary duration. This active phase came to an end in the Eocene after collision India and Eurasia and second phase of orogeny begun in Miocene, when underthrusting of India commenced. In the second phase zone of plate convergence progressively migrated towards the south by the formation of intracrustal thrusts on the subducting Indian plate. The earliest thrust was Indus Suture Thrust formed during the Eocene, followed by the MCT (Miocene-Pliocene), MBT (Pliocene-Recent) and the MFT (Late Quaternary). The MCT, MBT and MFT at present are showing the increasing order of seismic activity. According to this model present rate of convergences, 18±7 mm/year, is shared by the MCT, MBT and MFT in varying amount. In the steady state model (Seeber and Armbuster, 1984) all the active thrusts like MCT, MBT and MFT coincide with a subsurface detachment fault or Basement Thrust. The important components of this model are: a subducting slab i.e. Indian plate, an overriding slab i.e. the Tethyan slab and a sedimentary wedge located between and decoupled from Indian plate and the Tethyan slab. The underthrusting plate in this model extends up to Kunlun-Altyn Tagh ranges in the northern Tibet thus accounting the high elevation and double crustal thickness of the Tibet. The underthrusting Indian plate remains undetached and internally unaffected by the thrusting.

The evolutionary and steady state models are modified by a number of workers. Zhao and Morgan (1985) proposed that the underthrusting mass thrust into the lower crust – upper mantel underlying the Tibet and uplift of the Tibetan Plateau is hydraulically achieved. 'Extrusion Tectonics' based on experimental modeling by Tapponnier et al. (1982) believes in two phases of extrusion; in the first phase, Indo-China extruded along the Red River Strike-Slip Fault; and in the second phase Tibetan Plateau and SE China were extruded along the Altyn Tagh Fault in the eastern direction. Klootwijk et al. (1985) applied palaeomagnetic data in the evolution of the Himalaya. They proposed that the Indian plate collided with the Eurasian plate first in the west in the Palaeocene and convergence slowly moved in the east till Early Eocene (Fig. No.13.6).

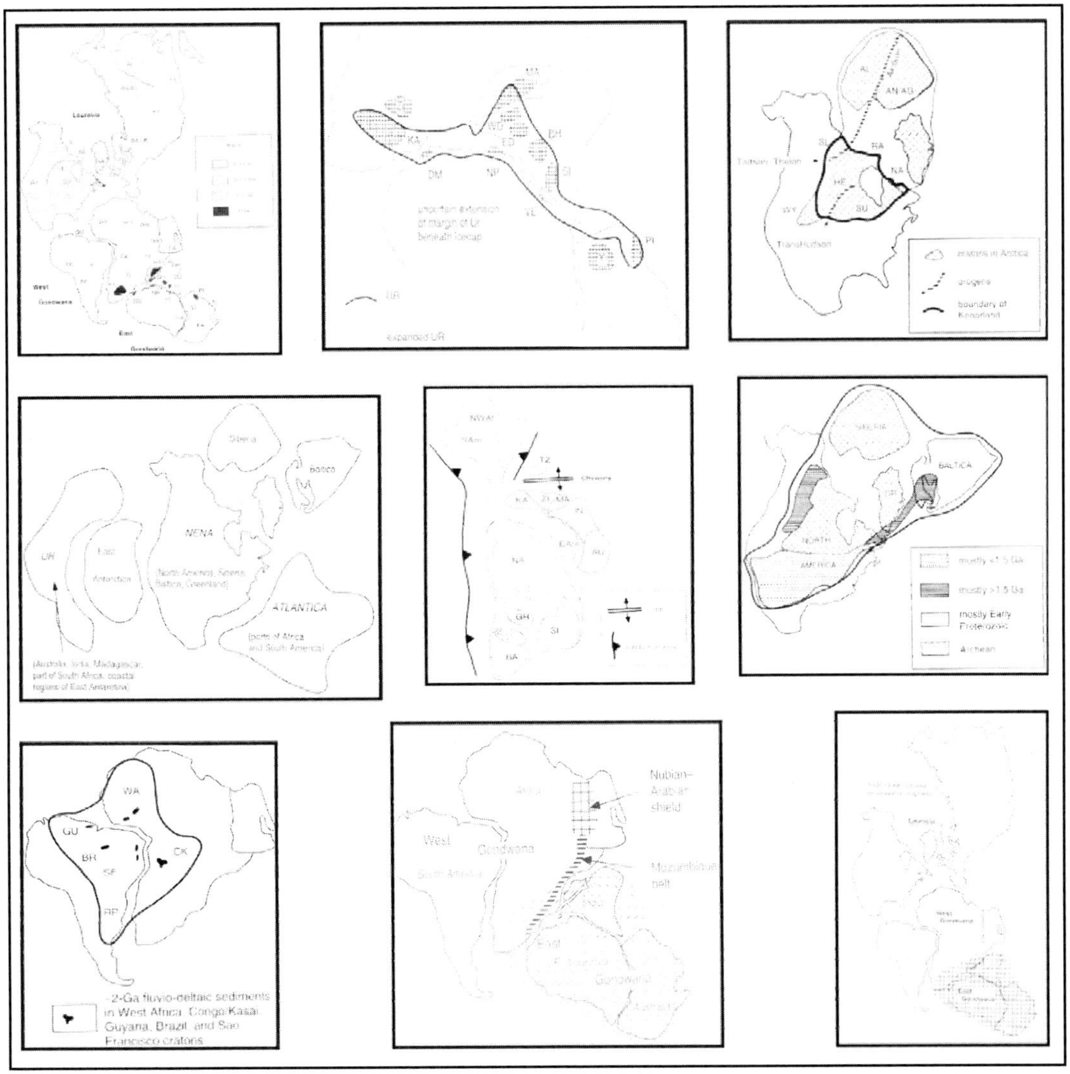

Figure 13.5. Assembly and super assembly of continents in different geological time. A- locations of cratons in Pangaea, abbreviations: AL- Aldan, AR- Aravalli, AN/AG- Anabar/ Angara, BA/UK- Baltic/ Ukraine, BH- Bhandara, BR- Brazil. BU- Bundelkhand, CA- Central Arabia, CK- Congo/Kasai, DH- DM-western Dronning Mud Land, EA-Eastern Australia, GA-Gawler, GU-Guiana, HE-Hearne, H-T-Hogger and Tibesti (N Africa), KA-Kappavaal, KI- Kimberley, KZ-Kazakhstan, MA- Madagascar, NA-North Atlantic, SC-South China, SF-San Francisco, SI-Singhbhum, SL-Slave, SU- Superior, TA-Tarim, TZ-Tanzania, VE-Vestfold, WA-West Africa, WN-West Nile, YI-Yilgarn, ZI-Zimbabwe; B-Ur and expanded Ur; C-Arctica and Kanorland; D-Rodinia; E-Columbia; F-Nena, G-Atlantica, H-Gondwana and I-Pangaea (after Rogers and Santosh, 2003)

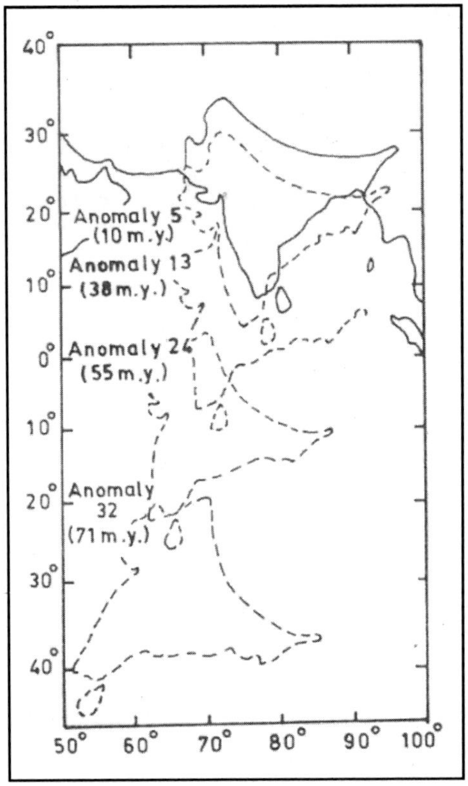

Figure 13.6. Drift of Indian Continent through Geological time (after Molnar and Tapponnier, 1975).

Once the entire plate started converging, eastward extrusion in the Tibetan region along the strike-slip faults occurred between Middle Eocene to Early Miocene and finally intracontinental rotational underthrust and formation of the syntaxial bends had taken place. Some workers have refuted the idea of extrusion or indentation tectonics.

Chengfer and Yusheng supported plate tectonic origin of the Himalaya in three stages. In first stage, oceanic subduction produced an ocean-continent or cordilleran type belt in the Tibet. In second stage, the oceanic domain vanished between Asia-India and a continent-continent collision (Eocene-Oligocene) produced Yarlung-Zangbo or Indus Suture. In the last stage, post-suturing, intracontinental convergence is characterized by the large scale downward (subduction), upward or sideways escape of continental lithosphere and formation of thrust like MCT (Middle Tertiary), MBT and MFT (Plio-Pleistocene) and leucogranites combinedly evolved the Himalayan range.

Behavior of the convergence or underthrusting Indian plate at lithospheric scale has been described as horizontal or very flat underthrusting (first proposed by Argand in 1924, he was also first proponent of idea of northward drift of greater India; Alfred Wegner and A. Holms in 1965), steep angle underthrusting, delamination the mantle or by convecting flow in the asthenosphere and downwelling of the mantle. Same models have been used to explain the altitude of the Tibet. (Matte et al., 1997).

13.3.7 Intra-Plate Evolution

The original proposal of plate model postulates the localization of orogeny at the plate boundaries or margins. Later findings revealed that orogenic belts could also be evolved within the plates (Miyashiro et al., 1982).

Crawford (1974) and Kaila and Narayan (1976) proposed intra-plate evolution of the Himalaya. Crawford (1974) visualizes a giant but weak and loose Gondwana plate comprising of India-Tibet, Australia and Antarctica on the basis of occurrence of Permo-Triassic vertebrate fossils of Gondwana affinity and of palaeomagnetic evidences. In the united Indo-Tibetan mass, India remained continental since Proterozoic but Tibet was submerged under the water. According to him, the Indus Suture represents a relic of the uplifted mantle. Opening and closing of the Indus Suture has taken place between Permian to Late Jurassic period because of counter clock wise rotation of Indian mass. The northern continent jammed in the Late Phanerozoic period was characterized by highly active sea floor spreading in the Indian Ocean and onward movement of the Indo-Tibetan block. As a consequence of jamming, overthrusting of rocks occurred along the weak planes. These overthrusts finally culminated into the Himalaya. Thus Himalaya is evolved without collision but by the deep-seated thrusts.

Kaila and Narayan (1976) on the basis of seismotectonic evidences considered the Tibet as a part of Indian plate and the northern boundary of Indian Plate run along the southern margin of the Tien-Shan and Nan-Shan mobile belts. Himalaya has formed due to block uplift along the parallel running steeply dipping faults in the intra-continental plate setting. Faulting is probably caused by excess loading of the mass in the Tethyan region.

Sinha (1990) described the evolution of the Himalaya from deep-seated faults. He stated that Himalaya is a separate mobile belt and renamed the Indus Suture Zone as the Great Himalaya Suture (GHS). The GHS is a deep fault zone showing evidences of collision and presence of ophiolites. To the south of the GHS, Himalaya shows two structural facies namely the northern facies (Tethyan Himalaya Zone) and the southern facies (Lesser Himalayan Zone). These two facies are separated by a transitional facies zone constituted by the Central Crystallines (Higher Himalaya) which is named as Main Axial Zone (MAZ). MAZ is the source of crystalline nappes covering a major part of the southern Lesser Himalaya. These facies are thrusts or faults bound i.e. by the MBT, MCT and the Tethyan Thrust and these thrusts have played main role in the evolution of orogen. He did not consider Himalaya as uplifted extension of the peninsular India.

Along the convergent margin particularly at the interface of Indian and Myanmar plate's ophiolite was emplaced during subduction (Dasgupta and Nandy, 1995)

13.3.8 Expanding Earth Hypothesis

Expanding earth hypothesis (Carey, 1982) describes earth is a continously expanding body in volume since its birth, however, expansion was more in Phanerozoic period and orogenesis is a primarily a dipiric process with secondary gravitational spreading. India was associated with Asia and Australia, Africa and Antarctica; these continents were situated in a small contiguous region before expansion. Post-Proterozoic expansion did not affect the position of the Asia and India but separated the Australia, Africa and Antarctica by opening of the Indian Ocean. The difference in expansion in the equatorial and the polar regions is responsible for generation of the shearing forces. This shearing force produces sphenochasm or great depressions or oceans on the surface of the earth as well as rotation of continents. The rotations of continental block along megashear produced Himalaya. Stacklin (1984) is of view that earth expansion is a result of excess generation of material in the Indian and Atlantic oceans in comparison to consumption or subduction along the Tethys. Crustal shortening, mantle diapirism and gravity gliding mechanisms are responsible for the Himalayan orogenesis.

Crawford (1982) proposed a concept of Turfan Tectonics on the basis of earth expansion. Turfan depression (42°30-N, 89°20-E in China) is at the center of a circle whose radius is 1600 km. Himalayan arc is a part of the periphery of this circle. He proposed that Turfan is a tectonic monopole about which earth's expansion has taken place in the post - Palaeozoic time. He also noted that due to vertical tectonics or vertical magmatism several major lineaments radiate from the Turfan center. He considered that formerly Asia and India constituting a single landmass and due to centrifugal force, which controlled the earth's expansion, India was separated from Asia through an annulus or a ring. In subsequent period this annulus was subsided and thick sedimentary rocks were deposited in it. Again due to centripetal forces India was attracted towards the Turfan center. A limited horizontal movement compressed the sedimentary rocks thereby produced folds and thrusts and uplifted the Himalaya. This horizontal movement of India is much less in comparison to what is proposed in the plate tectonics.

13.4 Vertical Tectonics and Gravity Tectonics

Vertical tectonics proposes vertical upliftment or subsidence of crustal blocks motivated mainly by the gravitational forces. The supporters of vertical tectonics do not deny the presence of horizontal or compressive forces but considers their role in the evolution of an orogen as peripheral one. Harmann led this school of thought (oscillation hypothesis) and subsequently idea is strongly supported by Van Bemmelan and Russian geologists such as Tetyayev, Beloussov and many others.

Hagen (1959) was perhaps first person who proposed that evolution of great nappes of the Nepal comprise of large crustal blocks have been uplifted along the deep fracture.

Qureshi (1969) argues that block upliftment, vertical tectonics and gravity gliding, are primarily responsible for upliftment of peninsular shield along the deep-seated faults into an intracratonic Himalayan mountain range. Polymorphic inversion of eclogite of higher density into lighter density gabbro is responsible for thickening of basaltic layer by 50 km. Some of this additional material moved up in the cracks to produce faults. Nappes and recumbent folds are produced by the gravity gliding as a consequence of vertical movements.

Ashgirei (1984) supported the deep-seated faults origin of the Himalaya. The Tethyan and the Lesser Himalayan rocks, separated by the crystallines, show divergent characters. MCT forms a deep Main Axial Fault Zone which is nearly 40 km wide and has been repeatedly metamorphosed during 260-250, 160-143, 79-73, 37, 13-7 Ma BP. Besides MCT, there in another deep axial fault i.e. the Main Boundary Thrust in the Himalaya. Himalaya is evolved by upliftment along these faults.

Beloussov (1981) and Beloussov et al. (1982, 1984) rejected the mobilist view of the origin of the Himalaya. He concluded that Himalaya is uplifted along deep-seated faults. The melting of ultrabasic rocks of mantle activates these deep-seated faults. Indus Suture Zone is a highly tectonized belt along which ultrabasic rocks risen to the surface. This suture zone was situated along the northern margin of Indian platform. All the deep-seated faults are situated to the south of the Suture Zone and movements in these parallel or sub parallel faults are responsible for the evolution of the Himalaya.

13.5 Conclusion

A reading of large number of contradictory views, hypotheses or theories dealing with the evolution of the Himalaya is sufficient to baffle or confuse a beginner. What we have is a well-developed and still growing mountain i.e. the Himalaya; the mountain is fairly approachable, having good exposures in deeply cut valleys. Despite these views have been expressed based on contraction to expansion and also from fixit to mobilist points of view, which are diametrically opposite to each other. This is enough for an impartial reader to have frustration or confusion on the origin of the Himalaya and one may even doubt about the validity of geology as science. In such a situation it is not easy to convince a new entrant. However, one must accept that differences in opinions are not unique in the geology alone but as a matter of fact same situation is also prevails in the higher realms of the other branches of science. May be it is more in the geology and less elsewhere. Truly, differences in opinions are based on observations and experimentations and this is the essence of science. Contradictory opinions are outcome of several factors in geology. Firstly, while dealing with a large region of the earth, heterogeneity in geological composition and settings in sub-regions is a normal phenomenon. Secondly, unlike other branches involvement of time in the various geological processes occur on stupendous scale and the common unit of time in geology is million year (Ma); while in other fields time required for completion of many processes is very small, therefore, actual operations of most of the processes can be easily monitored there. The laboratory scientists like of physics and chemistry cannot properly visualize this concept of time easily. This situation is akin to the concept of huge distances, which is measured in light years in astronomy. Thus in fields, geology and astronomy, probity is always influenced by the personnel dogmas and belief. Secondly, on the other hand in geology, except catastrophic events, in most of the cases we observe and note the final result. With this final product in hands, we make inferences about a series of sequences and those processes that are involved or might have taken part in the completion of that product or object of study. Here, when question of making inferences comes; interpretations are bound to vary from observer to observer. Having flexibility of geological scale, composition, setting and time some clues can readily be obtained which can support a particular point of view with further elaboration through search of similar line of evidences. This is sign of evolution of knowledge and perhaps a unique character of geology alone. Fourthly, our understanding of the subject is still not on matured level and as a fully established and mature science, geology has yet to cover a huge distance. Still happening of a geological phenomenon is controlled by an ill-defined or inadequately understood set of parameters and even these parameters are not well constrained yet. As a result of this leading practitioners of geology also behave at times in a highly partisan way; they prefer that explanation whlch they do believe or want to believe and ignore or severe criticize other explanations. Many have not taken incorporation of evidences or thought given by opponents kindly. Therefore, we have totally divergent opinions even on a readily examinable object. Multiplicities in the views have also eluded the agreements in most of branches of geology. Geology has enough space for every view to test its own strength in the course of time and also provides ample opportunities for a practitioner to advance or discard his preferences. This is a unique character of geology that it is a highly accommodative branch of science. However, scientifically sound views received validity with passage of time. Even God has proved himself wrong in the Himalaya. Lord Krishna while delivering famous discourse 'The Geeta' told being the Lord he presides over all physical actions of the nature but pronounced Himalaya as 'sthavar' Himalaya means among all immobile, Himalaya is the greatest. That is not the case, as we know very well. Or does Lord support contraction hypothesis?

Such dilemmas are best explained again in the Geeta, where Lord Kishna explains in response to a querry raised by Arjuna

Karmano hyapi boddhavyam boddhavyam ca vikrmanah

Akarmanasca boddhavyam gahana karmano gatih (Geeta, 4.17)

The truth about actions must be known and the truth about forbidden action must be also known; even so the truth about inaction must be known; for mysterious is the nature of action.